Structural Materials in Nuclear Power Systems

MODERN PERSPECTIVES IN ENERGY
Series Editors: David J. Rose, Richard K. Lester, and John Andelin

ENERGY: The Conservation Revolution
John H. Gibbons and William U. Chandler

STRUCTURAL MATERIALS IN NUCLEAR POWER SYSTEMS
J. T. Adrian Roberts

Library of Congress Cataloging in Publication Data

Roberts, J. T. A. (J. T. Adrian), 1924-
Structural materials in nuclear power systems.

(Modern perspectives in energy)
Includes bibliographies and index.
1. Atomic power-plants—Materials. 2. Nuclear reactors—Materials. I. Title. II. Series.
TK9185.R6   621.48′332   81-1883
ISBN 0-306-40669-1   AACR2

© 1981 Plenum Press, New York
A Division of Plenum Publishing Corporation
233 Spring Street, New York, N.Y. 10013

All rights reserved

No part of this book may be reproduced, stored in a retrieval system, or transmitted, in any form or by any means, electronic, mechanical, photocopying, microfilming, recording, or otherwise, without written permission from the Publisher

Printed in the United States of America

Structural Materials in Nuclear Power Systems

J. T. Adrian Roberts
Electric Power Research Institute
Palo Alto, California

PLENUM PRESS • NEW YORK AND LONDON

Preface

In recent years the effort devoted to assuring both the safety and reliability of commercial nuclear fission power reactors has markedly increased. The incentives for performing this work are large since the resulting improvement in plant productivity translates into lower fuel costs and, more importantly, reduced reliance on imported oil.

Reliability and availability of nuclear power plants, whether fission or fusion, demand that more attention be focused on the behavior of materials. Recent experiences with fission power indicate that the basic properties of materials, which categorize their reliable behavior under specified conditions, need reinforcement to assure trouble-free operation for the expected service life. The pursuit of additional information continues to demand a better understanding of some of the observed anomalous behavior, and of the margin of resistance of materials to unpredictable service conditions. It is also apparent that, next to plasma heating and confinement, materials selection represents the most serious challenge to the introduction of fusion power.

The recognition of the importance of materials performance to nuclear plant performance has sustained a multimillion dollar worldwide research and development effort that has yielded significant results, both in quantification of the performance limits of materials in current use and the development and qualification of new materials. Most of this information appears in the open literature in the form of research reports, journal articles, and conference proceedings.

For some time now, members of the nuclear materials community have realized that the number of such reports is staggering and that technical meetings on the various specialized topics are proliferating alarmingly. The survival solution for many, including myself, has been specialization in a narrow subset of nuclear materials issues. However, students or engineers just entering this field, and who are searching for background information, are faced with an overwhelming task. This book is an attempt to respond to their needs by providing a coherent treatment of materials in nuclear power systems and the main issues currently being addressed, within a single volume of reasonable length. Also, I hope that it will serve as a useful source of reference for the experienced engineer.

The scope of the book is confined to materials that comprise the key components in the fission and fusion reactor cores (i.e., nuclear steam supply system or N.S.S.S.), and in the so-called balance-of-plant (B.O.P.), which is assumed similar for both nuclear systems.

Following a general introductory chapter, subsequent chapters are sequenced to follow the change in material property/design priorities as one moves from those components that are exposed to the neutron irradiation field, to those outside the reactor vessel whose performance is dominated by the chemistry of the reactor coolants. A detailed description of each major component and materials of its construction is provided, reliability limiting problems are described, and recent research into understanding the material property/design aspects of the problem(s) is summarized against the background established in Chapter 1. If remedies have been proposed, each chapter generally closes with a status report on their evaluation.

Chapters 2 and 3 deal, respectively, with LWR and LMFBR core materials, and the effects of radiation damage on material properties and design considerations are emphasized. In Chapter 4, the treatment of the primary pressure boundary components of the fission reactor reveals a much stronger interaction between design and property considerations due to the need to demonstrate absolute integrity of the pressure boundary over a 30- to 40-year lifetime. In an attempt to show what lies ahead for fusion reactor development, Chapter 5 uses the background on irradiation effects generated in Chapters 2 and 3, and the structural integrity analysis methodology described in Chapter 4 to evaluate the probability of various candidates for first-wall materials meeting the requirement of absolute structural integrity for periods of years. Chapters 6 and 7 treat components that operate essentially outside the neutron radiation field, but which have in common an environment of time-varying stress and aggressive chemicals. Chapter 6 describes corrosion-related problems (and in some instances their solutions) in the various heat exchangers, i.e., steam generator, intermediate heat exchanger and condenser, and Chapter 7 examines the state-of-the-art in steam turbine material performance. The final chapter, Chapter 8, explores the likely future trends in materials for nuclear power systems from a personal viewpoint, but, at the same time, using the earlier chapters as retrospective background. Appendixes are included for the reader's convenience, to: (a) convert SI units to conventional nuclear industry units; (b) define unfamiliar terms and jargon; and (c) provide nominal compositions of the many alloys used in the nuclear systems.

I am indebted to many people in the industry for their cooperation in providing me with reports, data, original figures, etc. In this regard special mention should be given to A. Boltax, R. Conn, C. Cox, J. Holmes, L. James, C. Spolaris, and K. Zwilsky. Several friends and colleagues gave freely of their time to review early draft chapters and provide valuable comments. I am grateful to F. Gelhaus, T. U. Marston, R. O. Meyer, L. A. Neimark, H. Ocken, G. Thomas, G. Trantina, R. Vaile, and B. J. Wrona for their efforts. A particular debt of gratitude is owed to Floyd Gelhaus for his assistance in structuring the book and his continual encouragement, and to Che Yu Li, who had the patience to thoroughly review the entire draft manuscript. Mrs. A. Hodges and Mrs. S. Simkin typed the several versions of the manuscript, and I thank them for their resilience. The last helper whom I want to acknowledge is my wife, Susan. Not only is Susan responsible for most of the artwork in this book, but her enthusiasm for the project, her ability to find the right word when I could not, and her continued encouragement over the last year really made this book possible.

Most of the book was written during a seven-month sabbatical leave at Cornell University. I am grateful to the Management of the Electric Power Research Institute for giving me time off to complete the project, and to Professor Ruoff, Director of the Materials Science Department at Cornell, for providing me with a visiting faculty position. Also, I acknowledge with thanks receipt of a grant from Battelle Memorial Institute that partially covered the cost of preparing this book.

Sunnyvale, California   J. T. Adrian Roberts

Contents

1. Introduction and Overview 1
   1.1. Reactor Systems and Materials 2
      1.1.1. Boiling Water Reactor 3
      1.1.2. Pressurized Water Reactor 6
      1.1.3. Liquid Metal Fast Breeder Reactor 7
      1.1.4. Fusion Reactor Systems 9
          1.1.4.1. Magnetic Confinement Concepts 10
          1.1.4.2. Inertial Confinement Concepts 11
   1.2. Past Performance of Nuclear Plants 13
   1.3. Materials and Design Considerations 16
      1.3.1. Traditional Design Approach 17
      1.3.2. Structural Integrity Analysis 19
      1.3.3. Methods of Analysis 21
          1.3.3.1. Stress Analysis 21
          1.3.3.2. Fracture Mechanics 25
      1.3.4. Materials Properties and Phenomena 29
          1.3.4.1. Plastic Flow Properties 29
          1.3.4.2. Fracture Toughness 30
          1.3.4.3. Fatigue 33
          1.3.4.4. Environmental Fatigue 36
          1.3.4.5. Radiation Effects 38
          1.3.4.6. Corrosion 45
   References 48

2. LWR Core Materials 53
   2.1. Fuel and Core Designs 54
   2.2. Fuel Performance 60
      2.2.1. Fuel Reliability 60
          2.2.1.1. Pellet–Cladding Interaction 64
              2.2.1.1a. Results of Experimental and Theoretical Studies 65

     2.2.1.1b. PCI Failure Models and Criteria 77
     2.2.1.1c. PCI Remedies 82
   2.2.1.2. Waterside Corrosion of Zircaloys 85
     2.2.1.2a. Corrosion in BWRs 86
     2.2.1.2b. Corrosion in PWRs 91
  2.2.2. Fuel Operating Margins (or Fuel Safety Considerations) .. 93
   2.2.2.1. LOCA Criteria 95
     2.2.2.1a. Zircaloy Embrittlement Criteria 96
     2.2.2.1b. Fuel Pellet Densification 98
     2.2.2.1c. Fission Gas Release 100
   2.2.2.2. Heat Flux Limits 105
     2.2.2.2a. Departure from Nucleate Boiling 106
     2.2.2.2b. Fuel Rod Bow 107
   2.2.2.3. Core Damage Assessment in Three Mile Island
      Accident 109
 2.3. Plutonium Recycle Fuel Performance 112
 2.4. Stainless Steel–UO$_2$ Fuel Experience 114
 2.5. Control Materials 119
  2.5.1. Control Rod Materials 119
  2.5.2. Burnable Poisons 122
 2.6. Uranium Conservation Measures 124
 References 129

3. LMFBR Core Materials 137
 3.1. Fuel and Core Designs 140
 3.2. Performance of Current Mixed Oxide Fuel Designs 143
  3.2.1. Mixed Oxide Fuel Behavior 145
  3.2.2. Cladding and Duct Mechanical Behavior 150
  3.2.3. Fuel–Cladding Interactions 158
  3.2.4. External Cladding Corrosion 162
 3.3. Advanced Oxide Fuel Development 164
  3.3.1. Optimization of Mixed Oxide Fuels 165
  3.3.2. Optimization of Cladding and Duct Materials 167
 3.4. Advanced Fuel Development 174
  3.4.1. Carbide Fuel Development 174
  3.4.2. Nitride Fuel Development 179
 3.5. Control Rod Material Development 180
 3.6. Proliferation-Resistant Fuel Cycles 185
  3.6.1. Modified Mixed Oxide Fuel Cycles 186
  3.6.2. Alternative Fuel Cycles 187
   3.6.2.1. Thorium-Based Ceramic Fuels 187
   3.6.2.2. Metal Alloy Fuels 190
 References 192

4. Fission Reactor Pressure Boundary Materials 199
 4.1. Design and Materials of Construction 200
  4.1.1. LWR Vessel and Piping 201
  4.1.2. LMFBR Vessel and Piping 204
 4.2. Developments in Fracture Mechanics 205
  4.2.1. Elastic Fracture 206
  4.2.2. Elastic-Plastic and Fully Plastic Fracture 211

4.3. Material Characteristics 214
    4.3.1. LWR Materials 216
        4.3.1.1. Fracture Toughness 217
        4.3.1.2. Radiation Embrittlement 221
        4.3.1.3. Fatigue Crack Growth 226
        4.3.1.4. Stress-Corrosion Cracking 231
            4.3.1.4a. Stress-Corrosion Cracking of Piping ... 231
            4.3.1.4b. Stress-Corrosion Cracking in Nozzle
                    Safe-Ends 243
        4.3.1.5. Closing Remarks 243
    4.3.2. LMFBR Materials 244
        4.3.2.1. Elevated-Temperature Low-Cycle Fatigue 246
        4.3.2.2. Fatigue Crack Growth 255
        4.3.2.3. Closing Remarks 263
4.4. Materials Improvements 264
    4.4.1. Improvements to Vessel Steels 265
    4.4.2. Remedies for BWR Pipe Cracking 266
        4.4.2.1. Modifications to Type 304 Stainless Steel .. 266
        4.4.2.2. Alternative Materials 268
    4.4.3. Improvements to Castings 271
References 272

5. Fusion First-Wall/Blanket Materials 279

5.1. First-Wall/Blanket Designs 281
    5.1.1. Magnetic Fusion 281
    5.1.2. Inertial Confinement Fusion 283
5.2. Materials and Structural Integrity Considerations 285
    5.2.1. Candidate Metals and Alloys 287
        5.2.1.1. Helium Embrittlement 288
        5.2.1.2. Swelling Behavior 293
        5.2.1.3. Fatigue Behavior 295
        5.2.1.4. Closing Remarks 302
    5.2.2. Candidate Ceramics 304
        5.2.2.1. Brittle Fracture Characteristics 306
        5.2.2.2. Erosion Characteristics 311
        5.2.2.3. Irradiation Effects 313
        5.2.2.4. Closing Remarks 316
References 317

6. Heat Exchanger Materials 321

6.1. Design and Materials of Construction 321
    6.1.1. PWR Steam Generator 323
    6.1.2. LMFBR Heat Exchangers 324
        6.1.2.1. Steam Generator 325
        6.1.2.2. Intermediate Heat Exchanger 328
    6.1.3. CTR Steam Generator 330
    6.1.4. Condenser 332
6.2. PWR Steam Generator Experience 332
    6.2.1. Corrosion Damage Processes 335
        6.2.1.1. Denting 337

6.2.1.2. Stress-Corrosion Cracking 340
6.2.1.3. "Phosphate Thinning" 343
6.2.2. Vibration and Mechanical Problems 344
6.2.2.1. Fretting and Wear 344
6.2.2.2. Fatigue Cracking 344
6.2.3. Avoidance or Mitigation of Steam Generator Damage 345
6.2.3.1. Chemical Control 345
6.2.3.2. Design and Fabrication Changes 348
6.2.3.3. Modified or Alternative Materials 349
6.2.3.3a. Inconel 600 349
6.2.3.3b. Alloy 800 351
6.2.3.3c. Inconel 690 353
6.2.3.3d. SCR-3 Alloy 354
6.2.3.3e. Ferritic (Martensitic) Stainless Steels .. 354
6.3. LMFBR Steam Generator and IHX Development 356
6.3.1. Steam Generator Materials Properties 358
6.3.1.1. Sodium Corrosion 360
6.3.1.2. Water–Steam Corrosion Performance 368
6.3.1.3. Tube Wastage by Sodium–Water Reactions 373
6.3.2. IHX Materials Properties 375
6.3.3. Closing Remarks 379
6.4. Condenser Experience 381
6.4.1. Copper-Base Alloys 384
6.4.1.1. Admiralty Brass 385
6.4.1.2. 90–10 Copper–Nickel (CL and CA706) 385
6.4.1.3. 70–30 Copper–Nickel (CN and CA715) 387
6.4.1.4. Aluminum Brass and Bronze 387
6.4.1.5. Chromium-Modified Copper–Nickel Alloys
(IN-838 or CA-722)..................... 388
6.4.2. Iron–Chromium Base Alloys 391
6.4.3. Titanium-Base Alloys 393
6.4.4. Closing Remarks 397
References 398

7. Steam Turbine Materials 405

7.1. Design and Materials of Construction 407
7.1.1. Major Components 407
7.1.1.1. Rotors 407
7.1.1.2. (Shrunk-on) Discs 410
7.1.1.3. Blades 410
7.1.2. Material Characteristics 412
7.2. Turbine Damage Mechanisms 414
7.2.1. Stress-Corrosion Cracking 416
7.2.2. Thermal-Mechanical Crack Growth 422
7.2.3. Moisture Erosion 431
7.3. Improvements in Turbine Materials 432
7.3.1. Low-Alloy Ferritic Steels 433
7.3.1.1. Impurity Reductions 433
7.3.1.2. Elimination of Temper Embrittlement 438
7.3.2. Alternative Blade Materials 441
References 448

8. Future Trends in Nuclear Materials 451
  8.1. Overview of the Materials Problems 452
  8.2. Specific Materials Developments 453
    8.2.1. Clean (Low-Alloy) Steels 453
    8.2.2. "Stainless" Steels 457
    8.2.3. Nickel-Base Alloys 460
    8.2.4. New Materials—Titanium 461
    8.2.5. Barriers and Coatings 463
      8.2.5.1. Fuel Rod Cladding Barriers 463
      8.2.5.2. Fusion First-Wall Coatings 464
      8.2.5.3. Tritium Barriers 465
  8.3. Related Technologies 465
  8.4. Closing Remarks 467
  References 467

Appendixes 469
Appendix A 470
Appendix B 471
Appendix C 476

INDEX 481

1

Introduction and Overview

Materials degradation in service represents one of the major technological factors that can limit the efficiency and viability of nuclear power. Extensive experience with commercial thermal reactors has demonstrated the need for improved understanding of materials phenomena (principally related to corrosion and irradiation) and better analytical procedures for transferring test information to the real problem. While fast breeder reactor experience is limited, it is already evident that currently used materials may not be sufficient for a completely economical breeder economy, and so advanced materials are being investigated for long-term application. The materials problems are much more diverse and severe for fusion reactors than for fission reactors, and no material considered to date possesses all the characteristics required for long-term operation in the fusion environment. Clearly, without intensive long-range development programs in this area, fusion may never transcend the engineering feasibility stage.

This book has its origins in the research and development work being undertaken worldwide by industry and government organizations to improve the nuclear option through either quantification of material performance or development of alternative materials. To keep the book to a reasonable length, some selectivity has been exercised. First, although aspects of both fission and fusion nuclear power will be treated, the former will be emphasized inasmuch as it is a commercial reality and trends are more readily apparent; only one chapter is devoted to fusion materials, primarily to illustrate the transfer of technology from the fission experience base. Second, of the different thermal and fast breeder fission reactors currently operating,† the light water reactor (LWR) and the liquid metal fast breeder reactor (LMFBR) are selected for discussion because the overall trend worldwide seems to be in the direction of the LWR–LMFBR mix. Third, and perhaps most important, attention is focused on certain component problem areas and the associated materials damage processes: For instance the prestressed concrete containment and certain reactor core internals, although important components, are excluded from discussion.

By way of introduction, this chapter introduces important materials design considerations, both analysis methodology and properties, that will be referred to frequently throughout the book. To place this discussion in perspective, the first two sections provide a brief background on the power generation systems being considered, the materials utilized in the major components of the systems, and the effect that these materials and their performance have on overall plant performance and availability.

1.1. Reactor Systems and Materials

All current and planned nuclear reactor designs use the steam-cycle conversion system. Thus, as in fossil fuel power plants, steam (produced in this instance by nuclear heat) drives a conventional turbine generator to produce electricity. Thermal efficiency of the LWR is ~30% while that expected of the LMFBR is higher, in the range of 40%. The first-generation fusion (CTR)‡ systems are also expected to be in the 30 to 40% efficiency range. It may be noted, in passing, that one of the basic limits on thermal efficiency is the capability of materials to operate reliably in high-temperature environments.

Approaches taken to produce good-quality steam (i.e., steam containing <0.2% condensed water at the turbine inlet) differ with the reactor concept. The *direct cycle* boiling water reactor (BWR) produces steam in the reactor core, like a fossil-fueled boiler, but the *dual cycle* pressurized water reactor (PWR) utilizes a steam generator. On the other hand, the LMFBR (and probably the CTR systems) will employ both an intermediate heat exchanger (IHX) and a steam generator.

In the subsections that follow, a *simplified* flow diagram is presented for each reactor system, the essential features of reactor operation are described, and the *primary* materials of construction for *key* components

---

† The Canadian heavy water reactor (CANDU) and the UK gas reactor (AGR) systems currently provide a significant fraction of the world's nuclear electricity. Also, the high-temperature gas reactor (HTGR) and gas-cooled fast reactor (GCFR) are in advanced stages of development.
‡ Controlled thermonuclear reactor.

are introduced. These components and materials are also summarized in Table 1-1,† which is referred to frequently in subsequent chapters. It is important to note that the materials included in this table are, naturally, in use in BWRs and PWRs, are either planned or in use in LMFBRs, but are only under evaluation for CTR concepts. Also, only one or two materials are identified for each component inasmuch as these have been identified as having the biggest impact on component reliability. Nuclear components are, however, typically comprised of many different materials to ensure compatibility with both the environment and the adjoining component, and most of these meet or exceed design expectations throughout their lifetime.

1.1.1. Boiling Water Reactor

The direct-cycle BWR system, outlined in Fig. 1-1, is a steam-generating system that consists basically of a metal-alloy–ceramic nuclear core assembled within a low-alloy-steel pressure vessel, and necessary auxiliary systems for safeguards, control, and instrumentation. For plants in the range of 1000 MWe the vessel is about 22 m high and 6 m in diameter and weighs about 400,000 kg. Water is circulated through the reactor core, producing saturated steam which is separated from recir-

† Alloy compositions are given in Appendix C.

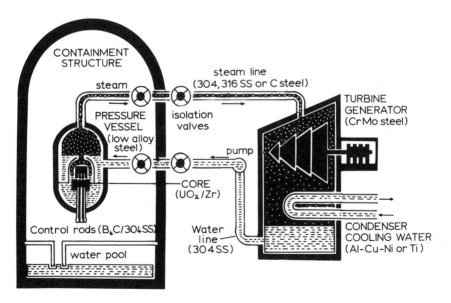

Fig. 1-1. Boiling water reactor (BWR).

Table 1-1. Major Components and Materials in Nuclear Systems[a]

Reactor component	BWR	PWR	LMFBR	Tokamak[c]	Inertial confinement[c]
Core					
Fuel assembly	UO_2/Zircaloy 2, 4	UO_2/Zircaloy 4	UO_2, 25 ± 5% PuO_2/ Type 316 SS[a]	D–T	D–T
Control materials	B_4C/Type 304 SS UO_2–Gd_2O_3/Zircaloy 2	AgInCd alloy B_4C–Al_2O_3; borosilicate glass	B_4C/Type 316 SS	NA[b]	NA
Breeding blanket	NA	NA	UO_2/Type 316 SS	Li/Li-containing material	Li/Li-containing material
Reactor Vessel					
Shell	Low-alloy steel (SA533 Gr. B)	Low-alloy steels (SA533 Gr. B, SA508)	Type 304 SS	NA	NA
Cladding	Type 308L SS (SA264)[e]	Type 308 SS; Inconel 617	NA	NA	NA
First wall/blanket	NA	NA	NA	Type 304, 316, 347 SS; 9–12Cr, 1–2Mo steels Nickel alloys (PE 16, Inconel 625 Titanium alloys, refractory metals (Mo, Nb, V, or alloys), or ceramics—C, SiC	Assume same as LMFBR
Steam generator					
Shell	Low-alloy steel (SA533 Gr. B)	Low-alloy steel (SA533 Gr. B)	2¼Cr–1Mo steel (SA 336)		
Support plates	NA	Carbon steel (SA515 Gr. 60)	2¼Cr–1Mo steel		

Component			
Tubes	NA	Inconel 600; Incoloy 800 (SB407, 513)[e]	2¼Cr–1Mo and 9Cr–1Mo steels (SA 213), Incoloy 800 (SB 407)
Intermediate Heat Exchanger (IHX)			
Piping	Types 304 SS or 316L SS (SA312, 358, 376, 403) or carbon steel (SA 106 Grade B)	NA	2¼Cr–1Mo steel or Type 304, 316 SS
Valves	Type 304 SS (SA351, 182); Stellite	Type 304 or 316 SS	Types 304 or 316 SS or 2¼Cr–1Mo steel
Pumps	Type 304 SS (SA351); Stellite	Type 304 SS; Stellite	Type 304 or 316 SS
Condenser			
Tubes	Al–Bronze; Al–brass (SB261); Cupronickel (SB111, 251); titanium (SB338); Type 304 SS (SA249)	Al–Bronze; Al–brass; Cupronickel; titanium; Type 304 SS	Type 304 SS or 2¼Cr–1Mo steel
Turbine Generator		Titanium	
Rotor		Cr-Mo-V or Cr-Ni-Mo steels (A469, 470, 471)[f]	
Disc		Type 403 SS;	
Blade		Ti-6Al-4V	

[a] SS = stainless steel.
[b] NA = not applicable.
[c] Candidate materials.
[d] Alloy compositions are given in Appendix C.
[e] SA and SB numbers are examples of specifications from ASME Boiler and Pressure Vessel Code and refer to both BWR and PWR.
[f] ASTM designations.

culation water, dried in the top of the vessel, and directed to the turbine-generator through large steel steam lines. The turbine has one high-pressure stage and, generally, two low-pressure stages, and employs a conventional regenerative cycle with condenser deaeration and condensate demineralization. Turbine materials are generally Cr–Mo based steels. Condenser materials vary widely depending on the source of cooling water. Typical materials include aluminum–bronze, stainless steel, cupronickel, and, more recently, titanium.

The reactor core, located in the lower portion of the vessel, is the source of nuclear heat. The core consists of cylindrical Zircaloy-clad uranium dioxide (UO_2) fuel rods, 12.5 mm diameter and 4 m long, and cylindrical Type 304 stainless steel clad boron carbide (B_4C) control rods which are contained in a stainless steel "cruciform" control blade. A modern 1220-MWe BWR core consists of 748 fuel assemblies, each containing 63 fuel rods in a Zircaloy channel, and 177 control blades, which form a core array about 5 m in diameter and 4 m high. Power level is maintained or adjusted primarily by positioning the bottom-entry control blades up and down within the core. Further adjustments to core power level can be made by changing the coolant recirculation flow rate (i.e., the water also serves as the neutron moderator) without changing control rod position. The BWR operates at constant pressure and maintains constant steam pressure, similar to most fossil boilers. Steam pressure is about 7 MPa and the steam produced is of high quality.

1.1.2. Pressurized Water Reactor

A PWR plant consists of two separate water systems, the primary and secondary, that meet in the steam generator (Fig. 1-2). The steam generator is a heat exchanger consisting of about 3000 nickel-based alloy steel tubes supported by carbon steel plates and contained in a steel shell. It is a large component, being about 20 m high by 4 m wide and weighing about 310,000 kg. PWRs may have two, three, or four coolant circuits, each with its own steam generator and one or two circulation pumps (called two, three, and four "loop" plants, respectively).

The PWR primary system, therefore, operates at conditions under which the water passing through the reactor does not boil. Pressure in the low-alloy-steel reactor vessel (12 m high by 4 m wide) and steel piping loop connected to it is about 15.5 MPa, which permits the water to be heated to 590 K without boiling. The heated water passes through the tubes of the steam generator, transferring its heat to the secondary side to produce steam at 560 K and 7 MPa pressure, which drives the turbine. Turbine and condenser design and materials are the same as for the BWR.

The core of a modern 1000 MWe PWR consists of about 190 Zircaloy-

Introduction and Overview 7

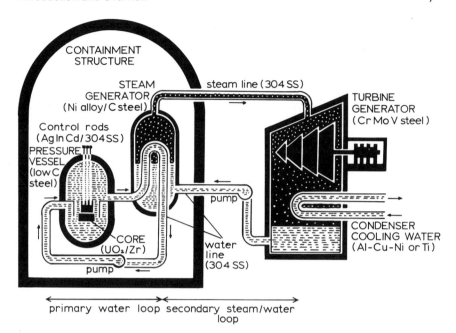

Fig. 1-2. Pressurized Water Reactor (PWR).

clad UO_2 fuel assemblies, 3.7 m in length, arranged in a 3.4 m equivalent diameter. The PWR contains about the same weight of fuel and cladding as the BWR. However, the number of assemblies is one-third that of the BWR because the number of rods per assembly is greater (i.e., 14 × 14 to 17 × 17 square fuel-rod arrays versus 7 × 7 or 8 × 8 fuel-rod arrays in the BWR). The fuel rods are also smaller in diameter than those of the BWR, being only 9.7 mm. Reactor power control is provided by approximately 50 top-entry AgInCd alloy or B_4C control rod cluster assemblies, each having 20 control rods, and a soluble chemical, boric acid, in the primary coolant. The control rods provide rapid control for startup and shutdown, while the concentration of boric acid is varied to control long-term reactivity changes, such as fuel depletion and fission product buildup.

1.1.3. Liquid Metal Fast Breeder Reactor

As the name suggests, the LMFBR utilizes a liquid-metal coolant, sodium, to transfer heat from the reactor core. Since this sodium becomes radioactive, an intermediate heat-transfer loop is employed between the reactor coolant system and the turbine water–steam system to eliminate any possible leakage of this radioactivity to the environment (Fig. 1-3). Both the primary reactor coolant system and the intermediate loop use sodium. A large, 1000 MWe LMFBR could have as many as six intermediate heat-transfer system (IHTS) loops, each with an intermediate heat exchanger (IHX) and steam generator. Both components are large: the IHXs can be more than 16 m high and 2.6 m wide, while the steam generators can be as much as 20 m high and 1.3 m wide. Each contains hundreds of heat-transfer tubes.

In operation, the liquid sodium in the reactor is heated to about 800 K, then flows to the IHX, where it transfers its heat to the sodium of the intermediate loop. The sodium in the IHTS passes to the steam generator where it heats water to produce steam at about 750 K. The turbine–condenser system uses LWR and fossil technology. Austenitic stainless steels (Types 304 and 316), high-nickel-based alloys (e.g., Alloy 800), and/or $2\frac{1}{4}$Cr–1Mo steel are used in IHX and steam generator construction. The piping is either Type 304 or Type 316 stainless steel.

The core of an LMFBR, contained in a large (17 m high by 6 m wide) austenitic stainless steel vessel or tank weighing nearly 500,000 kg, is essentially composed of two zones: the "active" core which, in a 1000 MWe plant, can contain about 100,000 Type 316 stainless steel clad mixed uranium–plutonium oxide fuel rods (6 mm diameter by 2.8 m long), 271 rods per assembly; and the "breeding blanket" region which surrounds the core and contains the fertile depleted UO_2 pellets. Reactor control is provided by top-entry, Type 316 stainless steel clad B_4C control rods.

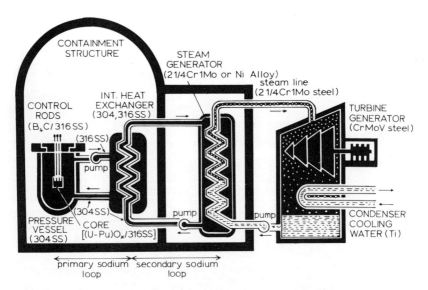

Fig. 1-3. Liquid metal fast breeder reactor (LMFBR).

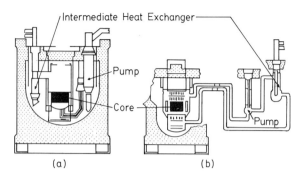

Fig. 1-4. Comparison of pool (a) and loop (b) LMFBR designs.

The LMFBR core is considerably smaller than that of the LWR, being of the order of 1 m high; but with the plenum region, end fittings, and upper and lower blanket regions the overall height is increased to ~3.0 m.

Two LMFBR concepts are being developed; Fig. 1-3 shows the conventional loop-type; but pool-type LMFBRs, in which the reactor vessel also contains the IHXs and primary sodium pumps, are also being constructed. Figure 1-4 compares the two concepts. The loop approach has been adopted by the U.S., Japan, and Germany for demonstration plants, whereas the pool concept has been selected by the U.K., France, and Russia. Both design concepts have a number of advantages and disadvantages which roughly balance each other. Only future experience with operation and licensing of large LMFBRs can show the more favorable design concept.

1.1.4. Fusion Reactor Systems

The most likely fusion reaction that will be exploited commercially in the future is that between deuterium and tritium, the so-called D–T reaction. Discussion of the emerging CTR Systems is therefore restricted to those employing the D–T plasma. Plasma confinement is accomplished by one of two methods: magnetic or inertial. The former, and specifically the magnetic tokamak reactor, is the most advanced. In the latter concept, electron beams, lasers, or ion beams are in the early stages of development. The total neutron flux to which materials are subjected is greater in the fusion reactor than in the typical fission reactor; more importantly, the spectrum is very different, with a much larger proportion of high-energy neutrons (14 MeV versus 1 MeV or less). The concern for material durability under such extreme radiation conditions is reflected in the CTR conceptual designs, which are quite different from the fission reactor.

1.1.4.1. Magnetic Confinement Concepts

The first and most dominant fusion approach in the world today is called a tokamak. A first-generation tokamak reactor is shown linked to a typical thermal cycle conversion system in Fig. 1-5. In the reactor, which is shaped like a torus or doughnut, the plasma is confined using the toroidal and transformer coils to produce the necessary magnetic fields, the plasma is heated by neutral beam injection, and fueling is accomplished via injection of D–T pellets. Operating characteristics of the many conceptual tokamak designs vary widely; plasma burn times range from 100 to nearly 6000 s and net electrical power ranges from 600 to 2000 MW.

The reactor, or torus, is both a vacuum chamber and heat transfer system. The "first wall," which surrounds the plasma and acts as the vacuum chamber, will most probably be constructed from a metal (Type 316 stainless steel, 9–13 Cr ferritic stainless steels, or titanium alloys appear to be the leading candidates) but will likely have a protective ceramic barrier (a low-atomic-number material such as graphite or silicon carbide) on the surface facing the plasma. Typical diameters are 5 to 10 m for low β† tokamaks and 2 to 4 m for high β reactors. The moderating breeding blanket serves several functions; it slows down the neutrons, provides a reflector to reduce the leakage of neutrons, provides a container for a coolant which carries the heat away, and contains a tritium breeding material so that the D–T reaction can be continued. One material that would satisfy three of these requirements is lithium (or lithium-containing material). The high-energy neutrons escaping the plasma enter the blanket, heat the lithium, and also react with the lithium to produce tritium which is removed to refuel the reactor. The heat deposited in the lithium is transferred to the intermediate coolant, sodium, via the IHX, and from this point on heat transfer is accomplished in the same manner as for the LMFBR, and component materials are expected to be the same (Fig. 1-5). However, other coolants such as other liquid metals or helium could be satisfactory. For example, a leading concept involves a molten-salt breeding blanket and helium cooling, which capitalizes on the molten salt-breeder and HTGR fission technology. Regardless of the design, a first wall plus a blanket of about 1 m depth is required to absorb about 97% of the energy from the plasma.

Other major components of the reactor include the magnet and personnel shield, usually consisting of iron, lead, and boron, to absorb gamma

---

† β is the ratio of the plasma particle pressure to the pressure due to the magnetic field and is, in effect, a measure of the efficiency of field utilization. For economic operation it is estimated that β would have to be larger than 10%, but current experimental machines come nowhere near this figure.

Introduction and Overview 11

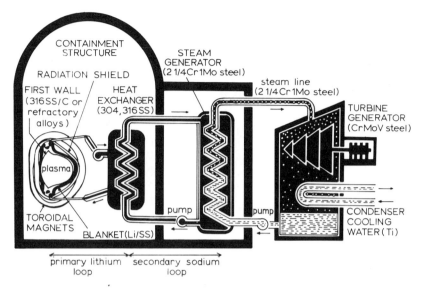

Fig. 1-5. Tokamak CTR and typical thermal cycle conversion.

rays and neutrons, and the superconducting magnets, which can be NbTi or Nb_3Sn alloys.

A fusion reactor as outlined above would have an extremely hot region in the interior (10^8 K in the plasma) and an extremely cold region on the exterior (4 K) at the magnets. Thus, thermal insulation will be one of the most difficult and challenging problems.

In closing, one should note the alternatives to the tokamak that are being pursued actively through experimental programs. These include the magnetic "mirror" reactor and the Theta Pinch reactor. Details on these and other magnetic confinement concepts are described in an EPRI† report.[1]

1.1.4.2. Inertial Confinement Concepts

Inertially confined fusion (ICF) systems characteristically have compact reactor cavities and employ laser or electron/ion beams to heat and compress the fuel pellet. The electron beam approach has fewer steps than the laser approach if a hot electron cloud can be produced directly. Since most work to date has been toward laser fusion, this device is chosen for this presentation of the unique features of ICF systems compared to the magnetically confined systems. The discussion is, nevertheless, general enough to cover all inertial confinement approaches.

† Electric Power Research Institute, Palo Alto, California, U.S.A.

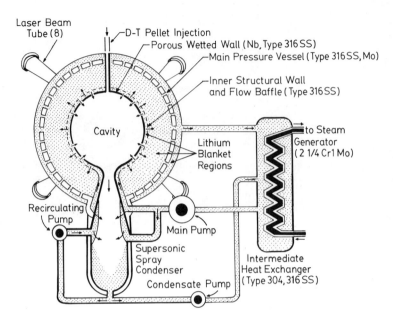

Fig. 1-6. Lithium wetted-wall laser-fusion concept.

A typical laser ICF reactor concept is shown in Fig. 1-6. In this example, the energy required for igniting the D–T fuel pellet is provided by eight laser beams focused on the center of the spherical reactor cavity. The basic reactor structure consists of two concentric spherical shells, of 1.5 to 3 m radius. Liquid lithium is contained in the region between the two shells. The inner shell is constructed from porous niobium metal or stainless steel, permitting the lithium to flow through the wall and form a thin film protecting the wall from pellet debris (i.e., the so-called "wetted wall" concept). Heat from the thermonuclear pellet reaction is deposited in the lithium film on the inner wall as well as in the lithium behind the inner wall. The latter serves to breed tritium and to transfer heat to the steam cycle system, as in the tokamak. Blanket structural materials might be austenitic stainless steel or ferritic stainless steel.

Laser- (or electron/ion beam-) driven fusion is presently less well demonstrated as a scientific possibility than is the magnetically confined fusion reactor. Nevertheless, a number of different design concepts are being actively pursued;[2] all but one (the SOLASE reactor) use liquid lithium for blanket cooling and tritium breeding, and the major distinction between them all is the method employed to protect the first wall from the short-range debris and x-rays. The combination of cyclic stresses with high-energy neutrons, charged particle, and x-ray pulses produces an environment never before faced by materials.

1.2. Past Performance of Nuclear Plants

The assessment of opportunities for improvement of the reliability of components and materials must necessarily start from an analysis of past plant performance.

Nuclear plant performance experience is limited to fission power plants, and to light water reactors in particular. At the end of 1979, the United States had 71 large commercial LWRs in operation with a combined output of over 50,000 MWe. Worldwide, a total of 137 LWRs are in operation, with the non-U.S. balance being primarily in Europe and Japan.

In 1979, electrical generation from this nuclear capacity was about 11% of all thermal generation in both the U.S. and Europe. With the additional nuclear plants firmly committed and under construction, this should rise to about 20% in 1985.

Liquid-metal-cooled breeder reactors are represented by four medium-scale prototypes totalling 1700 MWe, none of them in the U.S. The first LMFBR of truly commercial size, 1200 MWe, known as Super Phenix, is under construction in France. Germany, Japan, and Britain are also likely to move into commercialization soon with 800- to 1000-MWe units. In the U.S., LMFBR development has been interrupted by political decisions, and the status of the medium-sized prototype project, the Clinch River Breeder Reactor (CRBR), is uncertain at this time. The EBR II† continues to serve both as a test bed for fuel development and as an electrical power generating unit (~70% capacity factor over 10 years), and the FFTF‡ is in the startup stages. Also, the U.S. utility industry has completed a commercial LMFBR design study, but no commitments have been made for construction.[3]

Statistical performance data are available only for LWRs.[4, 5] Impressive gains in both plant capacity and plant availability§ have been made over the past few years. Figure 1-7 presents lifetime productivity data on U.S. LWRs collected through 1978,[6] which indicate an overall average capacity factor of 60.3% and an availability factor of 71.8%. However, the performance parameters over the past two years, July 1976 to June 1978, were 64.2% and 74.9%,[5] respectively, the improvement being due primarily to decreased forced outage rates as operating experience increases. Despite the plant productivity increases in recent years,

---

† Experimental Breeder Reactor II.
‡ Fast Flux Test Facility.
§ Availability factor (or plant availability) is the fraction of the time that the reactor is available to produce power. The capacity factor is then simply the availability factor multiplied by the output factor. In other words, the availability factor is always greater than the capacity factor, and the difference between the two is a measure of the total unit capacity lost from operating at less than full power.

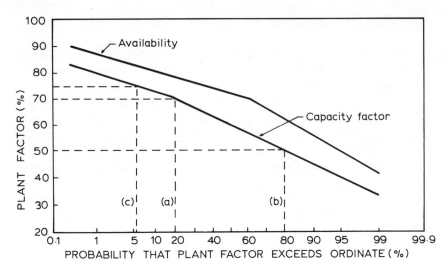

Fig. 1-7. Distributions of lifetime productivity for LWR nuclear plants (courtesy of J. James, Electric Power Research Institute).

there is still room for improvement. Although the upper half of the curves in Fig. 1-7 shows respectable plant performance [e.g., point (a) indicates that 20% of the U.S. nuclear plants exceeded 70% capacity factors], the lower half, particularly the lower 20% of the curve [point (b)], operates below 50% capacity factors, which indicates the need for corrective action. As a goal the 75–80% cumulative capacity factors now being achieved in some 5% of the plants [point (c)] in the long run should be achievable or exceeded in all plants. The concomitant savings in fossil fuel requirements are enormous (a 1% increase in capacity factor of a single nuclear plant is currently worth more than $1 million per year in differential replacement fossil fuel costs). Such savings emphasize that investments in improvement of the productivity of current and committed nuclear fission capacity probably represent the most highly leveraged expenditures in the energy field (on a cost/benefit basis) available to the world for at least the next decade.

Plant availability is impacted by all causes of forced and scheduled outage, including equipment failure, inspection, and repair time and refueling. A detailed analysis of plant data such as that in Fig. 1-7 reveals that "low-frequency, high-impact events" constitute the major sources of forced outage.[4, 5] In other words, materials failures in large key components such as the core (fuel-rod failures), primary piping (cracks), steam generator and condenser (tube leaks), and steam turbine (rotor or disc failures) do not occur often, but when they do, outages may extend to months. On the other hand, the regular inspection and repair activities associated with regulatory-based safety requirements, such as reactor pressure vessel inspections, appear to be of less consequence. The data in Table 1-2 taken from an EPRI report[4] indicate this result, and point to the components (and hence materials) that must receive detailed attention if plant performance is to be improved.

LMFBR experience, although limited, not unexpectedly shows some interesting similarities to LWRs. Numerous fuel failures were experienced in experimental fast flux reactors during development of the reference designs. However, recent experience with complete oxide-fueled cores is excellent. The 250-MWe French Phenix demonstration plant started full-power operation in 1973, and during its first year at full power it achieved greater than 80% availability, and no fuel failures were detected. Later, however, problems with the IHXs required shutdowns for repairs.[7] The British 250-MWe prototype, PFR, also showed good fuel performance but has been temporarily limited by steam generator problems. The Soviet program has included the 350-MWe BN-350, which has been operating since 1973, but it, too, has experienced steam generator problems.[7] Naturally, these problems, unless remedied, would also im-

Table 1-2. Representative Average Outage Durations and Partial Power Reductions in Nuclear Units through January 1, 1977 (Ref. 4)

Item	Duration (h/yr)	% Total
Forced outage (equipment malfunction)		
Turbine/generator	152	5.4
Condenser	88	3.2
Steam generator	131	4.6
Pumps	81	2.9
Valves	116	4.1
Vessel and core	91	3.2
Plant electrical distribution	77	2.7
All others	270	9.6
Subtotal	1006	35.7
Scheduled outage		
Maintenance	300	10.6
Refueling (including inspection and removal of failed fuel)	1360[a]	48.2
Training and administration	40	1.4
Subtotal	1700	60.2
Regulatory	115	4.1
Total	2821	100.0

[a] This figure represents hours for one refueling. Including it in this list, and in the totals, is tantamount to the tacit assumption that refueling occurs, on the average, on a 16-month cycle.

pact the performance of a CTR if it were based on a liquid-metal coolant system.

Research emphasis is obviously where the problems are: for example, in fuel rod reliability improvement, steam generator and condenser integrity improvement, and turbine reliability improvement. In addition, however, and also of great importance, is the continuing program to demonstrate the actual large margins of safety of present reactor vessel design.

1.3. Materials and Design Considerations

This section will provide introductory descriptions of current and developing design approaches and important material properties and damage processes that must be explicitly included in the design analyses before components can be considered truly reliable over the expected service life.

Examination of Table 1-1 and the accompanying diagrams, Figs. 1-1 to 1-6, reveals that nuclear systems employ a number of different materials that require joining to form an integral system. For instance, there are about 6000 welds just in the piping system of a modern BWR. Designers must account for the different properties of the component materials and welds, otherwise unacceptably high stresses or corrosion rates may well result. To assist the designer in this complex task, design rules have been developed specifically for nuclear components, and "nuclear grade" materials are fabricated to tighter specifications than, say, similar materials used in fossil plants or oil refineries. The safety record of the nuclear industry attests to the care and attention that goes into nuclear plant design, construction, and in-service inspection.

Nevertheless, as discussed above, uncertainties in either materials properties and/or the true in-service environment have led to unexpected material failures; consequently, improvements in both design analysis methods and materials are required. With respect to the latter, the improvements may take the form of just a better characterization of properties or, at the other extreme, replacement by an alternative material. The development of fission energy has already led to the evolution of new materials (e.g., the zirconium alloys known as Zircaloys) and the reoptimization of conventional materials (e.g., austenitic stainless steels) to meet the demands for reliability. This trend will no doubt continue, particularly as LMFBR operating experience grows. Fusion energy requirements are proving ever more demanding and unconventional material/design mixes are likely to be the order of the day when this energy source finally sees commercialization.

1.3.1. Traditional Design Approach

The American Society of Mechanical Engineers (ASME) Boiler and Pressure Vessel Code, Section III, Nuclear Power Plant Components,[8] sets forth rules for the design of the major nuclear components. Although enforceable only for the pressure-boundary components (i.e., reactor vessel, piping, steam generator, etc.) these rules are also used in the design of fuel rods.

The general conservative philosophy behind the code can best be illustrated by a direct quote:

> The potential failure modes and various stress categories are related to the Code provisions as follows:
> (a) The primary stress limits are intended to prevent plastic deformation and to provide a nominal factor of safety on the ductile burst pressure.
> (b) The primary plus secondary stress limits are intended to prevent excessive plastic deformation leading to incremental collapse and to validate the application of elastic analysis when performing the fatigue evaluation.
> (c) The peak stress limit is intended to prevent fatigue failure as a result of cyclic loading.
> (d) Special stress limits are provided for elastic and inelastic instability.
> Protection against brittle fracture is provided by material selection rather than by analysis. Protection against environmental conditions such as corrosion and radiation effects are the responsibility of the designer.

Except for the core region, materials in LWR components are designed to operate at temperatures in the subcreep regime ($<0.4T_m$, where T_m is the absolute melting point) and at stress levels that are, for the most part, below the yield stress. In some components (e.g., piping, steam generator) transient stresses, or local stresses at geometrical discontinuities (elbows, welds, etc.), might exceed yield for short times during the service life, which will result in small inelastic strains. However, stresses never attain the ultimate tensile strength of the materials. Therefore, fracture and fatigue below the general yield stress are the most important considerations in LWR pressure-boundary components and elastic stress analyses are generally applicable.

In contrast to the LWR, many of the LMFBR (and CTR) components will operate at temperatures and durations where metal behavior is time dependent. The ASME Code Case 1592[9] (and associated U.S. government standards such as RDT F9-ST) addresses high-temperature component design and establishes rules to prevent excessive inelastic damage (e.g., ductile rupture from short-term loadings, creep rupture from long-term loadings, creep-fatigue failure, and buckling and collapse). This extends Section III of the ASME Code to temperatures above 643 K for ferritic materials and above 698 K for austenitic materials.

Code Case 1592 provides time-dependent allowable stresses to limit the elastically calculated primary (load-controlled) stresses. These limits generally establish the preliminary structural design of a component, including the necessary wall thicknesses. In addition to the criteria for load-controlled stresses, the Case provides limits for deformation-controlled quantities. The principal of these are strain limits, to guard against excessive ratchetting deformation, and creep-fatigue criteria, to guard against excessive fatigue and creep-rupture damage.

Interestingly, Code Case 1592 currently contains stress (or deformation) limits, up to 1089 K, for only four alloys; Types 304 and 316 stainless, Alloy 800H, and $2\frac{1}{4}$Cr–1Mo. Inasmuch as CTR first-wall/blanket structures will likely operate in this regime, this situation will impose severe restrictions on the material selection process unless other materials become qualified soon.

Due to the high cost and the large number of man-hours and computer time required for rigorous inelastic analyses, extra efforts are made to allow demonstration of compliance with the Code design rules using only the less informative elastic analysis findings. These efforts involve (1) optimizing component configurations to reduce discontinuity and peak stresses, (2) removing welds from high stress regions, (3) selecting or changing material type for improved strength, (4) improving thermal-hydraulic analyses or system characteristics to remove undue conservatism in magnitudes of predicted thermal shocks and operating temperatures, and (5) developing simplified but justified conservative inelastic response by approximate analysis methods. Nevertheless, some components with envelope restrictions and/or with the more severe operating conditions have to be evaluated using rigorous inelastic analysis.

Every structure, no matter how carefully constructed, contains flaws. In metallic structures these flaws range in size from simple dislocations on the atomic scale, through slag inclusions and lack of penetration in welded joints, to large cracks originating during fabrication or from service-induced loading. In ceramic structures porosity and microcracks are the major flaws. Traditional design methods, embodied in the aforementioned Codes, do not explicitly account for the presence of such structural defects although a degree of control at the upper end of the size scale is exerted by exercising quality control [e.g., pass/fail nondestructive test methods (NDE)]. A safety factor applied to the analyses is used to account for the effects of flaws and any other uncertainties. Flaws that are encountered during the frequent in-service inspections, however, must be evaluated using procedures contained in Section XI of the ASME Boiler and Pressure Vessel Code.[10] The Section XI Code has also adopted a conservative approach. All observed indications (such as cracklike defects, slag inclusions, porosity, lack of weld fusion, laminations, and any

Introduction and Overview

combinations thereof) must be treated as planar cracks, irregularly shaped flaws are represented by idealized simple geometric shapes, and the flaw–structure interaction is assumed to be elastic and, therefore, analyzable by linear elastic fracture mechanics (LEFM). Clearly, these procedures cannot be applied to the assessment of defects in plants operating at high temperature.

It is also important to note that, in addition to ASME and RDT rules, there are, at least for the LWR, limits imposed on property degradation in service contained in the Code of Federal Regulations Related to Nuclear Power Plant Installation (i.e., 10CFR50) and in the Regulatory Guides issued by the U.S. Nuclear Regulatory Commission† (USNRC). The former is effectively the "law of the land" and must not be transgressed.

1.3.2. Structural Integrity Analysis

The traditional approach to design works very well for many engineering applications but can be improved by incorporating procedures for quantitative flaw tolerance assessment. This will be a key factor in improving component reliability and performance inasmuch as material flaws are unlikely to be eliminated entirely and, even if they were, the demanding environment of time-varying stress and thermal loadings, corrosive chemicals, and neutron flux may initiate new ones. Thus, we must live with flaws, and if the flaw assessment analysis is too conservative, time will be lost by unnecessary inspections and repairs; if it is nonconservative the component might fail unexpectedly in service, leading to long outages for replacement or even damage to other adjacent components.

Basic to quantitative flaw assessment is the Structural Integrity (SI) plan, schematically presented in simplified form in Fig. 1-8. The SI plan includes all elements that could lead to potential failure, including loading (either real or postulated), initial flaw size, material properties such as fracture toughness, fatigue life, and creep and plastic flow, and analytical techniques to determine flaw instability. Jones et al.[11] describe a general SI analysis as follows:

1. A flaw is found and sized by NDE techniques during a periodic inspection.
2. A cyclic crack growth analysis is performed to determine how much the flaw would grow by the end of the service life. The inputs to the analysis are the transient loading the component will undergo until the end of life, and potential crack growth mechanisms such as fatigue, corrosion fatigue, and stress corrosion cracking (creep would also be a consideration for high-temperature components in LMFBR or CTR, but these were not considered by Jones et al.).

---

† Other countries have similar regulatory bodies, but the laws tend to be very similar due to the international cooperation that exists.

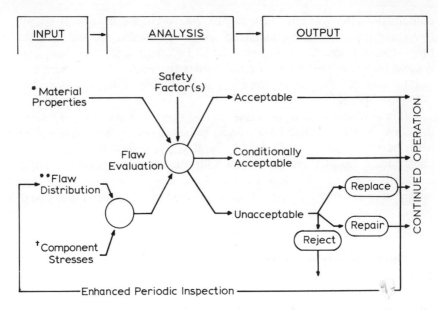

* Includes effect of service environment (irradiation, aging, corrosion, etc.)
** Includes expected crack growth due to fatigue, corrosion fatigue, etc.
† Includes normal and transient loadings

Fig. 1-8. A simple structural integrity analysis plan (courtesy T. U. Marston, Electric Power Research Institute).

3. Once the end-of-life flaw size is determined, a flaw evaluation analysis is performed to determine whether the structure is safe for continued operation or if it must be repaired. Inputs to this analysis include materials properties such as fracture or ductility characteristics; effects of environment such as radiation damage and corrosion; and the most severe accident loading condition that the structure is postulated to withstand during its lifetime.
4. If the structure cannot safely withstand the worst postulated accident condition, it is either repaired or replaced. If it can safely withstand this condition, it is returned to service and the flaw size is monitored by periodic inspections of the structure.

It is clear that the accuracy and reliability of the Structural Integrity or SI analysis will depend on the analytical procedures and material properties used in the analysis, and that any errors or inadequacies in either will be propagated through the analysis. A further consideration is the probabilistic nature of key inputs to the analysis such as inspection uncertainty, pre-inspection flaw distribution, duty cycle, stresses, and materials behavior.

The current Section XI procedure for LWRs represents a simple

Introduction and Overview

form of such a plan. Operating experience with the LWR pressure boundary is extensive and the incentives to improve the SI analysis are obvious. Accordingly a broad-based industry program has been established and this is discussed more fully in Chapter 4. A similar vigorous effort, though emphasizing a different set of concerns, has been mounted in the LMFBR program (also discussed in Chapter 4) and this approach has begun to be picked up by the CTR community (see Chapter 5).

1.3.3. Methods of Analysis

The currently used and developing analytical procedures are introduced here before the material properties inasmuch as the procedures have, to a large extent, determined which properties are emphasized. Some materials phenomena that are discussed later cannot yet be included in structural analysis due to limitations in analytical procedures; hence the application of safety factors in design.

There are two categories of structural analysis; in the first where defects are ignored, to satisfy the design codes three-dimensional stresses and strains are computed for startup, shutdown, and selected transient conditions. Elastic analyses, sometimes using finite-element techniques, are used wherever possible, because inelastic analyses are both expensive and time consuming. The second category of analysis is fracture mechanics which accounts for defects. Again, elastic analyses are emphasized (in fact LEFM is the only ASME Code approved fracture mechanics analysis) but elastic-plastic, plastic, and creep-fracture mechanics approaches are rapidly developing and may be included in the SI plan in the future.

1.3.3.1. Stress Analysis

Elastic stress analysis methods are widely understood and therefore will not be considered further. However, inelastic analysis is still a developing technology. Inelastic analysis methods have advanced in the last ten years to the extent that elastic-plastic-creep incremental load solutions to one-, two- and some three-dimensional structures are now available, and guidelines for inelastic analysis, material properties, and constitutive equations have been developed.[12-14]

Generally, when a combined time-independent elastic-plastic analysis and time-dependent creep analysis is required, which is capable of predicting stresses, total accumulated strains, and deformations as a function of time for specified load and temperature histories, the method of calculation is based on the assumption that the total strain at any instant of time consists of three parts: elastic, plastic, and creep. Figure 1-9 schematically represents the components in the form they are obtained experimentally, that is, tensile stress–strain data at constant extension (strain) rate and constant load (stress) creep data. In the analysis, small but discrete increments of time are considered in which elastic-plastic strains and creep strains are computed separately using simple constitutive laws shown in Fig. 1-9, and added to obtain the total strain, which is then accumulated during the expected operating lifetime.

The major drawback to this approach is that the laboratory data from which the constitutive laws are derived are obtained under conditions far removed from the actual service conditions, which typically involve temperatures and imposed loads that vary with time. Under such thermal and/or stress cycling conditions it is probable that creep and plasticity are not simply additive, even if they can be so distinctly separated (which is a controversial question in itself!). In any event, the situation has recently changed with the introduction of several models specifically designed to cover complex time-varying loading and thermal histories. Some are specific to phenomena, such as the creep-fatigue models that will be discussed later. Others, and two in particular are worthy of note, attempt

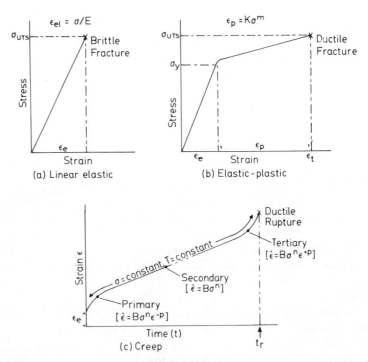

Fig. 1-9. Schematic representation of the different types of stress–strain relations (a,b), and the standard creep relation (c).

Table 1-3. Phenomena Modeled by MATMOD (Ref. 17)

1. Short-time plasticity, including
   (a) strain-rate sensitivity
   (b) temperature sensitivity
2. Long-time creep, including
   (a) primary creep
   (b) steady-state creep
   (c) proper variation of steady-state creep rate with stress
3. Cyclic stress–strain behavior, including
   (a) Bauschinger effect
   (b) cyclic hardening and cyclic softening
   (c) shakedown to a saturated condition of constant stress and strain amplitudes
4. Recovery
   (a) static recovery
   (b) dynamic recovery
5. Dynamic strain-aging effects, including
   (a) plateau in yield strength vs. temperature
   (b) negative strain-rate sensitivity
6. Complex histories
   (a) stress changes
   (b) strain-rate changes
   (c) temperature changes
7. Irradiation effects
   (a) irradiation hardening
   (b) irradiation-enhanced creep
   (c) channeling (strain softening)
   (d) swelling
8. Interactions of all of the above

the enormous task of providing a "unified" model for inelastic deformation.

The MATMOD† constitutive equations developed by Miller[15] attempt, with varying success, to model a wide variety of inelastic phenomena (Table 1-3) by blending physical mechanisms with phenomenology. Physical understanding of the controlling dislocation processes is used in the design of the overall format of the equations; for example, three structure variables, corresponding to three distinct types of obstacles to dislocation motion, are used. The specific algebraic expressions within the equations are, however, synthesized on a phenomenological basis.

At the center of the model is an equation for the thermal inelastic strain rate $\dot{\epsilon}_{th}$, which depends on current stress σ, on the current temperature T, and the three structure variables F_{def} (representing *isotropic* obstacles or deformation hardening), R (representing *directional* obstacles or kinematic hardening), and F_{sol} (representing *intrinsic, isotropic* obstacles such as solute strengthening). Thus

$$\dot{\epsilon}_{th} = B \exp(-Q^*/KT) \left\{ \sinh \left[\left(\frac{\sigma/E - R}{(F_{def} + F_{sol})^{1/2}} \right)^{1.5} \right] \right\}^n \quad (1\text{-}1)$$

where B and n are material constants, obtained by fitting to data, and Q^* is the activation energy which is allowed to vary with T. The quantities F_{def} and R, being history dependent, are governed by two differential

† *Materials Model.*

equations for F_{def} and R, respectively; both of these equations follow a work hardening–recovery format. The quantity F_{sol}, being independent of history, is governed by an equation which specifies F_{sol} as an instantaneous function of $\dot{\epsilon}_{th}$ and T. The various parameters and constants must be obtained from constant strain rate, constant stress (creep), and cyclic stress–strain (fatigue) tests.

The second model worthy of note is due to Hart[16] who, like Miller, employs basic dislocation theory to develop his constitutive law. The Hart "equation of state" model employs three basic elements; the $\dot{\alpha}$ element represents the barrier processes; the a element characterizes the dislocation pileups as a stored strain; and the $\dot{\epsilon}$ element represents the dislocation glide friction. The following constitutive equations result:

$$\ln(\sigma^*/\sigma_a) = (\dot{\epsilon}^*/\alpha)^\lambda \qquad (1\text{-}2)$$

$$\dot{\epsilon}^* = (\sigma^*/G)^m f \exp(-Q/KT) \qquad (1\text{-}3)$$

$$\dot{\sigma}^* = \dot{\alpha}\sigma\Gamma(\sigma^*, \sigma_a) - \sigma^*\mathcal{R}(T) \qquad (1\text{-}4)$$

$$\sigma_a = \mathcal{M}^a \qquad (1\text{-}5)$$

$$\dot{\epsilon} = \dot{a}^*(T)(\sigma_f/\mathcal{M})^M \qquad (1\text{-}6)$$

In these equations $\mathcal{M}$, M, m, λ, f, and Q are materials parameters to be measured for each material. Except for parameter f, they have been found to be insensitive to thermally induced structural changes. The quantity G is the shear modulus, and KT has the usual meaning. The parameters a^* (known as the "hardness"), Γ, and $\mathcal{R}$ are also determined experimentally. The latter two functions provide strain hardening and static recovery, respectively. These functions as well as $\dot{\epsilon}^*$ are sensitive to thermally induced structural changes. The particular advantage of the Hart approach is that all parameters may be measured from tensile and load relaxation tests conducted over short periods at various temperatures, thereby eliminating the need for lengthy creep and fatigue testing.

Both the Miller and Hart models represent a much more complex state of affairs than currently exists in Code-accepted inelastic analysis. Nevertheless, both models have been generalized to three dimensions and have been incorporated into conventional finite element computer codes.[17, 18] Further, the appropriate material constants have been obtained for Zircaloy[17, 19] and Types 304[15, 18] and 316[19] stainless steel over the temperature, strain, and strain-rate regimes of interest to fission reactors. Thus, in principle a more rigorous computation of the inelastic strains developed in nuclear components can be made. It will be interesting to see whether designers will pick up these new developments.

1.3.3.2. Fracture Mechanics

Currently the only method approved (by ASME Code and federal regulations dealing with nuclear power plants) for quantitative evaluation of the critical conditions for component failure to occur is linear elastic fracture mechanics (LEFM). In LEFM, all deformation is assumed to be elastic and structural instability is usually assumed to occur when the stress intensity factor K due to the applied load exceeds a critical value at which crack growth initiates. These assumptions are appropriate for perhaps 60 to 70% of all engineering situations of practical importance, and, accordingly, LEFM-based analyses are expected to gain increasingly wide acceptance by Code and regulatory bodies. In an additional 20% of the cases of interest, extensive plastic deformation occurs prior to structural instability. LEFM predictions can be excessively conservative in such cases and a new branch of fracture mechanics, elastic-plastic and general-yielding fracture mechanics, is being developed to provide an accurate predictive capability for ductile fracture.[20a,b,c] The remaining 10 to 20% of structural analysis problems involve time-dependent deformation and creep fracture processes. This is a much less well-developed area of fracture mechanics.

A detailed and complete discussion of the principles of fracture mechanics can be found in the work of Knott,[21] and recent developments are further discussed by other authors.[11, 20, 22–25] The presentation here, therefore, is only a brief summary of some major aspects of fracture mechanics.

To reiterate, linear elastic fracture mechanics (LEFM) considers only the linear elastic response of the material. Thus, for both loading and unloading, the stresses and strains are related linearly, as shown in Fig. 1-9(a). The elastic solutions for cracked bodies contain a stress field close to and surrounding the crack tip that lends itself to a one-parameter description. This field, identified as the K-field, is shown schematically in Fig. 1-10a, taken from Ref. 25. Under plane strain loading conditions, the elastic stresses and strains within this field are completely prescribed by the parameter K_I, the appropriate stress intensity factor. By definition, K_I is related to the applied stress σ and the crack size a by a relation of the form

$$K_I = Y\sigma(\pi a)^{1/2} \qquad (1\text{-}7)$$

where Y is a function of the component geometry and the way the component is loaded.

The ability of the material to support a K-field of a given intensity without crack extension is characterized by the critical value of K_I, which

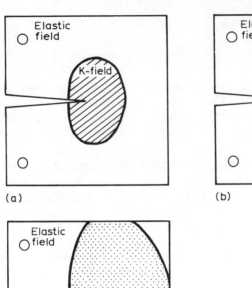
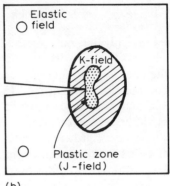
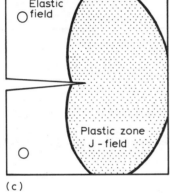

Fig. 1-10. Schematic representation of the stress fields of cracked bodies: (a) linear elastic behavior; (b) linear elastic behavior with small-scale yielding; (c) large-scale yielding (adapted from Ref. 25).

is denoted by a material property called the fracture toughness, K_{Ic}. Specifically, K_{Ic} is the value of K_I corresponding to the onset of crack extension in specimen thicknesses sufficient to give plane strain conditions. For thin structures where plain stress conditions occur, the appropriate toughness is designated K_c, and $K_c > K_{Ic}$.

The one-parameter description of the crack tip stress field provided by K has proved to be useful in characterizing subcritical crack growth processes as well as instability. For many materials in many environments it has been found that the rate of fatigue crack growth, da/dN, can be related over a wide range to the mode I stress intensity range ΔK_I by a simple power function of the type

$$da/dN = C(\Delta K_I)^n \qquad (1\text{-}8)$$

where the constants C and n depend on both the material and the environment.[26] Equation (1-8) has been used as the basis of reference curves for predicting fatigue crack growth below creep temperature in both LWR vessels[27] and LMFBR piping.[28] The extension of the power law to crack propagation at elevated temperatures has also been shown to be valid under cyclic loading conditions, presumably because the plastic zone remains small.[29] Also, the form of the da/dN vs. ΔK_I curve has provided useful insight into the mechanism of crack extension under corrosion fatigue conditions,[30] and it has been found that stress corrosion crack growth is often controlled by K_I rather than by the nominal applied stress.[31] For instance, Novak and Rolfe[31] showed that crack growth due to stress corrosion would not occur under linear elastic, plane strain conditions unless K_I exceeded a threshold value termed K_{Iscc}.

Thus, when the stress distribution ahead of the crack can be accurately characterized by K, LEFM can provide the basis for both the crack growth analysis and the flaw evaluation analysis steps in an SI plan of the kind shown in Fig. 1-8.

The elastic analysis must, however, be reconciled with the existence of a plastic zone, which is always present at the tip of a sharp crack, where the stresses exceed σ_y, the yield stress of the elastic-plastic stress–strain curve. The size of this zone, which is shown schematically in Fig. 1-10b, is related to σ, the normal net section stress at the onset of crack extension. As long as σ is less than about $0.6\sigma_y$, the plastic zone is smaller than the K-field. Under these conditions K_I describes the stresses and strains in the region adjacent to the plastic zone; i.e., K_I dominates the solution for the plastic zone. Consequently, the LEFM fracture toughness parameter K_{Ic} retains its significance as a geometry-independent material property and the LEFM criterion for crack extension, $K_I \geq K_{Ic}$, is valid.

When the material is tough and σ exceeds $0.6\sigma_y$, the plastic zone existing at the moment of crack extension is larger than the K-field. The LEFM parameter K_I then ceases to characterize either the elastic or the plastic zone. This case is shown schematically in Fig. 1-10c.

The J integral[21, 25] was introduced to account for a significant plastic zone at the crack tip. It is based on the nonlinear stress–strain curve shown in Fig. 1-9b and serves as a single-parameter description of stress and strain near the tip of the crack in the plastic zone.† This region is identified as the J-field in Figs. 1-10b and c. The J-integral value at the onset of crack extension provides a fracture toughness parameter, J_{Ic}, in the elastic-plastic deformation regime.

Although the nonlinear elastic representation of the J integral does not provide for crack extension, it can be shown that the J-integral continues to describe the stress–strain state of an advancing crack if the

---

† The quantity J is defined as $\int W \, dy - \overline{T}(d\overline{u}/dx) \, dS$, where W is the strain energy, $\overline{u}$ is the displacement, $\overline{T}$ is the traction vector, and dS is an increment along path S.

amount of extension is restricted to 5–10% of the remaining ligament.[25] In a similar manner, crack opening displacement (COD) is a viable crack initiation parameter and its variant, crack opening angle (COA), describes limited crack extension.[25]

While these parameters describe crack initiation and stable growth, it is also necessary to have a criterion that delineates stable growth from fracture instability. The principal candidates for instability criteria are the applied dJ/da,[25a] called the tearing modulus, and the applied COA. This area of general yielding fracture mechanics (GYFM) is covered in greater depth in Chapter 4.

For the creep crack growth phenomenon, we again discover limitations of the LEFM methodology and a situation somewhat analagous to the time-independent fracture analysis discussed above; that is, depending on the size of the "creep zone" at the crack tip, different load parameters determine the crack growth behavior. The earlier Figs. 1-9c and 1-10 will be used to briefly illustrate this trend. Stress analyses of macroscopically cracked bodies under creep conditions have included power law creep, power law creep and elastic deformation, and, more recently, elastic deformation and primary, power law, and tertiary creep.[24]

The short-time response of a cracked specimen is elastic and K_I determines the crack growth behavior (equivalent to Fig. 1-10a). This is the so-called creep-brittle regime which is to be avoided in nuclear structures. After a characteristic time, when the whole specimen creeps in primary creep, the integral[24] C_h^* determines the growth of the creep zone (equivalent to the J-field in Fig. 1-10b). For long times after load application, the whole specimen creeps extensively and the integral[24] C^* is the correct load parameter (equivalent to the large-scale yielding situation, Fig. 1-10c). The advantage of the C^* integral over the J integral, however, is that the former is truly path independent and hence a second parameter (such as dJ/da or COA equivalents) is not required to describe creep crack growth.

Not everybody has followed the J integral modification or GYFM approach to describing stable crack extension. Hart[32] recently introduced a theory for the growth of cracks in ductile materials that retains the usefulness of the stress intensity factor K. The theory predicts the steady-state crack tip extension velocity as it depends on the remote loading, geometry, and material parameters. It is based on a kinetic law for the crack tip velocity as it depends on the stress state at the crack tip, and the equation-of-state model of plastic flow described earlier.[16] Hart shows that with a plastic flow-rate law and for a finite crack tip velocity, the crack tip stress field is characterized by K. His theory predicts the crack velocity vs. K relationship generally observed in experiments.

1.3.4. Materials Properties and Phenomena

As was shown in Fig. 1-8, mechanical properties, together with flaw distribution and imposed stresses, determine the structural integrity of nuclear components. Accordingly, these properties and associated phenomena are emphasized throughout the book. As noted earlier in Section 1.3.2, mechanical properties are influenced to varying degrees by the service environment. Neutron flux and chemicals, in particular, result in "new" mechanical phenomena, such as irradiation embrittlement, irradiation creep, corrosion fatigue, and stress corrosion. The primary forms of damage encountered in nuclear systems, therefore, are cracking (and sometimes failure), wastage (corrosion), and distortion (creep).

Historically, mechanical property data have been obtained initially as a function of stress (strain) and temperature (we will call these basic data), and then as a function of either radiation or chemistry. It might be noted in passing that, unfortunately, very little data exist that show the combined (synergistic) effects of chemistry and irradiation, and therefore care must be taken when extrapolating properties to real component operating conditions. In any event, this scheme adequately serves the purposes of this introduction, and the following subsections will first define the basic properties before introducing the complicating effects of radiation and corrosion damage.

1.3.4.1. Plastic Flow Properties

Simple uniaxial, constant extension-rate tension tests and constant load (or stress) creep tests are widely used to characterize, respectively, the time-independent and time-dependent plastic flow properties of nuclear materials (Fig. 1-9). For the former, basic properties include the yield strength σ_y, ultimate tensile strength σ_{UTS}, and uniform and total strains to failure; while for the latter, primary and secondary or power-law creep properties are emphasized. Onset of tertiary creep in structures is to be avoided since it rapidly leads to rupture.

Plastic flow mechanisms are highly sensitive to stress and temperature (and irradiation, as is shown in Section 1.3.4.5). A convenient way of exhibiting the different deformation mechanisms is by the use of the deformation mechanism maps.[33] Figure 1-11[34] is such a map for Type 316 stainless steel, which shows the importance of creep deformation at stress levels below the normal yield stress, when temperatures exceed $\sim 0.4 T_m$. This domain is of relevance to the operation of some LMFBR (and CTR) ex-core components as is indicated by the shaded boxes. A point of interest, however, is that these components lie well inside the

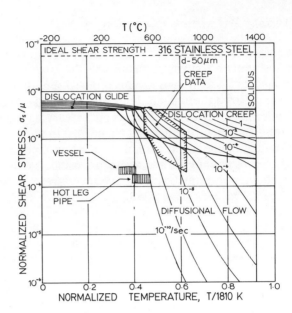

Fig. 1-11. Deformation map for 316 stainless steel showing the operating conditions of two LMFBR components; both deform by Coble creep (adapted from Ref. 34).

diffusional flow (or creep) field, in the regime controlled by grain boundary diffusion, whereas most of the laboratory data have been obtained in the dislocation or power-law creep regime. Thus, laboratory data are sometimes of no help in predicting component creep rates, and values must be obtained from theoretical models. Although the theory of diffusional creep is well developed it is perhaps fortunate that, in the case of these components, the creep strains will be negligibly small. However, as is shown later, creep strains at the higher temperatures and neutron flux environments of fission reactor cores and CTR first walls become life limiting.

1.3.4.2. Fracture Toughness

The ferritic steels used in nuclear components are characterized by a ductile-to-brittle fracture (or nil ductility) transition (Fig. 1-12). Below this temperature (so-called NDTT) the brittle fracture mode is usually cleavage. Above the transition temperature, the fracture mechanism is usually dimpled rupture, and large plastic strains are required for crack initiation and propagation; hence the name plastic fracture. Raising the temperature above the transition temperature first increases the toughness; then follows a region of relatively constant toughness—the so-called upper shelf (Fig. 1-12).

Although the ferritic steels used in the LWR vessel operate on the upper shelf, they might be fracture limited in some situations, such as transient loadings following long-term radiation exposures. (See Chapter 4). In contrast, the austenitic stainless steels utilized in the LMFBR pressure boundary, and most of the materials under consideration for the CTR first-wall/blanket structure, do not exhibit a ductile–brittle transition, and all indications are that fracture toughnesses are very high relative to required values.

Various measures of fracture toughness are in use. On the one hand, there are qualitative measures, such as the impact energy in the Charpy V-notch test or the nil-ductility-transition temperature (NDTT) in the drop-weight test. Materials specifications demand certain lower limits for Charpy "upper-shelf" energies and upper limits for nil-ductility temperatures, and such measurements, being relatively straightforward to make, are especially convenient for quality control purposes. The experimental study of the dependence of toughness on metallurgical factors (composition, structure, temperature, radiation dose, etc.) is also conveniently carried out by these means; in fact, most basic knowledge of this kind has been acquired through Charpy V-notch experiments.

On the other hand, the theory of fracture mechanics specifies the fracture toughness in terms of other quantities. As noted earlier, the

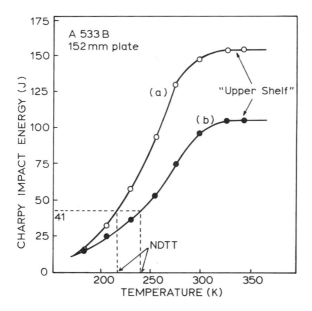

Fig. 1-12. Charpy impact energy as a function of temperature for A533B steel plate tested in longitudinal (a) and transverse (b) directions.

critical stress-intensity factor (K_{Ic}) for plane-strain conditions is widely used to predict the combinations of loading and crack size that will lead to the onset of crack growth in structural components. Once crack growth starts, the theory stipulates that it will continue and the structure will fail unless the stress-intensity factor at the crack tip falls below the crack arrest toughness, K_{Ia}, which is also considered to be a property of the material. Dynamic tests that yield a "dynamic" critical stress-intensity factor K_{Id} (as distinct from the "static" K_{Ic}) are also often undertaken to determine the fracture-toughness/strain-rate sensitivity of the material.

Ideally, reliable data are needed for K_{Ic}, K_{Id}, and K_{Ia} for all relevant material forms [plates, forgings, weld regions, heat-affected zones (HAZ), etc.] and for all appropriate conditions of temperature, strain rate, composition, orientation relative to the rolling direction, radiation and annealing treatments, etc. Compact tension specimens are commonly used to obtain the K values. Unfortunately, however, direct measurement of toughness parameters is difficult in the "upper-shelf" region for the very tough materials of interest to nuclear systems, mainly because impractically large test pieces are required to satisfy the condition for a "valid" measurement, i.e., specimen widths > 500 mm, approximately plane strain with plastic flow limited to a region around the crack tip that is small compared to the corresponding dimension of the specimen. This restriction not only places a limit on the maximum K value measurable, but also on the quantity of data obtainable.

Therefore, for design purposes the ASME Section III code recommends the use of a reference fracture toughness K_{IR}, which is actually based on the lower bound of K_{Ic}, K_{Id}, and K_{Ia} values obtained on the particular material. The relationship between the various parameters is sketched in Fig. 1-13. The underlying assumption here is that the temperature variation of toughness relative to the nil-ductility temperature is the same for all material forms meeting the specific requirments; thus toughness can be specified for the actual material used in construction simply from the measurement of the NDTT. More will be said on this approach in Chapter 4.

Inasmuch as the materials of interest are ductile, it would be more appropriate to characterize them using valid ductile fracture toughness parameters. As noted earlier, there are at least two theoretical developments in this area, and neither of them is Code approved. The parameters J_{Ic}, dJ/da, COD, and COA are material properties that can be measured experimentally, and data are currently being amassed and appropriate ASTM Code Committees are working on defining standard test procedures in preparation for code acceptance of plastic fracture methods.[25] The alternative approach[32] that uses a modified K would probably be more easily implemented if the theory were accepted. Critical reviews

Introduction and Overview

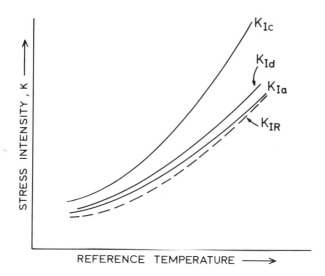

Fig. 1-13. Schematic comparison of fracture toughness parameters, plotted as stress intensity versus temperature, referenced to the nil ductility transition temperature (courtesy T. U. Marston, EPRI).

of Hart's theory have not yet begun. Clearly, much more work needs to be done in this area before final decisions can be made.

1.3.4.3. Fatigue

Generally, in actual structures fatigue is conceded to be the most widely encountered cause of failure; in other words, materials tend to be fatigue limited rather than fracture limited. Fatigue cracks can be initiated in a number of ways; however, the important fact is that they are usually nucleated at a free surface from small flaws, scratches, or other stress raisers. Crack initiation and early growth occurs above a characteristic "threshold" stress intensity K_{th} and is generally termed Stage I growth; fatigue propagation of a significant flaw ($\geqslant 3$ mm long) is termed Stage II growth; the crack grows a finite increment in each load cycle. Stage II continues until the crack becomes large enough for unstable fracture, Stage III (Fig. 1-14).[23]

Experimentally, fatigue failure is generally classified into high-cycle fatigue, low-cycle fatigue, and fatigue crack propagation. In high-cycle fatigue the net section stress is below the yield strength of the material, and the fatigue life is (roughly) 10^4 to 10^5 cycles. This phenomenon is of more interest to CTR first-wall life predictions due to the high frequency, pulsed nature of its operation. Low-cycle fatigue, which is more often

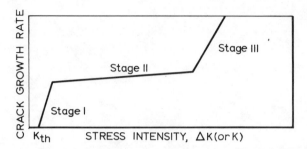

Fig. 1-14. Schematic of overall expected stress corrosion crack growth rate versus K.

encountered in fission reactor components, is characterized by the presence of macroscopic cyclic plastic strains, and the regime is generally limited to less than 10^4 cycles.

A common form for representing the fatigue behavior for a series of tests at different strain ranges is the "S–N curve" shown in Fig. 1-15, where the elastic, plastic, and total strain ranges are represented versus the cycles to failure. The time-independent or low-temperature (below creep) fatigue behavior has been successfully described by two relationships.[35]

Material behavior in the high-cycle fatigue, elastic strain range is describable by the empirical relationship proposed by Basquin,

$$\Delta\epsilon_e = \Delta\sigma/E = (B/E)N_f^{-\beta} \qquad (1\text{-}9)$$

where B and β are material constants, E is Young's modulus, and N_f is

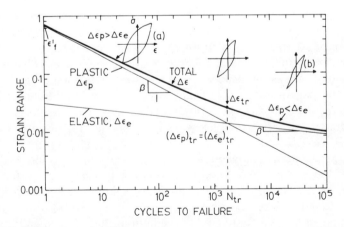

Fig. 1-15. Representation of elastic, plastic, and total strain ranges versus fatigue life and low- and high-cycle regimes (Ref. 23).

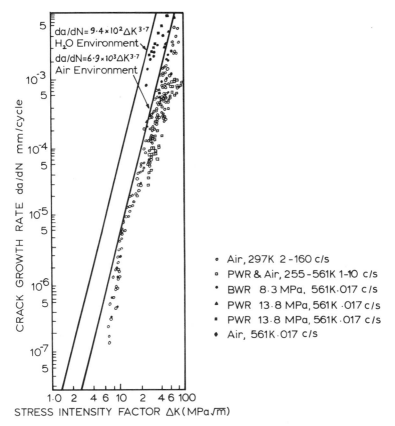

Fig. 1-16. Present Section XI fatigue crack growth reference curves, with supporting data (Ref. 30).

the number of cycles to failure. For inelastic strains, low-cycle fatigue damage obeys the Coffin–Manson equation

$$N_f^n \Delta\epsilon_p = C \qquad (1\text{-}10)$$

where C and n are constants, with $n \simeq \frac{1}{2}$. Low-cycle fatigue behavior of many nuclear materials can be characterized by this relation over a broad range of conditions of temperature, environment, stress history, and surface and microstructural condition. The total strain-range life relation is found by summing Eqs. (1-9) and (1-10).

Referring again to Fig. 1-15, the life at the point of intersection of the elastic and plastic lines (i.e., where $\Delta\epsilon_p = \Delta\epsilon_e$) is called the transition fatigue life, N_{tr}.[23] This reference point is very useful for distinguishing whether the failure in a given material is one of low-cycle fatigue or one of high-cycle fatigue. Implicit in the concept is the fact that high- or low-cycle fatigue is not determined by a specific cyclic strain or a specific number of cycles to failure, but rather by the geometry of the hysteresis loop for a given cyclic strain or cyclic life. Thus, for some given number of cycles to failure, one material may show a histeresis loop that is very wide [case (a) in Fig. 1-15 where $\Delta\epsilon_p > \Delta\epsilon_e$], while another material reveals a loop that is essentially a straight line [case (b) in Fig. 1-15 where $\Delta\epsilon_p < \Delta\epsilon_e$]. In case (a), failure is by low-cycle fatigue and the failure life is less than the transition fatigue life for the materials. In case (b), failure is by high-cycle fatigue and the life is greater than the transition fatigue life. The fatigue behavior of a given material can thus be defined by its transition fatigue life N_{tr} and by its loop width $(\Delta\epsilon_p)_{tr}$. In general, the lower the strength and the higher the ductility of a material, the higher the transition fatigue life.

The rate of fatigue crack growth da/dN can be correlated very effectively with the stress intensity factor range ΔK for many materials in many environments, as noted in Eq. (1-8) earlier. This equation has been used as the basis for ASME Code reference curves for predicting fatigue crack growth in a number of key nuclear component materials. By way of example, Fig. 1-16 shows "pre-1979" versions of the air and water fatigue crack growth curves for LWR reactor vessel steels; recent developments[30] are discussed further in Chapter 4. The power-law behavior of crack propagation also holds for time-dependent fatigue at elevated temperatures under cyclic loading conditions,[29] as is described in Chapters 4 to 7.

1.3.4.4. Environmental Fatigue

When either or both temperature and corrosive environment is increased, time-dependence enters fatigue life considerations, and this regime is referred to as either environmental or time-dependent fatigue. The result is more or less a continuous spectrum of mechanisms, as depicted in Fig. 1-17, and it is difficult to determine which mechanism—corrosion fatigue or creep fatigue—is actually controlling in the regime (0.2 to 0.5 T_m) where most nuclear component materials operate.

Only creep fatigue is recognized in the high-temperature Code Case 1592,[9] and the damage rule uses a semilinear cumulative creep and fatigue damage assessment based on the following empirical equation:

$$\sum_{j=1}^{p} \left(\frac{n}{N_d}\right)_j + \sum_{k=1}^{q} \left(\frac{t}{T_d}\right)_k \leq D \qquad (1\text{-}11)$$

where D is the total allowable creep-fatigue damage given in the Code

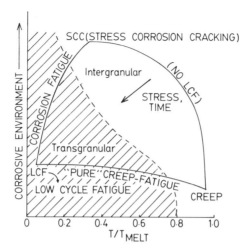

Fig. 1-17. Schematic interactions between creep, fatigue, and corrosion (adapted from Ref. 22).

Case, n/N_d is the ratio of the number of applied cycles for a given loading condition to the allowable number of cycles determined from a fatigue curve for that loading condition, and t/T_d is the ratio of the time duration for a given load to the allowable time at that load determined from a creep-rupture curve. This approach is intended to be conservative, but the growing data base indicates that this might not be the case.

The mechanistic explanations of high-temperature fatigue effects have either been corrosion or creep biased.[23, 35, 36–38] Coffin,[36] for example, has taken the view that time-dependent fatigue is essentially stress-corrosion cracking, and this approach led to the well-known "frequency-modified fatigue lifetime" correlation for crack initiation and growth rate. Thus,

$$(N_f \nu^{k-1})^n \Delta\epsilon_p = C_2 \qquad (1\text{-}12)$$

where ν is the frequency, k is a temperature-dependent constant, and n and C_2 are constants as before. Since development of this relation, however, the existence of a true creep-fatigue interaction has been proven by fatigue tests in a vacuum involving a hold time. Therefore, Coffin was later obliged to modify Eq. (1-12) to account for so-called wave-shape effects; this resulted in the frequency separation or FS method.[23]

An alternative creep- and plasticity-oriented approach to describing fatigue at elevated temperatures was proposed by Manson.[37] This so-called strain-range partitioning (SRP) method has as its basis the separation of total strain, first into elastic and plastic components (Fig. 1-15), i.e.,

$$\Delta\epsilon = \text{total strain range}$$
$$= \text{plastic strain range} + \text{elastic strain range} \quad (1\text{-}13)$$
$$= AN_f^a + BN_f^b$$

where A, B, a, and b are constants, and then a further division of the plastic strain into four components (PP, CC, CP, and PC) that have characteristic "life values" associated with creep (C) and plasticity (P). Higher temperatures and low strain rates tend toward creep, while lower temperatures and higher strain rates tend toward plasticity as the predominant deformation mode.

For further details on the frequency separation and strain range partitioning methods, the reader is referred to a detailed critique of these two approaches to describing elevated-temperature fatigue, which was prepared by the authors of the theories.[23]

More recently, Majumdar and Maiya[38] developed a phenomenological damage-rate approach to creep-fatigue interaction, which attempts to account for the damage mechanisms as they vary with plastic strain rate in tension and compression during creep-fatigue loadings.

Aspects of these approaches to modeling elevated-temperature fatigue are introduced in Chapters 4 to 7 during the discussion of the key fatigue data. It will be evident, however, that although a method can be shown to adequately explain one type of data or data set, a model that will explain the integrated effects of environment and loading history on fatigue behavior has not yet been developed. Indeed, data on fatigue behavior in such complex situations are still lacking; most of the data reported will be of the separate effects type.

In principle, all three of the "phenomenological" time-dependent fatigue models can be rewritten in terms of fatigue crack propagation rate. Maiya and Majumdar's model, for instance, explicitly accounts for fatigue crack growth that is assisted by grain-boundary creep cavitation damage. However, the most complete crack-growth algorithm, although based on Coffin's method, is empirical in nature. The so-called Carden parameter[23, 29] has been demonstrated to correlate fatigue data on Type 316 stainless steel to 923 K.[29]

1.3.4.5. Radiation Effects

Radiation damage is manifested in a variety of ways, none of which are conducive to long-term structural integrity. Detailed up-to-date treatments of radiation damage can be found in the books by Olander[40] and

Gittus[41] and in the paper by Nichols.[42] For the purposes of this work, three major types of radiation damage are of interest: effects of neutron irradiation (from either fission or fusion) on the bulk properties of materials, neutron irradiation damage from fusion source to the surfaces of materials, and fission-induced damage to nuclear fuels.

With respect to the first effect, which is perhaps of most direct concern to component integrity, neutron flux influences the mechanical and physical properties of surrounding materials through two processes that occur in the bulk. One process is a nuclear reaction resulting from neutron capture by a target atom and then decay of the excited nucleus. The transmutation reactions that yield helium, mainly (n, α) reactions, are of greatest importance. These nuclear reactions are dependent on neutron energy only and are unaffected by other components of the reactor environment such as temperature and stress state. Neutron energies follow the trend CTR (14 meV peak) > LMFBR (0.1 to 1.0 MeV peak) > LWR (high-energy peak at 2.0 MeV, but an important fraction below 0.7 eV), and the ratios of helium production to atom displacement (i.e., appm helium/dpa) are 100:4:1, respectively, for CTR, LMFBR, and LWR.[43] Thus, helium effects are a more important consideration for CTR first-wall/blanket structures than for LMFBR fuel rod cladding, and, due to lower temperatures, can be neglected entirely for LWR materials. In addition, these ratios indicate that, to the extent that useful life depends on bulk degradation of material properties, the life of materials in CTRs will be much shorter than in LMFBRs or LWRs. This fact is evident in the discussion in Chapter 5.

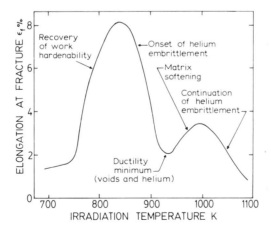

Fig. 1-18. Effect of irradiation temperature on the ductility of irradiated stainless steel. Tests at 323 K and fast-neutron fluence $> 10^{22}$ neutrons cm^{-2} · sec^{-1} (Ref. 40).

Fig. 1-19. Void formation in neutron-irradiated Type 304 stainless steel (courtesy P. Okamoto, Argonne National Laboratory).

The primary effect of helium on metal behavior is the so-called *helium embrittlement*, an effect that increases in severity beyond $0.5T_m$. As illustrated in Fig. 1-18, ductility in stainless steel is significantly reduced above 850 K, and intergranular failures result from helium bubble precipitation on grain boundaries.

The second process is atomic displacement, created when a neutron strikes a lattice atom and displaces it from its equilibrium position, forming a vacancy and an interstitial. The damage is further multiplied when the neutron transfers energy to the struck atom and it in turn displaces many other atoms. The damage production process again depends on neutron energy and is independent of temperature. However, the rearrangement of the damaged zone after the damage production event is strongly dependent on temperature, stress, and other variables.

The defect zones (either voids, as in Fig. 1-19, or dislocation loops) will act as barriers to the movement of dislocations and, therefore, will cause both an increase in the yield stress of the material, the *radiation hardening* effect (Fig. 1-20), and, for materials that exhibit a ductile-to-brittle transition, a shift in the transition temperature (ΔTT) and a reduction in upper-shelf energy (ΔUSE), the *radiation embrittlement* phenomenon (Fig. 1-21).

A volume increase in the material, termed (void) *swelling*, and an anisotropic shape change, termed *growth*, also results from this displacement damage. Swelling is a characteristic of most materials when subjected to fast neutron irradiation between 0.3 to $0.5T_m$ (Fig. 1-22). This is the temperature range where vacancies are fairly mobile, i.e., sufficient to coalesce, but not so mobile that they easily reach sinks and annihilate. The actual swelling is the result of the preferential trapping of interstitials by dislocations; the vacancies that remain coalesce to form the voids, shown in Fig. 1-19, that are characteristic of fast neutron-irradiated materials. The damage zones are large enough in LMFBR and CTR materials to produce significant swelling (volume changes of more than 10% are possible), but this phenomenon is not experienced in the LWR operating regime of lower fast fluence and temperatures.

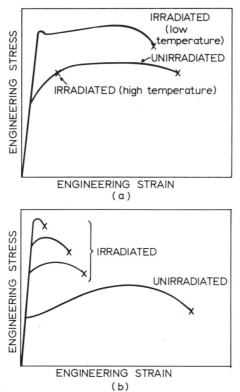

Fig. 1-20. Effect of fast-neutron irradiation on the tensile properties of reactor steels (Ref. 40). (a) Face-centered cubic structure; (b) body-centered cubic structure.

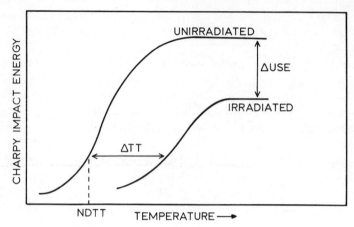

Fig. 1-21. Schematic of the transition temperature and upper shelf energy changes resulting from irradiation embrittlement of typical reactor vessel steels.

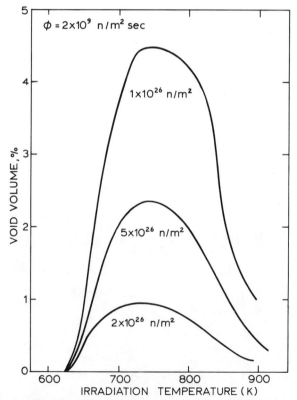

Fig. 1-22. Approximate curves of void swelling versus irradiation temperature in stainless steel at three fluence levels (Ref. 40).

Introduction and Overview

While growth occurs in the absence of applied stress, it is observed that when the material is stressed during irradiation its dimensional (shape) change is greatly increased through the phenomenon known as *irradiation creep*. The revised stainless steel deformation map in Fig. 1-23 (compare to Fig. 1-11) illustrates the general trends in irradiation creep exhibited by LMFBR core components. The LWR Zircaloys and CTR first-wall candidates would respond in a similar fashion under these conditions. The deformation is weakly temperature dependent, and is approximately linear in stress and neutron flux. With respect to stainless steel, as discussed in Chapter 3, LMFBR fuel assembly lifetimes are currently limited by irradiation creep/swelling distortions; but, as noted in Chapter 5, by comparison this phenomenon is of less importance to CTR first-wall performance.

To this point, the discussion of radiation damage has been limited to damage in the bulk. However, neutron flux can influence the surface characteristics of materials. The intense neutron flux of the CTR envi-

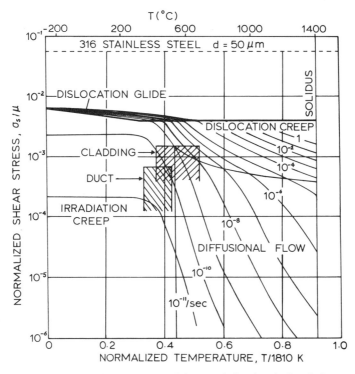

Fig. 1-23. Deformation map for Type 316 stainless steel showing the irradiation creep field and the operating regime for LMFBR and CTR components (courtesy H. Frost, M.I.T., 1979).

ronment, through a variety of surface damage processes, produces erosion of the first wall. These include vaporization, sputtering, and blistering.

Simulation experiments that bombard the material surface with beams of energetic helium and hydrogen ions have revealed the mechanism of blistering and erosion in metals. Because of the much lower solubility and slower diffusion of helium atoms in metals, the effect of helium on surface damage is greater than that of hydrogen. Accordingly more attention is being focused on helium effects. When the penetration range into a metal is on the order of 1 μm and the dose is such that the He concentration builds up to 10 atom% a plane of bubbles forms at a depth roughly equal to the range. These combine to form blisters; but at low and high temperatures, discrete, bulbous blisters form and rupture but their skin remains intact and attenuates the incoming beam and thus protects the metal below. At the intermediate temperatures ($\simeq 0.4 T_m$), the entire surface forms one large blister which peels off. Another blister forms beneath it, and this in turn peels off. This repeated peeling gives a pronounced maximum in the "erosion" or metal-loss rate (Fig. 1-24). Such a loss is undesirable on at least two counts: (1) the flakes would

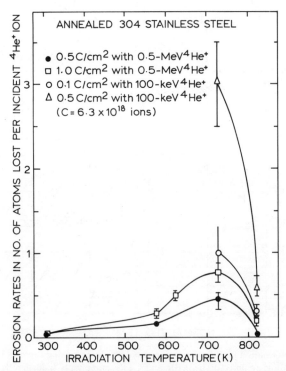

Fig. 1-24. Erosion rates for annealed 304 stainless steel versus irradiation temperature for varying helium-ion energies and doses (Ref. 44).

come off into the plasma and be damaging, and (2) the reduction in section is intolerable, sputtering rates of 10^{-2} or 10^{-3} atoms per incident helium atom being an upper limit for reasonable wall life.[44]

All metals studied seem to show this behavior with increasing temperature,[44] and it is not yet clear how or whether alloying will help. Also, the worst peeling occurs around 0.3 to $0.5T_m$, as with swelling. However, studies are still too preliminary to establish whether certain base metals or alloys may be used without peeling in the temperature range of the designers' ultimate choice.

The overriding consideration, however, is likely to be the introduction of impurities into the plasma which can cool and even quench the desired reactions. This concern has prompted examination of a series of low-activation (low atomic number or low-Z) materials, principal among these being the ceramics, silicon carbide, and graphite (discussed in Chapter 5).

In the fusion reactor, the fuel pellet is wholly converted to high-energy particle irradiation (neutrons, ions, neutral atoms and electrons, electromagnetic and x-ray radiation) by the process of fusion and, therefore, it is the products of fusion, rather than the pellet itself, that is the important factor in component integrity. On the other hand, in a fission fuel pellet, only a small fraction of the atoms actually fission, with the result that the pellet remains an important factor in component integrity throughout a fuel rod's operating life. The fissioning process controls the temperature distribution in the fuel rod and yields gaseous and solid fission products that result in fuel swelling (and possible straining of the cladding) physical, mechanical, and chemical changes in the fuel, and chemical attack of the cladding. Thus, the thermomechanical and thermochemical behavior of the fuel rod is to a large extent determined by the fissioning process.

Two basic differences between LWR and LMFBR cores must be considered in fuel rod design and materials specification: first, the neutron flux in the LMFBR is a factor of 100 larger than in the LWR, which causes the former to produce more power per unit volume and, hence, the fuel to run hotter; and second, the average fuel burnup in the LMFBR is about three times larger than in a LWR. Both differences are manifested in more fuel pellet swelling and larger fission product inventories in LMFBR fuel rods. Chapters 2 and 3, respectively, describe the evolution of the LWR and LMFBR fuel rod designs from the basis of these operational differences.

1.3.4.6. Corrosion

The chemical environment in a nuclear system is quite severe because of the many different metals used (electrochemical effects) and the existence of many crevices where chemicals can concentrate. Corrosion products derived from steam generator, piping, and vessel walls are transported through the core and become radioactive. Their subsequent deposition in the various components of the primary system constitutes a significant maintenance problem. In a few isolated incidents, deposition of this "crud" on the fuel in fission reactors has resulted in loss of heat transfer from the fuel rods and cladding overheating and accelerated corrosion.

Waterside corrosion effects are the most severe; air is next, followed by nitrogen (used as a cover gas in some parts of the LMFBR) and liquid sodium (and presumably lithium). In fact, the effect of liquid sodium on fatigue and crack growth properties appears to be similar to that of a vacuum[29] (see Chapter 4).

Chemical effects take two major forms (at least for the purposes of this discussion). In the first, schematically represented in Fig. 1-17, chemicals enhance mechanical processes such as fatigue and crack growth, that is, corrosion fatigue and stress corrosion. In the second, corrosion reduces component wall thickness, either locally (pitting, crevice attack, etc.) or more uniformly (wastage). Frequent focus on all these effects recurs in subsequent chapters inasmuch as corrosion is a major issue in nuclear plant reliability (as it is for any power system—fossil, hydroelectric, etc.).

With respect to environmentally assisted cracking phenomena, variously called corrosion fatigue (cyclic loading) or stress corrosion cracking (static or dynamic loading), the three-stage crack growth process alluded to earlier is well defined, and the characterization by the stress intensity factor K appears successful. Thus, as indicated in Fig. 1-14, the crack initiation stage (Stage I), slightly exceeding K_{th} (or K_{Iscc} for this situation), has a very strong stress dependence; the second stage (Stage II), covering a broad range of K_I values, exhibits a very weak stress dependence; at very high K_I values, in Stage III, a strong stress dependence is also found. The crack initiation and growth stages depend on the loading history. Generally, Stage I is prolonged under cyclic loads. Reducing the cyclic frequency or, in other words, increasing the time under constant load allows the crack to propagate by stress corrosion. In many components then, cracks initiate by corrosion fatigue, but propagate by stress corrosion.

The local or uniform corrosion damage that results in reduction of load-bearing area is the result of a chemical reaction, either between the environment and impurities or selected alloying agents (local attack), or between the environment and the base material (uniform attack).

The uniform oxidation process that is most prevalent follows classical laws.[45] In the example shown in Fig. 1-25, of Zircaloy corrosion, the

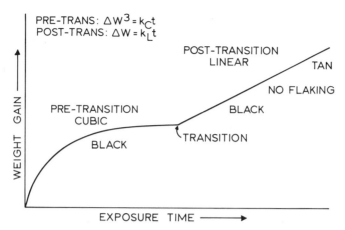

Fig. 1-25. Schematic representation of the corrosion of Zircaloy 2 and Zircaloy 4 in the temperature range 533 to 673 K (Ref. 45).

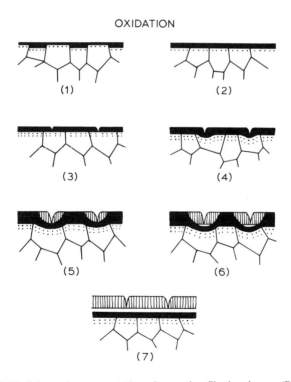

Fig. 1-26. Schematic representation of corrosion film breakaway (Ref. 46).

initial formation of a thin, black, tightly adherent corrosion film corresponds to an approximate cubic rate law behavior; further exposure produces a linear set of kinetics. Invariably, problems arise after long exposures in this posttransition region because the film becomes nonprotective and breaks off to reveal fresh surface for further reaction. In extreme situations, corrosion can lead to development of through-wall failures. The conditions leading to excessive wall thinning are schematically shown in Fig. 1-26.

The corrosion product can also stress an adjacent component; thus corrosion *per se* may not be the cause of failure but the stress imposed on an adjacent part by the volume increase of the oxide or other compound that forms may cause failure. A very good example of this effect, discussed in detail in Chapter 6, is the inward compression of steam-generator tubes that results from the oxidation of the carbon steel support plates (commonly referred to as "denting").[47, 48]

Clearly, the ideal situation to achieve is the formation of a stable, adherent oxide film, because a protective, self-healing oxide film renders the material immune to environmental cracking and consumes a minimum amount of "load-bearing" metal in its formation. Materials are generally selected for service in corrosive environments on the basis of their ability to form such a film (e.g., "stainless" steels, nickel-base alloys, titanium alloys, etc.). However, experience shows us that the long-term maintenance of a passivating oxide film in nuclear system environments is difficult, if not impossible. Flaws or weaknesses in the film, induced by stress or irradiation, quickly become crack initiation sites; and changes in corrosion potential can switch the oxidation process to nonprotective growth kinetics and accelerated metal consumption. Examples of such "failures" in oxide film passivation are described in just about every chapter.

References

1. Fusion Power: Status and Options, Electric Power Research Institute, EPRI ER-510-SR (June 1977).
2. J. A. Maniscalco, D. H. Berwald and W. R. Meier, The Material Implications of Design and System Studies for Inertial Confinement Fusion Systems, paper presented at First Topical Meeting on Fusion Reactor Materials, American Nuclear Society, Bal Harbor, Florida (January 29–31, 1979).
3. Pool Type LMFBR Plant 1000 MW(e) Phase A Extension 2 Designs, Electric Power Research Institute Reports NP-1014-SY (GE), NP-1015-SY (AI), and NP-1016-SY (W) (1979).
4. M. E. Lapides, Nuclear Unit Productivity Analysis 1976 Update, Electric Power Research Institute, EPRI NP-559-SR (October 1977).
5. R. H. Koppe and E. A. Olson, Nuclear and Large Fossil Unit Operating Experience, Electric Power Research Institute, EPRI NP-1191 (September 1979).

6. J. James, Private communication, Electric Power Research Institute, (May 1979).
7. L. Minnick and M. Murphy, The Breeder: When and Why, *Electr. Power Res. Inst. J.*, No. 2, 6–11 (March 1976).
8. Section III, Division 1 Rules for Construction of Nuclear Power Plant Components, ASME Boiler and Pressure Vessel Code, American Society of Mechanical Engineers, New York (July 1977).
9. Code Case 1592, Class 1 Components in Elevated Temperature Service, Section III, ASME Boiler and Pressure Vessel Code, American Society for Mechanical Engineers, New York (November 1977).
10. Section XI—Rules for In-Service Inspection of Nuclear Power Plant Components, ASME Boiler and Pressure Vessel Code, American Society of Mechanical Engineers, New York (July 1977).
11. R. L. Jones, T. U. Marston, S. T. Oldberg, and K. E. Stahlkopf, Pressure Boundary Technology Program: Progress 1974 through 1978, Electric Power Research Institute, EPRI NP-1103-SR (March 1979).
12. Z. Zudans, M. M. Reddi, H. M. Fishman, and H. C. Tsai, Elastic-Plastic Creep Analysis of High-Temperature Nuclear Reactor Components, *Nuc. Eng. Des.*, **28**, 414–445 (1974).
13. Guidelines and Procedures for Design of Nuclear System Components at Elevated Temperatures, Reactor Development and Technology (RDT) Standard F9-ST (March 1974).
14. D. S. Griffin, Inelastic Structural Analysis: Design Implications and Experience, *Nuc. Eng. Des.* **51**, 11–21 (1978).
15. A. K. Miller, An Inelastic Constitutive Model for Monotonic, Cyclic, and Creep Deformation (Part I. Equations Development and Analytical Procedures; Part II. Application to Type 304 Stainless Steel) *J. Eng. Mater. Technol.* 97–113 (1976).
16. E. W. Hart, Constitutive Relations for the Nonelastic Deformation of Metals, *J. Eng. Mater. Technol.*, 193–202 (1976).
17. A. K. Miller, Development of the Materials Code, MATMOD (Constitutive Equations for Zircaloy), Electric Power Research Institute, EPRI NP-567, Project 456-1, Final Report (December 1977).
18. V. Kumar, S. Mukherjee, F. H. Huang, and Che-Yu Li, Deformation in Type 304 Stainless Steel, Electric Power Research Institute, EPRI NP-1276, Project 697-1, Final Report (December 1979).
19. E. W. Hart, C. Y. Li, H. Yamada, and G. L. Wire, Phenomenological Theory: A guide to Constitutive Relations and Fundamental Deformation Properties, in: *Constitutive Equations in Plasticity*, A. S. Argon, ed., MIT Press, Cambridge, Massachusetts (1975), pp. 149–197.
20a. T. U. Marston, The EPRI Ductile Fracture Program, Proceedings of Seminar on Fracture Mechanics, April 2–6, 1979, ISPRA, Italy.
20b. P. C. Paris, CSNI Specialists Meeting on Plastic Tearing Instability, NUREG/CP-0010-CSNI Report No. 39, U.S. Nuclear Regulatory Commission, Washington, D.C. (October 1979).
20c. P. C. Paris and H. Tada, Further Results on the Subject of Tearing Instability. 1, NUREG/CR-1220, Vol. 1, U.S. Nuclear Regulatory Commission, Washington, D.C. (January 1980).
21. J. F. Knott, *Fundamentals of Fracture Mechanics*, Butterworths, London (1973).
22. M. Reich and E. P. Esztergar, Compilation of References, Data Sources, and Analysis methods for LMFBR Primary Piping System Components, Brookhaven National Laboratory, BNL-NUREG 50650 (March 1977).
23. L. F. Coffin, S. S. Manson, A. E. Carden, L. K. Severud, and W. L. Greenstreet, Time-Dependent Fatigue of Structural Alloys, Oak Ridge National Laboratory, ORNL-5073 (January 1977).

24. H. Riedel, Creep Deformation at Crack Tips in Elastic-Viscoplastic Solids, Brown University Rep. No. MRL-Ell4 (June 1979).
25. EPRI Ductile Fracture Research Review Document (T. U. Marston, ed.), Electric Power Research Institute, EPRI NP-701-SR (February 1978).
25a. P. C. Paris, H. Tada, A. Zahoor and H. Ernst, Instability of the Tearing Mode of Elastic-Plastic Crack Growth, NUREG 0311, U.S. Nuclear Regulatory Commission, Washington, D.C. (August 1977).
26. P. Paris and F. Erogan, A Critical Analysis of Crack Propagation Laws, *Trans. ASME*, 528–534 (December 1963).
27. Boiler and Pressure Vessel Code, Section XI, Appendix A, Article A-4000, American Society of Mechanical Engineers, New York (1977).
28. L. A. James, Fatigue Crack Propagation Analysis of LMFBR Piping, In: *Coolant Boundary Integrity Considerations in Breeder Reactor Design*, Series PVP-PB-D27, R. H. Mallett and B. R. Nair, eds.), ASME (1978).
29. L. A. James, Fatigue Crack Propagation in Austenitic Stainless Steels, *At. Energ. Rev.*, **14**, 37–85 (1976).
30. W. H. Bamford, Application of Corrosion Fatigue Crack Growth Rate Data to Integrity Analyses of Nuclear Reactor Vessels, paper presented at 3rd ASME National Congress on Pressure Vessels and Piping, San Francisco, California (June 1979).
31. S. R. Novak and S. T. Rolfe, Comparison of Fracture Mechanics and Nominal Stress Analysis in Stress Corrosion Cracking, *Corrosion*, **26**, 121–130 (1970).
32. E. W. Hart, A Theory for Stable Crack Extension Rates in Ductile Materials, Report No. 4111, Materials Science Center, Cornell University (July 1979).
33. M. F. Ashby, First Report on Deformation-Mechanism Maps, *Acta Metall*, **20**, 887–897 (1972).
34. H. J. Frost and M. F. Ashby, Deformation Mechanism Maps for Pure Iron, Two Austenitic Stainless Steels, and a Low-Alloy Ferritic Steel, in: *Fundamental Aspects of Structural Alloy Design*, R. I. Jaffee and B. A. Wilcox, eds.), Plenum Press, New York (1977), pp. 27–65.
35. L. F. Coffin, Jr., The Multi-stage Nature of Fatigue: A Review, *Met. Sci.*, 68–72 (February 1977).
36. L. F. Coffin, Jr., A Note on Low-Cycle Fatigue Laws, *J. Mater.*, **6**, 388–402 (1971).
37. S. S. Manson, Challenge to Unify Treatment of High-Temperature Fatigue—partisan proposal based on Strain-Range Partitioning, in: *Symposium on Fatigue at Elevated Temperatures*, June 1972, Storrs, Connecticut, Am. Soc. Test. Mater. Spec. Tech. Publ. 520 (1972) pp. 744–782.
38. S. Majumdar and P. S. Maiya, A Unified and Mechanistic Approach to Creep-Fatigue Damage, in: *Proceedings of the 2nd International Conference on Mechanical Behavior of Materials*, ICM-II, American Society of Metals, Metals Park, Ohio (1976), pp. 924–928.
39. A. E. Carden, Parametric Analysis of Fatigue Crack Growth, in: *International Conference on Creep and Fatigue in Elevated Temperature Applications*, Philadelphia (1973), Paper C324/73.
40. D. R. Olander, Fundamental Aspects of Nuclear Reactor Fuel Elements, Technical Information Center Energy Research and Development Administration, TID-26711-Pl (1976).
41. J. Gittus, Irradiation Effects in Crystalline Solids, Applied Science Publishers Ltd., London (1978).
42. F. A. Nichols, How Does One Predict and Measure Radiation Damage?, *Nucl. Technol.*, **40**, 98–105 (1978).
43. G. L. Kulcinski, D. G. Doran, and M. A. Abdou, in: *Properties of Reactor Structural Alloys after Neutron or Particle Irradiation*, Am. Soc. Test. Mater. Spec. Tech. Publ. 570 (1975), pp. 329–351.

44. S. K. Das and M. Kaminsky, Radiation Blistering of Structural Materials for Fusion Devices and Reactors, *J. Nucl. Mater.*, **53**, 115–126 (1974).
45. E. Hillner, Corrosion of Zirconium-Base Alloys—An Overview, *Zirconium in the Nuclear Industry, Proceedings of the 3rd International Conference*, A. Lowe and G. Parry, eds., Am. Soc. Test. Mater. Spec. Tech. Publ. 633 (1977).
46. D. L. Douglass, *The Metallurgy of Zirconium*, International Atomic Energy Agency, Vienna (1971).
47. R. Garnsey, Corrosion of PWR Steam Generators, Central Electricity Research Laboratories, England, RD/L/N4 79 (March 1979).
48. W. H. Layman, L. J. Martel, S. Green, G. Hetsroni, C. Shoemaker, and J. A. Mundis, Status of Steam Generators, paper presented to American Power Conference, April 1979, Chicago.

2

LWR Core Materials

The key component of the fission reactor core is the fuel pellet. The other core materials are used either for structural support or for nuclear control, and both their amounts and their geometries are carefully controlled to ensure maximum efficiency of the fission reaction. Experience with LWR cores is extensive and has been widely published in several important meeting proceedings.[1-3] Early problems with fuel rods such as hydriding, clad collapsing, and unanticipated growth and fretting, which were mainly caused by inadequate design or manufacturing, have been eliminated. In this chapter, therefore, the intent is to focus on those material limitations that are under current study, and to only briefly summarize past developments.

The LWR (and LMFBR) core comprises essentially only two components: fuel assemblies and control assemblies. Table 2-1, an expanded version of Table 1-1 (Chapter 1), is included here for the express purpose of providing a comparison of LWR and LMFBR fuel and core materials; the latter will be the subject of Chapter 3.

The goals of good fuel design are to achieve economical fuel cycle costs while meeting regulatory requirements for safe plant design and operation. The control rod system must be designed to ensure that adequate reactor shutdown capability is available at all times, and that an optimum power distribution can be maintained during fuel burnup. For about the first decade of LWR operation, attention centered on achieving design power and burnup of Zircaloy-clad fuel rods without breaching the cladding and releasing radioactive fission products into the primary coolant and into the off-gas systems. If the fuel achieved this goal, it had fulfilled its role in providing the power generation costs projected for the power plant portion of the fuel cycle. However, in the past five years or so, the fuel has also had to be designed to perform acceptably under even

Table 2-1. Fuel and Core Materials

Assembly	Reactor type		
	BWR	PWR	LMFBR
Fuel Assembly			
Fuel pellet	UO_2	UO_2	UO_2 25 ± 5wt.% PuO_2
Blanket pellet	NA^a	NA	UO_2
Cladding	Zr2	Zr4	Type 316 S.S.
Burnable poison	Gd_2O_3 in UO_2	$B_4C-Al_2O_3$ or borosilicate glass	NA
Skeleton	Inconel X, Zircaloy 2, Type 304 S.S.	Zr4, Inconel 718	Type 316 S.S.
Channel	Zr4	NA	Type 316 S.S.
Control assembly			
Neutron absorber	B_4C	AgInCd or B_4C	B_4C
Cladding	Type 304 S.S.	Type 304 S.S.	Type 316 S.S.

^{a} NA, not applicable.

more restrictive limits and under an increasing number of postulated accidents.

To meet the two goals of operating reliability and safe operation under hypothetical conditions, fuel designs have regularly been modified, with a definite trend toward increasing conservatism and design standardization. Current fuel and core designs are described in Section 2.1.

Extensive property data exist for LWR fuel rod materials, much of which is contained in the MATPRO[4] handbook, whose development has been supported by the U.S. Nuclear Regulatory Commission (USNRC). To avoid unnecessary duplication, therefore, descriptions of specific fuel and cladding properties will be limited to those of particular relevance to the fuel performance issues discussed in Section 2.2. In Sections 2.3 and 2.4, respectively, the performance aspects of the lesser used plutonium recycle and stainless steel clad fuels are reviewed. Section 2.5 reports on the status of control materials in LWRs, which have received far less attention than the fuel. The final section, 2.6, addresses the issue of uranium conservation and measures that are being taken to improve the efficiency of uranium utilization in LWRs. This latter discussion provides an indication of the direction that LWR core materials development will take in the next two decades.

2.1. Fuel and Core Designs

As shown in Fig. 2-1, the fission reactor core is built up from its basic component, the ceramic fuel pellet. Slightly enriched (with U-235)

LWR Core Materials 55

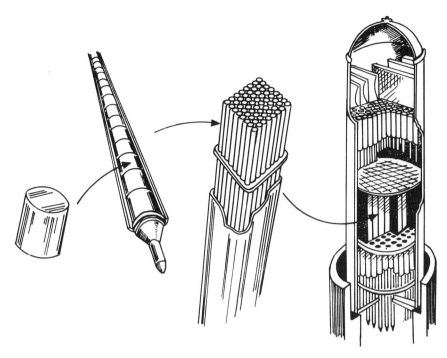

Fig. 2.1. Components of water reactor core (BWR is the example).

UO_2 pellets are manufactured by compacting and sintering powders into cylindrical pellets and grinding to size. The nominal immersion density of the pellets ranges from 93 to 95% TD.† A LWR fuel rod is made by stacking pellets into a zirconium alloy cladding tube. The BWR uses Zircaloy 2 cladding, and the PWR uses Zircaloy 4 cladding. Following loading of the fuel pellets, the fuel rod is then evacuated, back-filled with helium to some specified pressure, and sealed by welding end plugs in each end of the tube. A space called the plenum is provided at the top of the fuel column to contain fission gases, and a spring inserted between the fuel column and the top of the rod keeps the fuel column integral during handling. Typical fuel rod parameters are summarized in Table 2-2, which is again used to compare the LWR and LMFBR approaches.

The fuel assembly design differs markedly for the BWR and PWR, and for a particular reactor type, the fuel vendors employ minor design differences. The example in Fig. 2-1 is of a General Electric BWR assembly, which is comprised of 64 rods spaced and supported in a square (8 × 8) array by lower and upper tie plates fabricated from Type 304 stainless steel (refer also to Fig. 2-3a). Rod-to-rod separation is maintained

---

† % TD = percent of theoretical density (10.96 g/cm^2).

Table 2-2. Typical Fuel and Core Parameters[a]

Parameter	BWR	PWR	LMFBR[b]
Pellet diameter (mm)	10.5	9.0	4.9
Pellet l/d	0.9–1.2	1.0–1.3	1.0–1.5
Pellet density (% TD)	94	94	95
Pellet O/M ratio	2.00 ± 0.01	2.00 ± 0.01	1.94–1.97
Cladding thickness (mm)	0.9	0.6	0.4
Fuel rod outside diameter (mm)	12.5	10.0	5.8
Fuel–clad diametral gap (mm)	0.25	0.2	0.17
Pellet column length (m)	3.7	3.7	1.6 (including two 0.3 m axial blankets)
Plenum length (mm)	250	200	1220
Prepressurization level (MPa)	0–0.5	2	0
Number of fuel rods in assembly	60 to 63 (8 × 8)	176 (14 × 14) to 264 (17 × 17)	217
Assembly width (mm)	139	270	116
Number of fuel assemblies	560	177 to 217	198
Core height (m)	4.4	4.0	1.0 (excludes axial blankets)
Core width (m)	5.0	3.4	2.0 (excludes outer blankets)
Number of radial blanket A/S	NA	NA	150

[a] Design parameters will vary slight depending on the particular reactor vendor.
[b] The U.S. Clinch River breeder reactor (CRBR) is the example used.

by seven grid spacers regularly distributed along the length. Eight of the rods screw into the tie plates to complete the framework or skeleton.

The fueled rods in the BWR assembly contain different U-235 enrichments to reduce the local power peaking. Low-enrichment uranium is used in the peripheral fuel rods while higher-enrichment uranium is used in the central part of the fuel assembly. In addition, selected fuel rods in each assembly contain UO_2–Gd_2O_3; the gadolinia (Gd_2O_3) is a neutron absorber which burns out during the first irradiation cycle and provides temporary reactivity control. One or two rods do not contain UO_2 fuel but have coolant flowing through them to facilitate moderator distribution. The fuel assemblies in an initial core have average enrichments that range from 1.7 to 2.1 wt.% U-235 depending on initial cycle requirements. Reload fuel will have an average enrichment in the range of 2.1 to 3.0 wt.% U-235. Exxon Nuclear Corporation, which provides some of the reload fuel batches, follows the GE design except for minor changes to pellet size and cladding thickness, the use of solid Zircaloy rods instead of water rods, and the selection of cold-worked and stress-relieved rather than fully annealed Zircaloy 2.

To complete the fuel assembly in a BWR, the fuel rod bundle is enclosed in a close-fitting Zircaloy 4 can or channel, which directs the coolant flow upward through the individual assemblies.

In contrast, the PWR fuel assembly is "canless" and is comprised of a larger square array of fuel rods, typically 14 × 14 or 16 × 16 (Combustion Engineering and Kraftwerk Union) and 15 × 15 and 17 × 17 (Westinghouse, Babcock & Wilcox, and Framatome) arrays. The example in Fig. 2-2 is a Westinghouse design. Control-rod guide thimbles

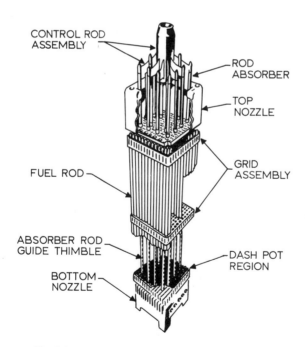

Fig. 2.2. PWR fuel assembly (Westinghouse design).

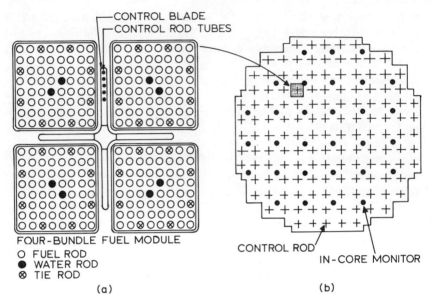

Fig. 2.3.(a) BWR core lattice; (b) typical BWR core arrangement.

replace fuel rods at selected spaces in the array and are fastened to the top and bottom nozzles. To complete the skeletal framework, as in the BWR design, grid spacers are fastened to the guide thimbles along the length of the fuel assembly to provide support for the fuel rods. The trend in the industry is to use "low parasitic" Zircaloy 4 skeleton components as far as possible, although both stainless steel and Inconel are used in some designs. Burnable poisons, generally B_4C–Al_2O_3 pellets contained in Zircaloy 4 tubes, or borosilicate glass rods in stainless steel tubes are installed either in the fuel-rod (i.e., lattice) positions or in the control-rod guide thimbles in some assemblies, depending again on the particular vendor design. Unlike the BWR design, a constant U-235 enrichment is maintained in the rods of a single assembly.

The core of a fission reactor is constructed of fuel assemblies interspersed with control rods. The BWR core (Fig. 2-3b) is made up of a repetition of modules, each module consisting of four fuel assemblies surrounding a cruciform control blade (Fig. 2-3a). The control blades are movable and control the reactivity and power distribution within the core. They are cooled by core leakage (bypass) flow through holes in the cruciform. Each blade or cruciform is comprised of 84 Type 304 stainless steel tubes (21 tubes in each wing) filled with boron carbide (B_4C) powder compacted to approximately 65% TD. Additional reactor control is provided by adjusting coolant flow rate.

A uniform-scatter refueling pattern is generally adopted in BWR fuel management. For annual refueling, approximately one-fourth of the fuel is discharged each cycle. For 18-month refueling cycles, approximately one-third of the fuel is discharged at the end of each cycle. The residual fuel, which is at various reactivity levels corresponding to the number of cycles of irradiation (one to four) can then be rearranged with respect to fresh fuel assemblies to give a nearly uniform reactivity across the core. Peak fuel burnups at discharge are typically 2250 to 2590 GJ/kgM.†

Although similar in shape and dimensions to a BWR core, the core of a PWR is designed on a multizone (usually three) basis (Fig. 2-4b). All fuel assemblies are mechanically identical, although the fuel enrichment is not the same in all assemblies. In the typical initial core loading, three fuel enrichments are used. Fuel assemblies with the highest enrichments are placed in the core periphery, or outer region, and the two groups of lower-enrichment fuel assemblies are arranged in a selected pattern in the central region. In subsequent refuelings, one-third of the fuel is discharged

† M is normally the SI symbol for mega; an exception in this book is the additional use of M in burnup to denote heavy metal. 1 GJ/kgM = 11.6 MWd/MTM.

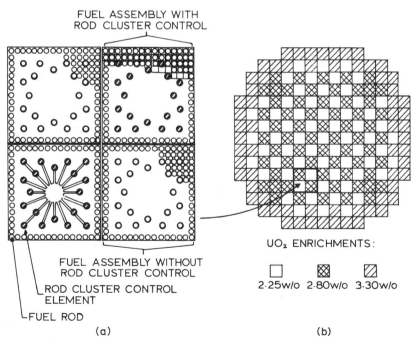

Fig. 2.4.(a) Cross section of PWR fuel assemblies with and without control rod clusters (Westinghouse design); (b) typical PWR core arrangement.

and fresh fuel is loaded into the outer region of the core. The remaining fuel is arranged in the central two-thirds of the core in such a manner as to achieve optimum power distribution. Peak fuel burnups are higher in PWRs, attaining 2850 to 3100 GJ/kgM.

Reactor control in a PWR is more finely dispersed than in a BWR. Rod Cluster Control (RCC) assemblies, which consist of clusters of cylindrical absorber rods (AgInCd alloy or B_4C contained in stainless steel or Inconel tubes) move within guide tubes in certain fuel assemblies (Fig. 2-4a). The control rod clusters are used to follow load changes, to provide reactor trip capability, and to furnish control for slight deviations in reactivity due to temperature. However, normally, a PWR operates with the control rods out of the core, and control "at power" is provided by a neutron absorber (boric acid) dissolved in the reactor coolant. The concentration of the boric acid is varied during the life of the core to compensate for changes in reactivity that occur with fuel depletion.

2.2. Fuel Performance

Performance of Zircaloy-clad UO_2 nuclear fuels in LWRs has generally been more than adequate, but with extremes that in some plants have required early displacement of certain batches of fuel, and conversely, extremes where fuel has performed flawlessly, even beyond design expectations.[1-3]

A variety of fuel-related problems have been experienced in the last decade. For plants up to eight years old, *total* capacity losses from fuel problems was about 7.0% for BWRs, 1.5% for PWRs, and 3.6% for all units.[5] Table 2-3, adapted from Levenson and Zebroski,[6] summarizes the main factors that lead to limited fuel performance, and remedies that have been applied to eliminate or minimize impact on plant output.

The two main aspects of fuel performance that impact on plant availability and capacity factor are (a) reliability and mechanical integrity, and (b) the margin between the fuel operating limits and the fuel damage limits, which is established to ensure acceptable fuel conditions under normal operations as well as during postulated accident events. Recent research into increasing both reliability and operating margin will be discussed in the following subsections.

2.2.1. Fuel Reliability

An LWR rod (as is an LMFBR fuel rod) is designed as a pressure vessel. The ASME Boiler and Pressure Vessel Code, Section III, is used as a guide in the mechanical design and stress analysis of the fuel rod. Complex computer models have evolved which treat analytically the syn-

Table 2-3. Main Factors That Have Limited Fuel Performance and Remedies[a]

Factors	PWR	BWR	Remedies
Hydriding of Zr	×	×	Elimination of moisture in fabrication; addition of getters
Scale deposition		×	Elimination of copper tubing from feedwater heaters
Enrichment errors	×	×	Gamma scanning all rods prior to shipment
Clad collapse	×		Prepressurized cladding, stable pellets
Pellet densification	×	×	Stable pellet microstructure
Other manufacturing and handling defects (e.g., faulty welds)	×	×	Improved quality control (now constitute <5% of defects found in reactor)
Clad corrosion or fretting		×	Rare; control of clad quality and cleaning; spacers
Fuel rod growth and bowing	×		Control of texture; axial clearances; spacer design
Channel bulging		×	Use of thicker wall channels and control of residual stress
Pellet–clad interaction (PCI) on power increases	×	×	1. Slow power rise (PWR), 2. Plus local power shape control (BWR), 3. Plus fuel "preconditioning" phases (BWR and PWR), 4. Plus fuel pellet design changes (BWR and PWR)

[a] Adapted from Ref. 6.

ergism between the thermal and mechanical responses of the fuel pellet and cladding. A detailed treatment of these so-called fuel performance codes is outside the scope of this chapter; the reader is referred to Ref. 7 for details. Progress in the development of key submodels of fuel-rod thermomechanical and thermochemical behavior will be presented as appropriate.

The fuel rod is designed to withstand the applied loads, both external and internal, that develop during irradiation. However, deleterious interactions between the Zircaloy cladding and the core environment (coolant, fast neutron flux, and temperature), other assembly components (grids or spacers), or the fuel pellet and associated rod internal environment, have resulted in fuel rod and assembly corrosion, distortion, and, sometimes, leakage. Clad breaching or failure invariably requires remedial action, but excessive distortion and/or corrosion may also require premature discharge because of their negative impact on plant operation and maintenance.

Only a very low fuel-rod failure rate can be tolerated because of rigid limitations on coolant-activity radioactive releases. Experience over the last decade, based on commercial reactor operation, suggests typical rates of fuel rod failures on the order of 0.05% or less (i.e., <15 rods out of 30,000) per reactor per year, but with extremes of up to ~1.0%, during early "epidemics." As indicated in Table 2-3, corrective design modifications and improved manufacturing techniques were introduced by fuel suppliers to eliminate the latter problems, such as faulty welds, Zircaloy hydriding, and fuel-rod fretting (wear). Reviews by Robertson[8] and by Garzarolli and Stehle[9] have described these developments.

Today's concerns are focused on the random, low-frequency fuel defects that appear to be strongly dependent on the details of core operation. Of current interest to both BWRs and PWRs (but having greater impact in BWRs) is the mechanical/chemical interaction between the fuel pellet and cladding tube known generally as pellet–cladding interaction (PCI). The incidence of PCI failures depends on absolute power, increase in power, duration of the power increase, previous power history, and burnup. Also, there is a power threshold below which failures do not occur. PCI failure rates are currently maintained at an acceptably low level by conservative reactor operations which are proving to be costly. For the two-year period 1975 and 1976, losses due to restrictions on the rates of power increases recommended by the fuel suppliers were about 0–2% (average <1%) for PWRs and about 2–8% (average 3.5%) for

Table 2-4. Parameters That Affect Cost of Fuel Failure[a]

Fuel costs
  Incomplete burnup
  Decreased capacity factor
    Reactor derate or shutdown
    Leak testing and inspection time
    Assembly reconstitution time
  Cost of replacement power
  Increased storage capacity for good and failed fuel
  Penalties, or restrictions, for shipping and reprocessing failed fuel
  Increased services
    Failure investigation
    Fuel management plan modifications
    Licensing of modified reload

Operation and maintenance ("O&M") costs
  General maintenance
  Radwaste System maintenance
  Increased personnel exposure

[a] Adapted from Ref. 11.

BWRs, with a total cost in replacement power estimated at $50–100 million.[10] Despite recent improvements in plant availability due to both increased experience with these operating recommendations and an increased fraction of improved fuel in the cores, the replacement power costs are still as high due to increased costs of oil and coal.

Nevertheless, the cost of fuel failure avoidance is more acceptable than the cost of fuel failures, which can have such broad ramifications, not the least of which is the damage to the public image of nuclear power. The costs of failure are independent of their cause, and affect both the fuel costs and the operation and maintenance (O&M) costs. The parameters that effect each are given in Table 2-4. Most of these are self-explanatory, the two of note being:

(1) Incomplete burnup, which is the result of premature discharge of failed assemblies. The additional costs incurred by not receiving full energy output from the failed fuel is often covered partially, but not totally, by warranties provided by the fuel supplier, which can be applied to the replacement fuel.
(2) Decreased capacity factor, which can result from a regulatory mandated reduction in power (i.e., a derate) or reactor shutdown due to high activity levels. It can also be the result of increased refueling time due to the need to leak-test fuel assemblies to identify those that failed, and subsequent disassembly, removal of the few failed rods, and reassembly using spare rods or rods from a companion assembly (so-called assembly reconstitution). In the event of high system demands, the purchase of replacement power may be necessary at considerable incremental cost.

The individual cost of the various parameters related to failures have been analyzed recently to establish the relative significance of each to the cost of failure.[11] The conclusions are given in Fig. 2-5 as a function of percent failed fuel rods in the core. The example shown is the PWR case, but the BWR case is similar. The electricity generating costs in M/kWh† for a U.S. plant are compared to the dollar cost of failure on the ordinate. The graph shows that as failure levels increase, the cost of derating and lost burnup are the most significant cost components. The current cost of fuel fabrication (i.e., fuel cycle cost) is <0.5 M/kWh. Figure 2-5 indicates that the cost of fuel failure would reach the fuel cycle cost if fuel failures exceeded the 0.10% level. Even though the majority of the reactors in the world are operating at <0.05% failure level, this result does indicate that a considerable increase in fuel fabrication costs might be permissible if the new fuel design(s) guaranteed uniformly low failure rates (hence improved capacity factors).

† Mills/kWh = 1/1000 of $1, or 0.1¢/kWh.

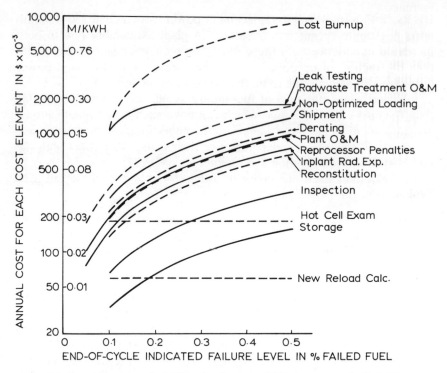

Fig. 2.5. Annual costs vs. fuel failure level using a PWR as an example (Ref. 11).

Therefore, despite the fact that fuel reliability has increased and failures have decreased significantly in the past few years, analyses such as the foregoing provide a continuing economic incentive to eliminate all known failure mechanisms through optimized fuel design and/or operations. In recent years, worldwide research emphasis has been on an effective solution to the PCI problem, but, in addition, concern about waterside corrosion of Zircaloys at high burnups has motivated a research effort that is growing gradually. The following subsections will review progress in these two areas.

2.2.1.1. Pellet–Cladding Interaction

PCI is recognized as the major, if not the only, potential generic mechanism for cladding rupture during normal operation.[9] It is also being evaluated as a possible clad failure mechanism during off normal power transients.[12]

To understand this phenomenon, which, by the way, is exhibited in some form or other in all metal-clad ceramic fuel rods, one must first consider the processes that occur in a fuel rod in a reactor.

When the power is raised initially the Zircaloy cladding is ductile and the as-fabricated gap between the fuel pellets and the cladding is sufficient to accommodate the larger thermal expansion of the fuel pellets. The pellets also crack and relocate outwards to contribute to gap closure. Later in life, however, the Zircaloy cladding becomes somewhat embrittled due to its exposure in the intense neutron environment, while the pellet–clad gap continues to decrease due to dimensional increases (fission product swelling) in the fuel pellets and, in the case of PWRs, inward motion or creepdown of the thin-walled cladding in response to the applied coolant pressure. Under such conditions, if the fuel rod power is increased, local power changes may be of sufficient magnitude to cause direct interaction between the UO_2 pellet and the Zircaloy cladding with the resulting straining and possible fracture of the cladding. A more detailed review of the precursors to PCI is provided by Baley et al.[13]

Early observations of such failures gave rise to the designation PCMI† failures because they were attributed to a combination of localized pellet-induced strain of the cladding and the low strain-to-failure capability of the irradiated Zircaloy. Recent information suggests, however, that volatile fission products generated in the fuel pellets as a consequence of the fissioning process, and subsequently released to the gap during steady-state operation or during an increase in power, are essential to the degradation of the cladding and its eventual failure. Most investigators now accept the view that both the presence of aggressive chemical species and high localized stresses (strains) are prerequisites for power-ramp-induced PCI failures, although the relative contributions of these two factors remains to be resolved.[9] This synergism between stress or strain concentrations at pellet crack and/or pellet–pellet interface locations and the harmful species is schematically illustrated in Fig. 2-6.

As noted in Chapter 1, failure of susceptible metals under combinations of both load and aggressive environments is designated stress corrosion cracking (SCC). Recent research has therefore proceeded from the assumption that PCI-induced fuel-rod failures are a consequence of the SCC of Zircaloy.

2.2.1.1a. Results of Experimental and Theoretical Studies. The principal basis for concluding that PCI fracture of Zircaloy cladding is the result of stress corrosion is the large body of data on the shape and morphology of the cracks found in both power reactor[14, 15] and test reactor[16, 17] fuel rods, which exhibit similar characteristics to those

---

† Pellet–clad mechanical interaction.

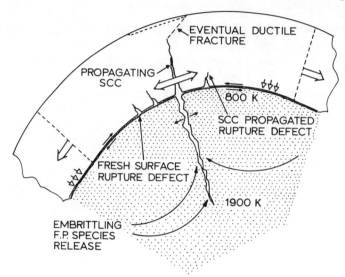

Fig. 2.6. Schematic of PCI Zircaloy fuel rod failure mechanism.

found for cracks induced in laboratory SCC experiments using iodine (a known fission product) as the embrittling agent.[18, 19] Roberts et al.[20] listed the key features in common as follows:

1. Cladding cracks are normal to the cladding surface (i.e., 90° to the hoop stress).
2. A number of partially penetrating (incipient) cracks are observed.
3. Incipient cracks are tight, i.e., openings are small compared to their length.
4. Fully penetrating cracks are characterized by an absence of general deformation or necking of the cladding.
5. Fracture surfaces are essentially macroscopically flat with little evidence of extended 45° shear fracture segments except perhaps near the outer cladding surface.
6. At the microscopic level, fracture surfaces exhibit flat cleavage planes and fluted regions.

Figures 2-7 and 2-8 illustrate some of these features.

The SCC phenomenon in Zircaloy appears to be the operative failure mode over a broad range of reactor operating conditions. PCI failures have occurred in Zircaloy 2 BWR and Zircaloy 4 PWR cladding, operated under quite different conditions but within the normal range for fuel rod linear heat generation rates in power reactors (26 to 43 kW/m); similar failures are also seen in test reactor fuel rods that are power ramped outside the normal power reactor domain (e.g., to powers >60 kW/m at

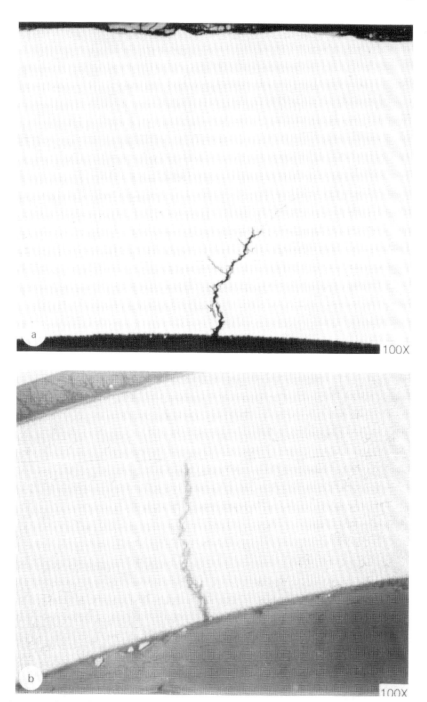

Fig. 2.7. PCI cracks in power reactor fuel rods. (a) Dresden 3 (BWR); (b) Maine Yankee (PWR) (Ref. 21).

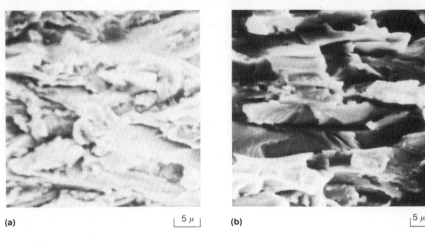

Fig. 2.8. "Cleavage and fluting" SCC features on the surfaces (Ref. 23) of cracks in Fig. 2.7. (a) and (b) are from Maine Yankee and (c) is from Dresden 3 cracks.

ramp rates up to ~0.16 kW/m/sec).[21] The implication of this conclusion is that stress corrosion cracks can form and propagate under a broad range of mechanical/chemical conditions, extending from high stress–low fission product (iodine) availability to low stress–high fission product availability situations. Thus, fission product availability and stress can be considered to be reciprocal in the Zircaloy PCI-SCC process.[22]

Zircaloy Stress Corrosion Cracking. Mechanistic studies of Zircaloy SCC have been, and still are, involved in (a) characterizing the steps involved in iodine SCC and the role, if any, of Zircaloy's metallurgical properties; (b) identifying the rate-limiting or critical steps in the process; and (c) establishing the mechanical/chemical conditions for SCC, in particular the existence of "thresholds." The majority of the data is from unirradiated Zircaloys 2 and 4, but sufficient data have been obtained on irradiated cladding to show the major effects of irradiation.[23, 24] Since a number of papers and reviews have been published on this subject,[13, 18-20, 22-25] it will suffice here to relate only those results that are germane to the development of PCI failure criteria (or models) for power reactor fuel reliability applications.

The current data base supports a threshold-stress concept; that is, a sustained critical tensile stress in a sustained environment of aggressive fission-product species.[23] The threshold stress is associated with the formation of a crack at inhomogeneities in the Zircaloy surface (e.g., FeNiCr or Al/Si rich particles).[19] If the ID oxide film is cracked, small intergranular cracks will form below the threshold stress, but these are nonpropagating.[19] The threshold stress is influenced, as expected, by iodine availability, and Peehs et al.[25] measured a well-defined iodine concentration threshold, of $1-3 \times 10^{-2}$ g I_2/m^2, below which no cracking was observed (Fig. 2-9). Times to failure above this threshold were reduced by about an order of magnitude for each order-of-magnitude increase in iodine concentration, and failure strains dropped to near zero. This work also indicates a temperature influence such that a temperature increase from 623 to 673 K produced two orders of magnitude reduction in time to rupture. Similar behavior was observed in tests on irradiated Zircaloy.[24]

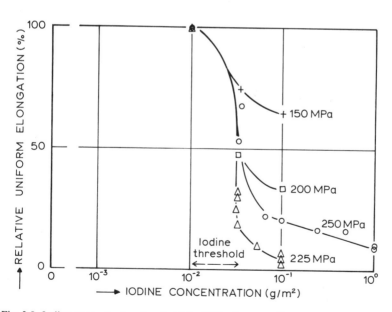

Fig. 2.9. Iodine concentration threshold for SCC of stress-relieved Zircaloy at 673 K (Ref. 25).

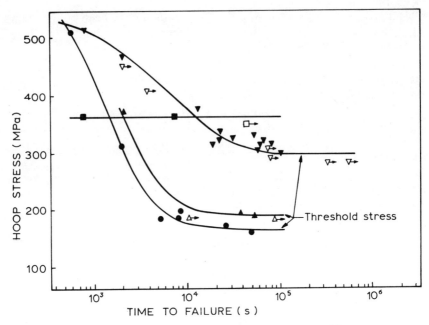

Fig. 2.10. Effect of irradiation on the resistance of internally pressurized Zircaloy tubes to iodine-induced failure at 620–630 K (Ref. 23).

The effect of metallurgical variables on Zircaloy SCC does not appear to be strong.[19, 23, 26] Data on unirradiated Zircaloys indicate a slight effect of microstructure (annealed is inferior to stress relieved), and there is an influence due to either Zircaloy surface condition, surface texture, or internal stress, or a complex combination of these features.[26]

By far the strongest effect on threshold stress, however, is that of irradiation, which tends to override any small differences in microstructure or chemical composition.[23, 24a,b] Results from pressurized-tube tests on power reactor irradiated Zircaloys (from which the fuel was removed before testing), shown in Fig. 2-10, clearly show the large reduction in threshold stress (when compared as a fraction of the yield strength) at long times to failure, which is observed after 5×10^{24} n/m^2 fast neutron fluence. SCC can therefore occur at stresses ranging from close-to-yield stress to a fraction ($\sim \frac{1}{3}$) of the yield stress (i.e., in the elastic range!). At the high stresses short, axial splits result, whereas at the low stresses pinhole failures were most common. Cracks of both type have also been observed in defected power reactor fuel rods.[14, 15]

A theoretical picture of stress corrosion crack formation and propagation has evolved from the experimental data.[27, 28] As schematically represented in Fig. 2.11a, environmentally assisted cleavage of appropriately oriented material (i.e., planes that make an angle of <15° with the basal plane) is presumed to occur in the immediate vicinity of the flaw. In many instances, the "flaw" is an impurity or alloy particle as noted above, but in theory it could be a fabrication defect, a small flaw arising from the extension of an oxide crack into a locally oxygen-em-

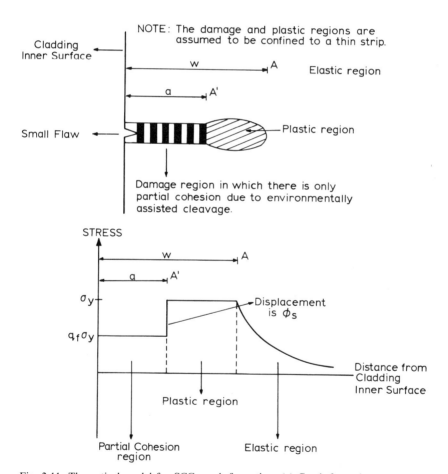

Fig. 2.11. Theoretical model for SCC crack formation. (a) Crack formation process; (b) tensile stress vs. distance from cladding inner surface (Ref. 27).

brittled layer of Zircaloy, or a small defect that is produced by dislocation channeling or plastic flow concentration. The size of the cleavage-damaged region then increases with time until the local strain at the flaw is sufficient to rupture, by plastic "fluting," material that is not oriented for cleavage. This latter event is the condition for crack formation, and the critical threshold stress is the lowest stress at which a crack can form by this process (refer to Fig. 2-10). If the cladding stress exceeds the critical value, the time required for crack formation is that needed for the damage region to spread sufficiently for plastic fluting to occur at the initial flaw, and this time obviously depends on the magnitude of the cladding stress. The boundary of the damage region (A' in Fig. 2-11) is therefore an interface which separates material that has plastically yielded from material where there is only partial cohesion, this partial cohesion being provided by material that is not oriented for environmentally assisted cleavage.

The critical threshold stress σ_{th} has been expressed in terms of four material parameters: the yield stress of the irradiated material σ_y; the local plastic strain ϵ_s [$= \phi_s/d$, where ϕ_s is the displacement (Fig. 2-11b) and d is grain size] associated with environmentally assisted cleavage of appropriately oriented material; the local plastic strain ϵ_f associated with rupture of the remaining material by the fluting process; and the fractional area q_f of this latter material (Fig. 2-11b). Very approximately, σ_{th} is given by the expression

$$\frac{\sigma_{th}}{\sigma_y} = 3q_f\left(1 - \frac{\epsilon_s}{\epsilon_f}\right) + \frac{\epsilon_s}{\epsilon_f} \tag{2-1}$$

With $\epsilon_s/\epsilon_f < 0.1$ and q_f in the range of 0.1–0.3, this relation shows that the threshold stress can be low compared to the yield stress (e.g., when $q_f = 0.1$, $\sigma_{th}/\sigma_y \cong 0.3$), which is in agreement with the experimental results in Fig. 2-10.

The effect of iodine availability on the threshold stress has also been considered in some detail with this model.[28] While it is unlikely to have a significant effect on q_f or ϵ_f, it should have a marked effect on ϵ_s, and this should be manifested in a marked effect of iodine availability on the threshold stress at low concentrations when an almost step-function variation is expected. Figure 2-9 tends to support this conclusion, lending credence to the crack formation mechanism.

The growth of the stress corrosion crack is also considered as occurring by the same two mechanisms.[27, 28] Environmentally assisted cleavage occurs in grains so oriented that the hoop stress resolved in the direction normal to the near basal cleavage plane is sufficiently large, and the remaining grains fracture by a plastic (ductile) process. The combination of these two failure mechanisms is what gives rise to the characteristic fluted fracture surface illustrated in Fig. 2-8. Fracture is therefore viewed as being a sequential process in which the cleavage grains first fail so as to overload the fluting grains, which then fail in turn and allow a true (i.e., cohesion-free) crack to propagate. The crack will propagate in this manner until the onset of plastic instability, locally at the crack tip, at which point the remaining ligament fails by ductile rupture.

Role of Zircaloy Plastic Deformation. Although there is a general consensus that plastic tensile strain is needed for SCC to occur, there is some doubt as to whether a macroscopic critical strain is a prerequisite for SCC. Rather, the data support a threshold stress concept as discussed above.

Due to a combination of anisotropic properties and varying axial and circumferential stress and temperature profiles, the strain distribution in a fuel rod cladding is irregularly distributed even on a macroscopic scale, as is evidenced by the observation of both creep ovalization (particularly in PWR cladding) and circumferential ridge formation. The former, as noted earlier, is partially responsible for closure of the fuel–clad gap—ovalization generally results in point contact between fuel and cladding. The latter form as a result of a "ratcheting" process—cyclic stretching of the cladding by the fuel—principally during power cycling.[13] There is clear evidence from test reactor data that ridges form or grow during power ramps.[29, 30] These ridges are the location of the PCI-induced cracks, but since this is also the region of maximum fission product concentration, it is not clear whether the high stress (or strain) concentration at ridges is responsible for crack initiation. As reported earlier, stress corrosion cracks can form at stresses in the elastic range.

What is clear, however, from fracture surface observations, is the fact that plastic deformation on a microscopic scale is instrumental in propagating the stress corrosion crack, once formed. As in the macroscopic situation, microscopic deformation in neutron-irradiated Zircaloys occurs in a highly inhomogeneous manner with deformation bands confined to localized regions within grains. In the theoretical picture described earlier, such grains would be fluting grains and it is thought that the stress (or strain) concentration that results on neighboring grains may encourage the cleavage process there. Thus the SCC process is one of interaction between slip and cleavage, with slip predominating on near-basal planes for the $\pm 30°$ Zircaloy texture commonly used in LWRs.[28]

Lee and Adamson[31] have evolved a dislocation picture of localized deformation in Zircaloys, which helps to explain the reduced SCC resistance of irradiated cladding. Localized deformation initiates through dislocation channel formation at stress concentrations, i.e., grain boundaries, triple points, surface flaws, etc., and irradiation damage is swept and annihilated. Channels start propagating from one grain to another, depending on availability of slip systems and orientation of grains. Some of the propagating channels link up to form slip bands, and with increasing displacement the channels strain harden over the region that has been sheared. This intense local strain concentration allows crack formation at fairly low macroscopic stresses. Thus, the flow concentration susceptibility in irradiated materials appears to be the critical factor responsible for the degradation of Zircaloy's SCC resistance by irradiation. Lee and Adamson[31] state that "crack initiation under iodine atmosphere is caused by local strain, not by the applied stress." On examination, this conclusion is not totally in conflict with the stress threshold concept since a macroscopic applied stress may be the indicator of SCC formation, although, at the grain level, it is local strain concentration that initiates cleavage.

A model of the macroscopic inelastic deformation of Zircaloy which implicitly accounts for this plastic flow localization behavior has recently been developed.[32, 33] The model, MATMOD,† introduced in Chapter 1, could independently predict the strain softening characteristic of irradiated Zircaloy that Hosbons et al.[34] had earlier shown was the manifestation of this localized deformation. Two additional features make MATMOD a useful model for predicting macroscopic stress (or strain) conditions during PCI; these are the effects of anisotropy and the development of the constitutive equations in a form appropriate for either the axisymmetric fuel codes or the general-purpose finite-element analyses.[35]

Role of the Fuel Pellet. The fuel pellet is both the source of clad loading and of fission-product chemical environment. In the former role, it is the cracking, and relocation of the cracked fuel segments, that provides the principal source of cladding stress, with fission-product swelling making a smaller contribution (in contrast to the situation in the LMFBR fuel rod—Chapter 3). At higher fuel rod powers and hence fuel temperatures, fuel pellet plasticity can of course relax part of these stresses, but, in general, operating conditions minimize this effect. Stresses (and hence strains) are concentrated over fuel pellet cracks at pellet–pellet interfaces, due to the "wheatsheafing" expansion characteristics of the fuel pellet. Figure 2-12, due to Gittus,[36] illustrates the "theoretical" picture of a cracked fuel pellet at operating temperature. Utilizing such a model, hoop stress concentrations of factors of 2 or more have been calculated in PCI situations.[13, 28, 36] The stress concentration is a sensitive function of the assumed coefficient of friction between fuel and cladding, typically $\mu = 0.5$, and the number of fuel pellet cracks, typically $N = 6$ to 10. Increasing μ or decreasing N raises the stress concentration; likewise low clad stresses are calculated when μ is low and N is high.

---

† *Materials Model.*

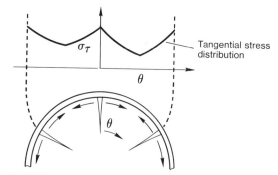

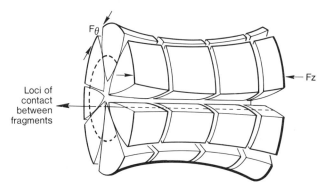

Fig. 2.12. Theoretical pellet condition during irradiation (Ref. 36).

Support for the hypothesized large diametral expansions due to UO_2 cracking and relocation has recently been obtained by out-of-pile simulations of the nuclear thermal environment.[37] The key results, summarized pictorially in Fig. 2-13, indicate that: (a) Pellets subjected to maximum equivalent nuclear power ratings in excess of 25 kW/m exhibited substantial diametral relocation after a single power cycle (Fig. 2-13b). Below this power threshold, virtually no relocation was observed despite extensive cracking (Fig. 2-13a); (b) in tests with ramp rates <185 kW/m-h, pellet relocation increased (up to 2% $\Delta D/D$ as the power rating increased; and (c) in tests with ramp rates >185 kW/m-h, several pellets exhibited fragmentation and particularly high relocation of up to 4% $\Delta D/D$ (Fig. 2-13c). Carried over to in-reactor behavior, these results could be quite significant to the understanding of the PCI phenomenon since they point to power and ramp-rate thresholds for pellet relocation which are within the range of power reactor operations. For instance, PWR fuel has historically operated below the 25 kW/m threshold and is observed

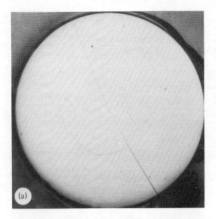
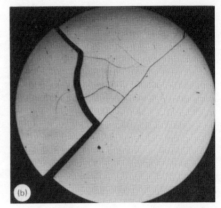
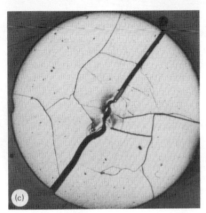

Fig. 2.13. Effects of power and power ramp rate on UO_2 pellet fragment relocation as observed in direct electrically heated tests (Ref. 37). (a) Pellet tested below power and ramp-rate thresholds shows no relocation. Maximum equivalent nuclear power 25 kW/m. Ramp rate 33 kW/m-h. (b) Pellet tested above power threshold, below ramp-rate threshold shows 1.9% $\Delta D/D$. Maximum equivalent nuclear power 48 kW/m-h. (c) Pellet tested above ramp-rate threshold, close to power threshold shows 4.1% $\Delta D/D$. Maximum equivalent nuclear power rating 31 kW/m. Ramp rate 225 kW/m-h.

to be highly reliable in general.[1-3, 10] In contrast, a significant fraction of early (7 × 7) BWR fuel has operated above this threshold power and is observed to have been PCI-failure-prone.[1-3, 10]

The second key aspect of the PCI process is the availability of aggressive fission products at the cladding inner surfaces. Current information about fission product availability in fuel rods is very limited and is only sufficient to indicate a correspondence between the occurrence of PCI cracks and the presence of fission products at the cladding inner surfaces.[38] Specific information about fission product availability under various regimes of fuel rod operation is lacking.

Earlier work established that the species most aggressive to Zircaloy are iodine, cesium, and cadmium (in that order) which are released in vapor form from regions of the fuel that operate at high temperatures.[18, 20, 38] These vapors then migrate and form compounds (e.g., CsI) that condense on the cooler inner surface of the cladding. In calculating the release of the aggressive volatile fission products (VFP) the assumption is typically made that the fraction released is equal to the fraction of inert fission gases released by the fuel. The advantage of this approach is that reasonably sophisticated models are available to calculate the latter quantity (see Section 2.2.2.1c).

A recent "first generation" model for the behavior of VFP provides a more detailed picture of their behavior.[39] As schematically represented in Fig. 2-14, several different regimes, and hence behavior patterns, are expected:

(1) For low fuel temperatures, the VFP do not redistribute.
(2) For steady-state operation at higher temperatures or for slow temperature increases, the VFP migrate readily toward the fuel periphery and for the most part remain in the UO_2 (a small fraction find crack surfaces and hence an easy path to the rod free volume).
(3) For low burnups or small increases in burnup after the most recent temperature rise, the VFP will migrate radially to the fuel periphery and remain in the UO_2 even for fast temperature increases.
(4) For fast temperature increases after some burnup (larger than that of Item 3) at low temperature, the VFP will be released and deposited on cladding inner surfaces and will be available to cause SCC or PCI failures.
(5) The fractional release of VFP during the operational mode identified in Item 4 above will be approximately equal to the fractional fission gas release.
(6) The fractional release of VFP during operational modes 1, 2, and 3 above is expected to be smaller than the fractional fission gas release.

The interesting feature of this model is that it can explain the effects of duty cycle on PCI failure probability strictly in terms of the availability of VFP at the cladding inner surface, and hence downgrades the importance of mechanical loading of the cladding. Wood reached a similar conclusion with respect to PCI failure characteristics of CANDU fuel;[40] failure threshold power was highly correlated to the onset of fission gas release.

2.2.1.1b. PCI Failure Models and Criteria. Some form of model of the PCI failure phenomenon is a prerequisite for initiating remedial actions. The conservative operating recommendations introduced early on by the fuel suppliers to minimize PCI damage constitute an empirical model of the process, i.e., failure probability is a function of maximum power, Δpower, burnup, etc. This empirical philosophy has to a large

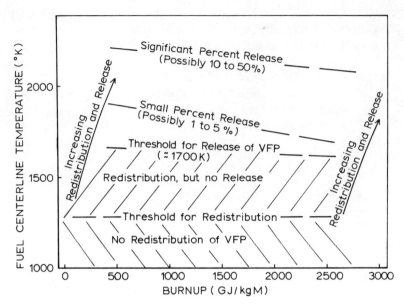

Fig. 2.14. Schematic representation of redistribution and release behavior of VFP predicted by state-of-the-art model (after Ref. 39a).

degree dominated PCI model development ever since. Attempts to become fully mechanistic have been confounded by the complexity of the problem, whereas empirical models have been regularly "tuned" as more and more "field" experience is gained.

Thus the situation today is that PCI failure predictions are based primarily on empirical or semiquantitative models, which statistically treat a combination of power reactor, test reactor, and/or laboratory data. Theoretically based models, derived from fracture mechanics concepts[41, 42] or dislocation theory,[43] and utilizing the basic research discussed in Section 2.2.1.1a, are still being pursued but are not yet ready for field application. Nevertheless, they are providing confirmatory support for decisions made as to the directions that various remedies should take.

Empirical or data-based statistical models utilize a broad range of data and generally yield predictions in terms of failure probabilities.[12, 23, 44, 45] Best known is a Canadian model, FUELOGRAM, developed from and for CANDU fuel operations.[44] Recently, this concept of defining failure probability in terms of maximum power, Δpower, and burnup has been extended to LWR performance, with mixed results.[12] Similar models, derived from LWR data, are PROFIT[12] and POSHO.[45]

The PROFIT model[12] is the result of applying standard statistical regression methods to the available PCI fuel failure data and an analysis of the environmental and strain-rate-dependent stress–strain properties of the Zircaloy cladding. It incorporates pretransient power, transient increase in power, and burnup parameters, and introduces a strain-rate dependent, *s*train-*e*nergy-*a*dsorption-to-*f*ailure (SEAF) concept as the mechanistic corollary of the power ramping rate. Predictions of the PROFIT model have been compared to predictions from other empirical fuel failure models but, to the author's knowledge, no independent predictions have been reported to date.

The POSHO model, on the other hand, received intensive evaluation, using both European and U.S. fuel operating experience. The original equations were based on test reactor experience (specifically the Halden test program), with parameters subsequently tuned to the growing commercial reactor experience. Due to its basis on purely mechanical interactions between fuel and cladding, one of the major problems encountered with POSHO was its inability to predict failures in the low-stress, high-fission-product PCI-SCC regime. It did, however, provide reasonable predictions of failure probability in situations where a single large "power shock" was imposed on the fuel.[45]

There is reasonable evidence to conclude that the PCI damage mechanism is cumulative; that is, cracks that might initiate under one power increase might only propagate as a result of subsequent power increases. Thus a large number of small power ramps might lead to the same result (i.e., failure) as a small number of larger power increases. This cumulative nature of PCI damage requires that detailed knowledge of the fuel rod duty cycle be used in the analysis of integrity. Failure to include all power shocks will result in underprediction of PCI defects, for example.[45] The treatment of damage accumulation in the empirical models has also caused difficulty. Generally, simple linear damage laws have been used. However, recent laboratory measurements do not support this approach.[46] The data suggest that the use of a linear damage accumulation rule may result in underestimates of failure probability under rising stress, and overestimates under falling stress situations. This is one problem area that has motivated the mechanistic modeling activities.

The probabilistic nature of PCI damage is equally evident; the low incidence of failures attests to the fact that clad breaching is a consequence of the overlapping of the "tails" of several distributions—stress, chemistry, material characteristics, etc. Experimental evidence supports the view that a threshold stress exists that is characteristic of the fission product iodine availability and below which cladding cracks are unable to form. The cladding hoop stress must exceed the threshold for a finite time at any particular fission product iodine concentration for the crack(s) to form. Zircaloy can sustain stresses greater than the threshold value, without failing, provided that the time spent above the threshold value is not excessive.[19, 23, 24] Hence, in probability terms, the probability of fuel rod failure increases as either the stress or time spent above threshold, or both, increase. Stress and time are related through the stress relaxation properties of the Zircaloy; for example, if the cladding relaxes rapidly, time above threshold will be short unless the initial cladding stress was well above threshold. However, the time component must also include a contribution to allow for the conditions at the inner surface to become ripe for cleavage to begin. Thus, one expects (and observes) a wide variation in failure times in in-reactor PCI situations, even when test conditions are considered identical. Also, the basis for a "damage" accumulation law has been provided, since a number of short-time excursions through the threshold should have a similar effect as a single long "hold time" above the threshold.

Because of this "sensitivity" of PCI to a number of operational and material variables, a deterministic treatment of the mechanism is an essential ingredient for a probabilistic failure model if one wants to extend the regime of application of the model beyond that covered by real data. This fact was recognized in the development of the SPEAR fuel reliability code system[47] which will be described in some detail below by way of example of PCI model developments.

Fuel failure predictions from a mechanistic model (virtual data generator) are combined with predictions from a separate (real data) empirical model; the relative weights assigned the two models in reaching a final prediction will depend on their relative regime-by-regime prediction capability. The two approaches are further combined by using some intermediate variables from the mechanistic model (e.g., predicted cladding stress) as independent variables in the empirical model. The deterministic submodel is called *C*ladding *F*ailure *M*odel (CFM), and currently version 1, CFM 1, is in SPEAR. It represents a somewhat simplistic view of the failure process but contains the essential ingredients of (a) a threshold stress, (b) a requirement for fission product attack, and (c) damage accumulation.

Details of CFM 1 and its application are provided by Roberts *et al.*[23] and Miller *et al.*[28] The MATMOD Code is used[33] first to compute the stress-versus-time evolution (including stress concentration) in the cladding and, from this history and information on corrosive fission product (I_2) availability obtained from a fission gas release model, CFM 1 is used to generate failure predictions, applying a linear damage accumulation rule. In CFM 1, iodine availability at pellet–pellet interfaces is assumed to be proportional to total burnup and fractional fission gas release. Average time to failure, t_f, is obtained as a function of failure (threshold) stress σ_f and texture from a fit to the laboratory data described earlier (Figs. 2-9 and 2-10, for example). Thus,

$$\frac{\sigma_f}{\sigma_y} = 1.2 - 1.96 \cos\psi \sin\psi \left\{ 1 - \exp\left[-\left(\frac{x}{490 \text{ sec}}\right)^{1/3} \right] \right\} \quad (2\text{-}2)$$

where σ_y is the irradiated yield strength, ψ is the texture angle, and $x = t_f \cdot I_2/(3 \times 10^{-2} \text{ g/cm}^2)$; here t_f is the average failure time in seconds and I_2 is the iodine concentration in g/cm² of Zircaloy Surface.

Equation (2-2) is applicable only for fluences greater than 3×10^{20} n/cm² and for $0 < \psi < 30°$.

The linear cumulative damage rule for failure in CFM 1 (based on time to failure at σ and I_2 present during any given time step Δt) is

$$D = \sum \Delta D = \sum \frac{\Delta t}{t_f(\sigma, I_2)} \quad (2\text{-}3)$$

where the average D for failure is equal to 1.0, and the standard deviation on $\log_{10}(D_{\text{failure}})$ is equal to 0.5 (i.e., 95% of specimens fail when $0.1 < D < 10$). Failure probability predicted by CFM 1 applies to a cladding I.D. surface area of 0.009 m² (i.e., equivalent to the test areas in laboratory samples).

A more satisfactory model would allow one to predict crack length versus time and even the failure mode (intergranular, transgranular cleavage plus fluting, or ductile rupture), rather than simply fail/no fail. However, to achieve this, the time-varying local conditions (i.e., stress, strain, and iodine concentration) must be tracked, as opposed to using global parameters such as stress intensity or average hoop strain. The theoretical picture outlined earlier in Section 2.2.1.1a has provided the basis for development of a quantitative mechanistic treatment of the SCC process in Zircaloy which is usable in fuel rod failure codes. To achieve this goal for the second-generation Model CFM II, Miller[43] employs a mathematical simplification to avoid the complexities and expense of finite-element analysis and yet allows one to track the progress of the stress corrosion crack on a grain-by-grain (and hence texture-dependent) basis. CFM II will be included in the second-generation SPEAR Code[47] SPEAR β.

The empirical model component of SPEAR is a statistical analysis of actual fuel rod data derived from power reactor and test reactor experience.[47] For example, the SPEAR α version contains PCI fuel failure data from one PWR, one BWR, and two test reactors, representing 599 fuel assemblies or ~56,000 fuel rods. A computer "pattern matching" procedure is used to develop patterns or correlations between the dependent and independent variables. The two best failure indicators from the current version are high creep strain at maximum fuel rod power and high fission gas release, which is consistent with our understanding of the PCI/SCC mechanism. As more and more data are included in the pattern matching analysis, the regime of application of the empirical model will broaden and the uncertainties on predictions will reduce. The SPEAR β version will likely contain more than twice the α version's data set. Preliminary results from SPEAR are encouraging.[47b] In one test case, SPEAR's predictions were better than POSHO's by a factor of 2 in error magnitude; but perhaps more important, SPEAR appears to predict well for both BWR and PWR fuel, whereas POSHO predicts well for BWR fuel but poorly for PWR fuel.

2.2.1.1c. PCI Remedies. As one can see from the foregoing discussion, the understanding of PCI is advanced but incomplete. Nevertheless, design and operating remedies have been introduced in an attempt to alleviate the impact of PCI on plant capacity. Design remedies have addressed the mechanical-chemical nature of PCI, namely reducing clad stresses, fuel temperatures, and/or fission product release. Operation remedies have emphasized reduced severity of the duty cycle through less use of control rods for achieving power maneuvers and power shaping (i.e., more use of flow control in BWRs or boron in the coolant of PWRs).

One of the most significant changes to the LWR fuel design that is expected to reduce PCI damage was the increase in the number of fuel rods per assembly to reduce the power per rod. For PWRs, however, the principal reason for this change was not PCI alleviation, but to reduce the consequences of hypothetical accidents to acceptable conservative levels (see Section 2.2.2.1). The number of rods per assembly for PWRs was increased from a 14×14 or 15×15 array to a 16×16 or 17×17 array, respectively. These new PWR designs are not generally retrofittable to older 14×14 or 15×15 plants. In contrast, the BWR changeover from a 7×7 to 8×8 array could be retrofitted into existing plants. The reduction of PCI failure incidences in BWRs since the introduction of 8×8 fuel has been significant, but the contribution of the design change is difficult to quantify since the reactors have also been operating under vendor guidelines during this period.[2,3,10] Nevertheless, at the request of a major German utility, Kraftwerk Union has fabricated 9×9 BWR and 19×19 PWR assemblies in an attempt to further capitalize on the PCI resistance that results from low fuel-rod power levels.

In broad terms three parallel paths are currently being pursued to achieve PCI-resistant or "nil capacity loss" core operations.[21] This procedure is necessary since full-scale demonstration programs can take up to ten years—through test reactor, lead test assemblies in a power reactor, and, finally partial or full reloads. These three efforts have been categorized as near-term (1979), mid-term (1982), and long-term (1986) time frames.

In the near-term time frame the emphasis is on relaxation of the conservative vendor operating recommendations which have most impact on BWRs.[5,10] Two approaches are being pursued. In the first approach, adopted by General Electric, "field" experience is being used as a basis for relaxing the limitations on power ascension rates; as PCI failure incidences at one ramp rate are seen to have been reduced to a statistically acceptable minimum, the ramp rate is increased on a plant-by-plant basis until the operating statistics again provide confidence for a further relaxation. In the second approach, being developed by EPRI under the name power shape monitoring system (PSMS), a fully computerized site-based system for tracking the mechanical integrity of the fuel on a near real time basis is being evolved.[48] The software in the system is state-of-the-art computer codes that calculate the power-exposure of fuel on a nodal basis and the associated failure probabilities due to PCI. The POSHO code is being used to compute failure probabilities in the first-generation system currently under test at the Oyster Creek BWR plant.[48] Since the system is modular, improved models based on current developments can be inserted when available. The hardware is a minicomputer that is linked to the process computer to obtain all core parameters essential for the subsequent calculations. A color CRT display provides the engineer with a graphic display of the fuel failure situation in the core, thus assisting decision analysis both in near real time (e.g., during power ascension) or in an anticipatory mode (e.g., planning reloading patterns to avoid PCI failures in future cycles).

The design changes being pursued over the mid- and long-term time frames involve changes to the fuel rod. Early changes adopted by fuel vendors included fuel pellet shape, clad thickness, and metallurgical structure, all of which were intended to reduce clad stresses. Subsequent and current changes being evaluated address either or both (a) the thermal, and hence fission product, environment in the fuel rod, and (b) crack formation at the cladding inside surface.[21,49]

In the mid-term time frame only minor modifications to the fuel pellet and cladding, which fall within the capabilities of current fabrication technology (if not within current specifications) and do not raise new safety or licensing questions, could be introduced into fuel reloads. One such "simple" change recently introduced by BWR vendors is prepressurization (typically 0.3 to 0.5 MPa helium) of the fuel rods, which is the current practice in PWR designs. This step should stabilize the internal thermal environment (i.e., fuel-to-clad gap heat transfer), thereby eliminating the possibility of the sharp temperature rise and high fission product release that is sometimes observed in unpressurized fuel rods.[14,20]

Modifications to the fuel pellet itself that would reduce (or eliminate) fission product release also appear practical. Three that are being pursued are the annular pellet,[50] a large grain size fuel pellet (fabricated using Nb_2O_3 as an additive) in which the large grains delay the grain boundary saturation and release process,[21] and a dual enrichment (or duplex) pellet which, by virtue of an outer enriched UO_2 region and an inner natural UO_2 core, operates at lower average temperatures and flatter temperature gradients for the same equivalent powers.[21,49]

The mechanical properties of Zircaloy have been modified progressively over the years to respond to various concerns (creep collapse, fuel rod bow, etc), and they seem to be optimized for overall performance in LWRs—this writer does not foresee radical changes to creep strength or ductility values, for example. However, texture could perhaps be modified to make the I.D. surface more resistant to crack formation and early-stage propagation. Theoretical analyses[28] and limited experimental data[31] indicate that a 0° basal texture exhibits substantially higher SCC resistance than the typical ±30° texture. Thus, in principle, a 0° texture, either at the inner surface or through the clad wall, should impart improved resistance to SCC, although any texture change will have to be compromised to meet other physical and mechanical property requirements.

A longer-term approach to modifying the inner surface of the Zircaloy cladding would involve altering fabrication schedules to eliminate (or disperse) the locally high concentrations of alloying impurities (Fe, Cr, Ni), or impurities (Al, Si) at or near which cracks have been found to initiate.[19] Suitable heat treatments, combined with a general "clean-up" at the ingot, and pickling and grit blasting stages could possibly achieve some significant improvements.

Of the "remedies" that fall into this general mid-term category, none alone, or in combination, appears likely to lead to a nil-capacity loss fuel. A modified design combination, such as a prepressurized fuel rod containing an annular, large grain size fuel pellet, when combined with controlled core operations using a power shape monitoring type of system, may well be more than adequate for PWRs in which power maneuvers are made by chemical shimming (with boron), or for base-loaded BWRs.

For load-follow capability in BWRs, however, which would make extensive use of the control rods, a more forgiving design must be sought. Since Zircaloy and UO_2, in contact, are basically incompatible, a barrier between the two fuel-rod components seems a logical remedy. There is some evidence which indicates that the I.D. surface oxide film itself could be an effective barrier to SCC if it remains intact,[51] but most vendors have opted for a barrier material. The Canadians have developed a graphite coating on the Zircaloy I.D., called CANLUB fuel, for their CANDU reactors;[52] extensive data indicate an improvement in PCI resistance which is due, primarily, to a fission product "gettering" mechanism, rather than to relaxation of interface stresses.[40] The Canadians are now pursuing a siloxane coating which will likely provide improved chemical gettering properties.[40,52] In the U.S. the Department of Energy (DOE), in cooperation with Exxon Nuclear, Battelle Northwest Laboratory, and Consumers Power Company, is pursuing this concept for LWR application.[50] In addition, this same project is evaluating particle fuel (e.g., Vipac or Spherepac†) as a PCI-resistant design for LWRs. Early LMFBR (see Chapter 3) and more recent European LWR operating experience indicates increased PCI failure threshold powers and power ascension rates.[53]

For LWR application a more acceptable barrier would be a metal on the cladding surface, since it is unlikely that a ceramic coating could survive a LWR fuel cycle (e.g., the discharge burnup of CANDU fuel is 1037 GJ/kgM compared to 2000 to 3000 GJ/kgM). It is generally believed that the BWR duty cycle will necessitate that the barrier be both impervious to iodine attack and sufficiently plastic to relax interfacial stresses in the cladding. Two candidates that are receiving worldwide attention are copper (plated as a 5–10 μ layer on the inner surface) and zirconium (coextruded with the Zircaloy).[54,55] Both laboratory PCI simulation tests and test-reactor ramp tests have confirmed the advantages of these concepts over conventional Zircaloy cladding in resisting cracking. The Cu- and Zr-barrier fuel rods survived ramps to ~60 kW/m and 6–24 h holds at peak power, after 864 GJ/kgM burnup, whereas conventional rods failed by PCI-SCC in an hour or less.[54b] Figure 2-15 shows the typical appearance of the (a) copper and (b) zirconium layers after ramping, compared to a (c) conventional cladding; these data are from the General Electric and Commonwealth Edison study sponsored by DOE. Power reactor demonstration programs are underway both in the U.S. and in Europe. First full reloads are expected for selected plants in 1986.

2.2.1.2. Waterside Corrosion of Zircaloys

The characteristic water corrosion (oxidation) behavior of Zircaloy was described in Chapter 1 (refer to Figs. 1-25 and 1-26). Zircaloy oxidation rates are increased, however, under irradiation in the reactor coolants of BWRs and PWRs. For the purposes of this discussion the external (i.e., waterside) in-reactor corrosion can take three major forms, viz. a thin, uniform cohesive layer, 1–10 μm thick, which will not exceed ~10 μm even after 10 years exposure, nodular corrosion that is a localized attack sometimes building up to $\gtrsim$ 200 μm thickness, and increased uni-

---

† Trade names for various particle fuel processes.

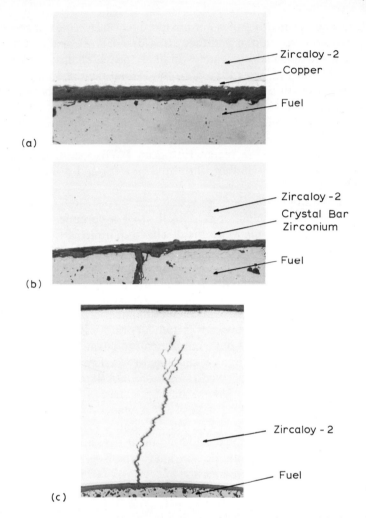

Fig. 2.15. Typical appearance of (a) copper barrier, (b) Zr liner (in as-polished condition), and (c) reference fuel rod after ramp testing (courtesy of S. Armijo, General Electric Corp.).

form oxidation that starts to spall at $\gtrsim 70$ μm thickness. Waterside corrosion of fuel rods has not proved to be life limiting to this point, but BWR channels have been removed prematurely due to spalling corrosion.

2.2.1.2a. Corrosion in BWRs. The corrosion of Zircaloys in BWRs is strongly influenced by the fast neutron flux. Both nodular corrosion and spalling corrosion have been observed in BWR Zircaloy components, with the former being a typical observation on cladding, grids, and channels.[21]

LWR Core Materials

The nodules are small white pustules that are apparent to the eye (Fig. 2-16), the pustules protruding outward from a black background-type oxide. The nodules protrude into the metal as well as above the surface and are lenticular in cross section (Fig. 2-17). While it is well established that the appearance of nodular corrosion is a product of the free oxygen (and possibly nitrogen) from the radiolysis in the coolant, and may be aggravated by the presence of dissimilar metals, the mechanism of nucleation is not clear. Various workers[56,57] have suggested that galvanic effects localized within the oxide film, and associated with particles, may be important. In addition, since nucleation does occur readily at sites of large curvature (e.g., edges and scratches), mechanical failure of the protective oxide may also be involved. It is postulated (but cannot yet be verified) that subsequent growth is controlled by radiation-enhanced anion diffusion through a continuous protective layer of nominally constant thickness maintained at the base of each nodule throughout the oxidation period.[58] Such an effect could arise by stress-induced cracking of thicker oxides and rapid growth of thinner ones; extensive cracking is a characteristic feature of nodules, as shown in Fig. 2-17.

In principle, nodular corrosion could result in (1) excessive thinning of the cladding wall thickness; (2) loss of corrosion product to the coolant stream; (3) enhanced hydrogen pickup due to the oxidation reaction and/or (4) increased cladding temperature due to the insulating effect of the thicker oxide. Experience to date, however, indicates that only the first might prove detrimental, that is, excessive wall thinning. Typically the

Fig. 2.16. View of Class III nodular corrosion, BWR rod at end of second irradiation cycle in Peach Bottom-2 (Ref. 37).

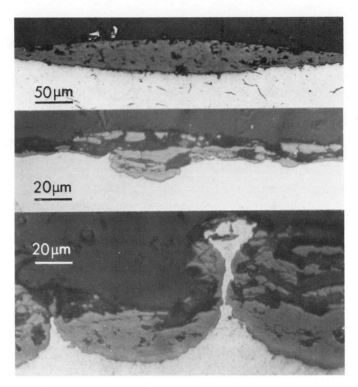

Fig. 2.17. Sections through nodules on irradiated cladding at grid positions (Ref. 54).

maximum wall thinning observed in BWR fuel rods never exceeds 10%, and even the thickest nodules have proved to be remarkably tenacious and adherent. A few exceptions to this rule occur each year, and point to the as yet incomplete understanding of this phenomenon. For example, a very recent incidence of fuel rod failures in the Vermont Yankee BWR might be the result of excessive wall thinning due to accelerated nodular corrosion.[59] A preliminary view is that the insulating effect of tenacious crud on the fuel in this plant resulted in increased cladding temperatures and hence accelerated oxidation. Contributing effects such as cladding contamination or unusual second-phase particle distribution in the cladding could account for the accelerated corrosion being limited to certain rod batches, rather than distributing uniformly throughout the core.

Notwithstanding the very few occurrences of this type, nodular corrosion on BWR fuel rods does not appear to be life limiting since there is sufficient conservatism in the wall thickness to allow up to a 10% wall reduction without loss in the required load-bearing cross section. However, the spacer grids, which also exhibit nodular corrosion, are thin enough to be influenced mechanically by the effective reduction in thickness caused by nodule growth, and this factor is an area of current attention.[21]

By far the most significant Zircaloy waterside corrosion problem in BWRs is the spalling corrosion observed on some Zircaloy 4 channels. Although it is known that Zircaloy 4 is not as resistant to oxidation as Zircaloy 2 in the BWR environment, it is not clear whether the accelerated corrosion results from growth and coalescence of nodules or simply from increased uniform oxidation. Channel corrosion has, however, in some instances resulted in discharge of channels after only one irradiation cycle (they are intended for up to two fuel lifetimes, or 8–10 years, and many have survived), and the oxide flakes that spall off during both operation and handling have caused radioactivity buildup in the reactor vessel, which increases vessel maintenance problems.[58]

In some plants channel corrosion *per se* is not life limiting but enhances the channel creep effect (outward deflection or bulging) by effectively reducing the load-bearing wall thickness, which is of concern because of potential interference with control blade movement (refer to Fig. 2-3).[60] An interim measure in this case is to use thicker wall channels, and many utilities have opted for 2.54-mm and even 3.05-mm wall channels to replace the earlier 2.03-mm design. Older plants do not have this option open to them, however, because of the smaller tolerances in the core design.

From the available evidence, therefore, it appears that either creep deflection or corrosion or a combination of both can limit channel life to less than a fuel bundle lifetime (4 to 5 years). The contribution of each effect varies from reactor to reactor, presumably because of differences in core design (i.e., pressure drop, fast flux, and water chemistry), and characteristics of the channel material. This "plant dependence" of waterside corrosion has motivated utilities to mount their own channel surveillance programs.[60]

Thus, improvement of channel corrosion resistance (to a 10-year lifetime) is the primary goal of R&D in this area. However, both the fuel suppliers and utilities also regard the nodular corrosion on fuel rods as a problem, albeit "cosmetic," and, accordingly, in conjunction with channel improvements, are developing various modified Zircaloys (different heat treatments and surface preparations) for cladding. The basis for much of the research is the galvanic nature of the corrosion process. Beta-treated materials appear more resistant than conventional alpha-treated materials, which is explained by the effect of the different heat treatments on the distribution of second-phase particles in the microstructure.[57,61] Evidence exists which shows that the intermetallic particle provide sites

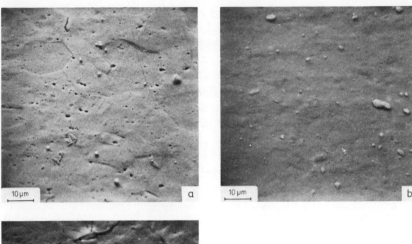

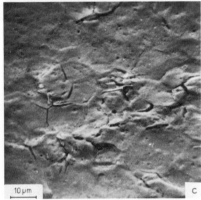

Fig. 2.18. Comparison of anodized Zircaloy surface (b) with conventional Zircaloy surface treatments (a and c) after 700 days reactor exposure (Ref. 62).

for easy electron transport through the oxide film. The particles are larger in size than the film thickness initially, in which case the particles extend through the film, and the conduction is electronic. However, as the film thickens, the particles become isolated from the substrate as they are incorporated into the film and thus cannot provide easy conduction paths. Once the particles are insulated from the metal, localized attack can occur, leading to nodule formation. The beta-treated samples have intermetallic particles in very close contact along grain boundaries, and thus they never become insulated from either the metal or oxide surface. In other words, the easy conduction paths remain even though the film thickness is greater than the particle size.

Anodizing is also being evaluated as an alternative to autoclaving for pre-oxidizing the Zircaloy surfaces.[62] Figure 2-18 illustrates the improvements in oxide integrity after exposure in-reactor. Channel vendors are currently irradiating anodized channels in commercial BWRs.

Alternative zirconium-base alloys being re-examined for application as future channel materials include Zr–2½%Nb, Zr–3%Nb–1%Sn, Zr–1%Sn–1%Nb–0.5%Fe, Zr–Nb–Cu, and ZrCrFe alloys. All seem to have corrosion properties either equivalent to or slightly better than the Zircaloys, and some, such as Zr–2½Nb, have better high-temperature strength also. The general consensus is, however, that a new class of alloys will not be subject to a detailed development program unless the aforementioned modifications to Zircaloys provide no real improvement.

2.2.1.2b. Corrosion in PWRs. Most early data showed that the corrosion rate of Zircaloy 4 under normal conditions (i.e., hydrogen overpressure and crud inhibitors such as lithium hydroxide) in PWRs was only enhanced slightly over ex-reactor rates.[58] Dalgaard[63] argued that the oxidation rate in PWRs does not show any enhancement. This difference in opinion undoubtedly arises because of the difficulty in unambiguously detecting an enhancement factor of 2 to 3 when both material and local environmental conditions (e.g., temperature) are imprecisely defined.

Data recently obtained from both nondestructive and destructive measurements of oxide film thicknesses in commercial PWRs confirm that uniform oxidation is enhanced in-reactor, Fig. 2-19.[64] There is large scatter in the data, especially at higher burnups, where typical oxide layer thickness of up to 50 μm at a rod average burnup of 3268 GJ/kgM are observed. The thickest oxide layers are formed in the upper one-third of the fuel rods due to the increasing coolant temperature with axial position. In-reactor corrosion rates appear to be enhanced by up to 3.6 times that expected from thermal corrosion alone.

Apart from the neutron flux, oxide film thermal conductivity and coolant chemistry (lithium hydroxide in PWRs) are major factors which can influence corrosion behavior.[9,21] Poor heat transfer characteristics of the oxide films developed in PWR environments could result in an increase in the oxide-cladding interface temperature and further oxidation, which is autocatalytic in nature.[9,21] Alternatively, since data show that the Zircaloy corrosion rate is accelerated in the presence of lithium hydroxide when the pH exceeds 11.3,[64] concentration of lithium hydroxide in the thicker oxide films could lead to the observed corrosion enhancement. In this mechanism the lithium increases the anion vacancy concentration when introduced into the ZrO_{2-x} lattice, and so if oxidation depends on the concentration of anion vacancies, the presence of lithium should increase the oxidation process.

Indirect evidence from the study noted above[64] suggests that there may be a decrease in film conductivity of thicker films as the corrosion film changes in structure and circumferential cracks develop (refer to Fig. 1-26 earlier). However, the effect of lithium hydroxide on corrosion behavior was not seen in this study. In one reactor the lithium hydroxide

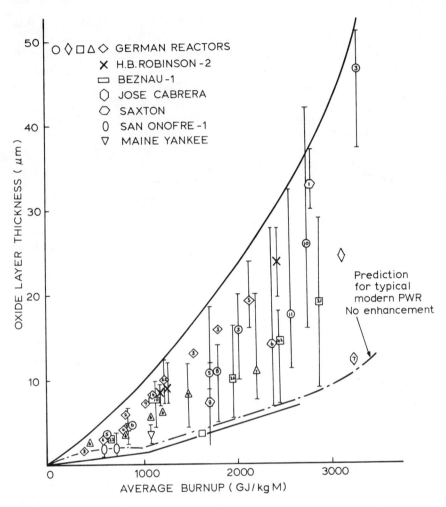

Fig. 2.19. Oxide layer thickness of commercial PWR fuel rods, at the axial position of maximum corrosion, vs. burnup. Numbers within symbols refer to number of data points (Ref. 64).

content was intentionally reduced from an average of 1.4 ppm in the first cycle to 0.8 ppm in the second cycle, but the corrosion enhancement factor only decreased from 2.3 to 1.9, which is probably within the variability of the data. Therefore, it must be concluded at this point that the Zircaloy corrosion enhancement in PWRs is primarily a function of local thermal conductivity reductions in the oxide films caused by cracking, and is more a function of cladding temperature than of fast neutron fluence.

Although this phenomenon has never restricted operating strategies or impacted design limits on current PWRs, there is concern for the future operations where incentives for minimizing uranium requirements and spent fuel storage facilities have led to considerations of fuel rod designs and/or duty cycles that would result in higher life-average linear heat ratings, higher coolant temperatures, and/or longer in-reactor residence times (see Section 2.6). Inasmuch as any or all of these factors will accelerate the corrosion rate, there is a growing need to develop the statistical data base necessary to determine operating limits and margins under these new conditions (e.g., maximum allowable fuel rod power and surface heat flux versus exposure).

2.2.2. Fuel Operating Margins (or Fuel Safety Considerations)

Thermal limits on fuel rod components, such as maximum cladding temperature, fuel temperature, or fuel stored energy; cladding strain limits; and thermal/hydraulic limits, such as departure from nucleate boiling (DNB), are established to ensure acceptable fuel conditions even under hypothetical accident situations. These so-called fuel damage limits have been developed over a period of years in response to three reactor licensing needs: viz., to assure no fuel rod failures during normal operations and moderate frequency transients; in the event of a severe accident, to determine the numbers of failed rods for accident dose calculations; and to assure that "coolability" is maintained even in the worst accident situation. The term coolability means that the fuel assembly retains its rod-bundle geometry with adequate coolant channels to permit removal of residual heat.

These material performance limits are employed in three primary operational categories: (1) loss-of-coolant accidents (LOCA); (2) reactivity-initiated accidents (RIA), covering events such as control rod ejection for PWRs and control rod drop for BWRs; and (3) steady-state operation, and moderate frequency transients.[65] For the purpose of reactor operation the limits are translated into core parameters. For example, the maximum linear heat generation rate (LHGR) represents the PWR LOCA limit and the BWR RIA limit, while the maximum average planar linear heat generation rate (MAPLHGR) serves as the BWR LOCA limit, the departure from nucleate boiling ratio (DNBR) is the PWR RIA and moderate frequency transient limit, and the minimum critical power ratio (MCPR) is the BWR transient limit.[10]

Operating margin (sometimes called "thermal" margin) can be defined as the difference between these established limits and the actual condition prevailing in the operating reactor core. Operating margins depend on a number of variables, including the fuel and core design, accumulated irradiation, actuation criteria for engineered safety features, and reactor operating mode. Accordingly, these margins vary from plant to plant, depending on many design features and the mode in which the unit is operated. Generally, the maximum margins to design limits are experienced when the core is in steady-state condition, that is, when the unit is base loaded. Margins are diminished during operational transients such as load-follow maneuvers.[10]

To illustrate the concept of operating margin, consider the relationship between LOCA operating limits and actual values of maximum LHGR which prevail under load-follow and base-load operating modes. These are sketched in Fig. 2-20 for a modern PWR.[10] LOCA limits on fuel rod power range from ~43 to 57 kW/m for 1200 MWe PWR designs, depending on core elevation. Typical projected values of base-load margins are in the range of 13 to 20 kW/m (i.e., maximum fuel rod powers for base-load operation range from ~20 to 37 kW/m). For load-follow operation the margin is reduced to the range of 3 to 13 kW/m. Given uncertainties in instrumentation and less than optimum operator control strategies, these margins are not particularly large. It should be noted in passing that, for the latest BWR designs, LOCA criteria are not expected to be limiting; instead, the moderate frequency transients will be the limiting accidents and MCPR the controlling core parameter.[10]

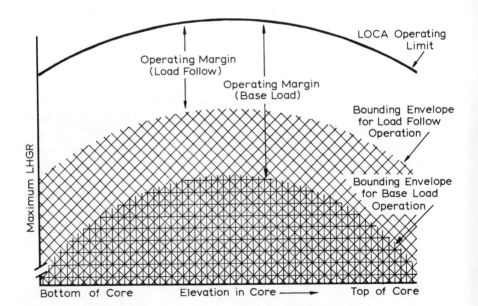

Fig. 2.20. Example of PWR operating margin (Ref. 10).

It is prudent design practice to maximize this operating margin at the outset subject to other physical and economic constraints, because a large margin means improved operational flexibility for the plant and, as a result, probably improved capacity factors. Also, there is an intangible value of a more-than-adequate operating margin as a hedge against more restrictive regulatory changes and new "phenomema" or data that might adversely effect the models used in accident analyses. For instance, in the following subsections it will be noted that conservative treatment of data on Zircaloy embrittlement under LOCA conditions, and uncertainties in fission gas release and fuel rod bow behavior have resulted in progressive erosion of some of the operating margins. In the worst case, of course, thermal output must be reduced (i.e., plant derates) to preserve some margin. This has happened for short periods in the past, while new analyses or new operating procedures were developed that would allow a return to normal operating conditions.[66]

Clearly, in this general area of maintenance or improvement of thermal margin, the better quantification of those limiting materials phenomena plays an important role. However, the relative importance of the Zircaloy–steam reaction, fuel rod bow, and fission gas release varies from reactor to reactor and with fuel design, and, consequently, research is proceeding on all fronts. Progress in this area of fuel performance improvement will now be reviewed.

2.2.2.1. LOCA Criteria

During a LOCA, although the fission reaction has been terminated by "scramming" the reactor (i.e., inserting all control rods), the fuel pellets continue to transmit stored heat and radioactive decay heat for some time. Inasmuch as the heat transfer efficiency has been reduced by the loss of coolant, the cladding temperature rises rapidly and the normally very slow Zircaloy oxidation process (Section 2.2.1.2) is now accelerated enormously in the steam environment that remains. Additionally, the concomitant pressure increase inside the fuel rod causes the cladding to dilate.

For the purposes of radiation dose calculation, it is assumed that, under such severe oxidation/deformation conditions, all the fuel rods fail, and so emphasis is placed on maintaining coolability of the core. The idea is that when the emergency core cooling system (ECCS) is activated a few seconds after accident initiation, the water entering the top of the core should have an easy path through the core to achieve effective heat removal. This is achieved by establishing a limit on both peak cladding temperature, 1478 K (2200°F), and on total cladding oxidation, 17% of the wall, attainable during the accident. These criteria are spelled out explicitly in Appendix K to Part 50 of the Code of Federal Regulations (i.e., 10CFR50).

It must be shown by analysis that steady-state operating conditions do not cause the plant to exceed these limits during the postulated LOCA. Any uncertainties in fuel rod behavior (cladding or fuel properties) that impact fuel stored energy (temperature), rod internal pressure, or oxidation rate calculations might result in LOCA limits being exceeded. One classic past example was the issue of fuel densification, in which uncertainties in the amount of fuel shrinkage in-reactor produced large uncertainties in fuel stored energy, and hence in cladding temperature calculations. The consequence of this problem was the ordering of power reduction (derates) or operating restrictions for several plants in order to regain the required thermal margin.[66] A current, lesser, but still important, issue, is the uncertainty in the rate of fission gas release from the fuel pellet to the fuel rod interior during normal and transient operations.

In the case of the LOCA criteria impact, therefore, materials research was required for two major aspects of the problem. In the first instance, it was essential to either eliminate the phenomena or remove the uncertainties from the data, which impacted the steady-state fuel thermal performance; secondly, it was recognized that the LOCA limits themselves were conservative (although the degree of conservatism was unknown), and so additional data on Zircaloy properties were required to quantify conservatism there. Some of the key results from the rather intense government- and industry-sponsored research undertaken over the past five years are presented below.

2.2.2.1a. Zircaloy Embrittlement Criteria. The LOCA criteria contained in 10CFR50, Appendix K, reflect conservative positions with respect to Zircaloy properties because the scatter or nonstatistical nature of much of the data available at the time required use of conservative upper bounds on the existing data. In particular, the need for more information on the oxidation, embrittlement, and deformation of Zircaloy cladding at high temperatures and its physical and mechanical properties at high strain rates was identified. Johnston[67] has summarized the research projects initiated worldwide in response to this need.

The results of this research have since been extensively reported.[4,37,68–76] Zircaloy oxidation behavior in steam has been measured by several laboratories. A recent critical review of the data[71] showed that the specimen heating procedure must be taken into account in comparing the various data sets. The "internally heated" data set, which is more appropriate for the fuel rod situation, yielded a pre-exponential term and activation energy some 20% smaller than the "externally heated" data set. At 1478 K these data predict oxidation levels almost a factor of 2 lower than computed by the Baker–Just equation

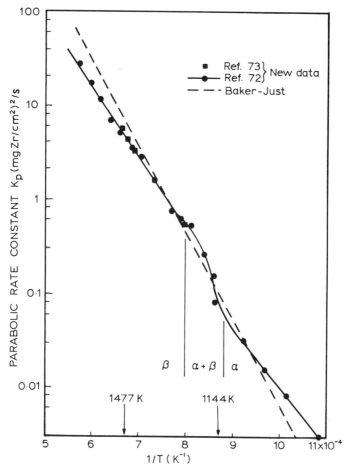

Fig. 2.21. Comparison of new isothermal oxidation kinetics data for Zircaloy in steam with Baker–Just correlation used in Appendix K (adapted from Ref. 71).

adopted in Appendix K (Fig. 2-21). Pre-oxidation effects provide an additional element of conservatism in calculations of oxidation over the course of the LOCA transient. For example, results from prefilmed specimens suggest that the total amount of oxidation measured over a simulated PWR-type LOCA transient is less than that calculated from isothermal reaction kinetics data obtained from measurements on as-received tubing.[72]

Following cladding rupture, Appendix K requires that the rate of oxidation for both inside and outside surfaces of the cladding be considered equal (i.e., I.D./O.D. ratio = 1) over an axial distance of 38 mm from the rupture, and the maximum oxide thickness must not exceed 17%. Measurements of the I.D./O.D. ratio were made as a function of initial internal rod pressure, final hold temperature, time, and distance from the rupture position.[73] Although a great deal of scatter was found in the I.D./O.D. oxide ratios, the average value even at the rupture position was less than unity. About 10% of the measurements were larger than unity, the maximum value being 1.4. The strongest trends in the data were a decreasing I.D./O.D. ratio with decreasing test time and increasing distance from the rupture site. The average I.D./O.D. ratio at a distance 51 mm from the rupture was less than one-half of the value at the rupture site, and at a distance of 102 mm, the ratio was about one-fourth that at the rupture site.

The 17% total oxide thickness criterion was proposed to ensure no loss of cladding integrity upon cooldown (i.e., the so called "reflood" stage utilizing the emergency core cooling system or ECCS) from thermal shock loads.[65] Recent results, summarized in Fig. 2-22, indicate that the present criterion is conservative with respect to both thermal shock and impact loads.[74] Interestingly, hydrogen embrittlement has been observed in addition to oxygen embrittlement in these Zircaloy–steam experiments. No explanation for this effect is available, but the data in Fig. 2-22 implicitly account for all forms of steam embrittlement and therefore can be used as a basis for revised criteria. A new criteria would probably be based on a maximum content of oxygen in a minimum thickness of wall material.

From the foregoing, it is possible that the principal LOCA cladding criteria could be replaced by less conservative limits, which would lead to increased thermal margins. In addition, other data on cladding dilation (or swelling) under LOCA conditions[75a] and improved physical property values[75b] have resulted in refined calculations of heat transfer and stored energy, both of which represent potential penalties in LOCA analyses. Further, the concerns over cladding ballooning and the potential for so-called localized coplanar flow blockage have been put to rest. The USNRC[76] has confirmed that uniform ballooning of cladding depends on uniformity of stresses and temperatures which are unlikely to occur in practice. This view is supported by the results of many prototypic single- and multirod LOCA simulation tests.

Thus, to conclude, in this instance no fuel rod design modifications are necessary; revised analyses using the improved data base would be sufficient to regain lost margin.

2.2.2.1b. Fuel Pellet Densification. Fuel densification (i.e., fuel pellet shrinkage that resulted in formation of fuel stack gaps and clad collapse) is no longer a licensing issue, but is worthwhile mentioning here because the solution to the problem resulted from a unique utility and

LWR Core Materials

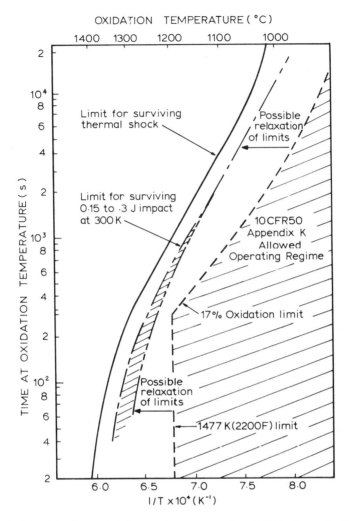

Fig. 2.22. New Zircaloy embrittlement data compared to LOCA Appendix K criteria (adapted from Ref. 74).

vendor industry cooperative R&D effort. Investigations into the physical mechanisms of densification by several groups involved hot cell examinations and both in-reactor and out-of-reactor parametric tests on a variety of UO_2 fuel types.[66,77,78] Particular emphasis was placed on those microstructural factors, such as grain size and porosity characteristics, which had been analytically identified as important to densification in a radiation environment. By comparing pre- and postirradiation pellet structure, it was found that the major difference in densified fuel was the initial absence of fine porosity (Fig. 2-23). This observation led to the present theory which attributes densification to annihilation of fine porosity by fission fragment action.[9] With this theoretical basis in hand, it became clear that control of the final pellet density, fine porosity, and grain size was the key to stable fuel. Extensive studies of the pelletizing process were carried out to determine how process variables influenced the important structural characteristics. Increased grain size and reduced fine porosity levels, both desirable in achieving stable fuel, were found to be correlated with higher sintering temperatures. For a specific final density, control of sintering time and temperature provided good control of grain size and fine porosity. Use of pore formers provided an additional degree of freedom, allowing large grain size and large porosity, but lower fuel densities, to be obtained with high sintering temperature. Appropriate process revisions have been introduced by all fuel manufacturers, and fuel that is essentially stable to in-reactor densification (i.e., $\leq 1\%$ density change) is used in all new core loadings. Thus, for the densification problem, thermal margin was regained through a straightforward fuel pellet modification.

2.2.2.1c. Fission Gas Release. Gaseous fission products are produced in quantity during the irradiation of LWR (and LMFBR) fuels. The total yield of three gases—krypton, xenon, and iodine—in the LWR is greater than 125 atoms per 100 fissions. These gases are essentially insoluble in the solid fuel matrix and, hence, tend to precipitate out as bubbles. During the fairly long fuel lifetime at high operating temperatures, the gases can, therefore, migrate within the fuel and, upon reaching free surfaces, such as grain boundaries and cracks, find easy paths for release to the fuel rod's free space.

This release of fission gases to the fuel rod atmosphere affects the performance and safety of nuclear plant operation in two fundamental ways. Firstly, by contributing to the internal gas pressure in the fuel rod, they affect both thermal and mechanical performance during normal and transient operation. Secondly, many of the fission gases have short half-lives and emit penetrating radiation when they decay. Since they are gaseous, and, therefore, very mobile, they constitute an important class of isotopes that must be taken into account in the source terms for radiological releases.

In contrast to fuel densification, fission gas release behavior represents a current unresolved issue. The uncertainties lie in the conditions (e.g., temperature, burnup, microstructure) that are necessary for large gas releases to occur from fuel pellets. Several reviews[79–84] of the data base and prevailing models of fission gas release published over the past five years have suggested that increased rates of fission gas release should be expected in LWR fuels that operate at either high LHGRs, ≥ 40 kW/

Fig. 2.23. Changes in UO_2 pellet density following irradiation versus percent of pellet volume occupied by small-diameter pores when fabricated (Ref. 51). (a) Fine porosity typical of early fuel, (b) large pores typical of later fuel fabricated with pore formers.

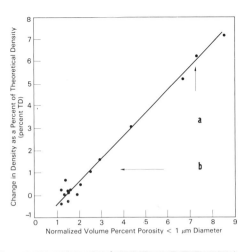

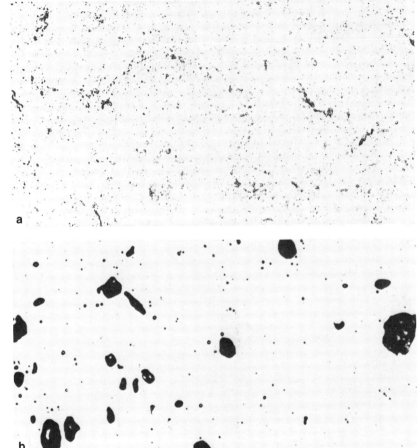

m, or to high burnups, above ~2600 GJ/kgM. However, the data in this high-burnup regime is very limited for modern fuels operating at typical LWR power levels (≤30 kW/m).

The lack of a significant burnup dependence in most U.S. fuel vendor models caused the USNRC,[80] in 1978, to introduce a correction (or enhancement) factor that must be applied to vendor models to account for the apparent burnup dependence of fission gas release. A slightly more refined fission gas release model, since developed by an ANS Standards Committee,[84] uses the same bases and predicts essentially the same trends.

For licensing purposes, the enhancement factor must be applied to fission gas release at burnups greater than 1730 GJ/kgM.[80] The enhancement factor is an exponential function containing three constants that were obtained from a treatment of high-burnup mixed (U-Pu) oxide LMFBR data. These data are the only published source of well characterized high burnup gas release data, and the assumption was made that mixed oxides and UO_2 produce the same gas release under identical conditions. In calculating fission gas release at high burnup, an existing vendor model is first used to calculate gas release at the appropriate fuel temperature. This calculated value is then weighted by the enhancement factor, which is evaluated at the burnup of interest above 1730 GJ/kgM. The enhancement factor is not imposed on low burnup (i.e., ≤ 1730 GJ/kgM) release calculations since current vendor models were derived from data in this range. Some degree of conservatism is provided for fuel that operated at low temperatures by requiring a minimum value of 1% be used for the fission gas release at 1730 GJ/kgM.

Although the impact of this new correction factor on fuel damage limits (i.e., fuel temperature and stored energy, and cladding strain) is considered to be not too significant for current design burnups,[85] with the trend to higher fuel burnups (see Section 2.6) there is a growing industry concern that the requirement to assume increasing fission gas releases above 1730 GJ/kgM burnup will result in thermal margins being reduced and, possibly, in the event of failure, radiological release limits being exceeded. Either eventuality would severely impact the cost/benefit of high-burnup fuel cycles, Accordingly the industry is gearing up to obtain the necessary high-burnup LWR UO_2 gas release data.

As a preliminary step in this direction, fission gas release data have been obtained from fuel rods irradiated in seven commercial LWRs.[14,15,37,86,87] Although the burnup range is low to medium, the data, together with accompanying fuel performance code analyses,[7] are of relevance to the high-burnup gas release issue, because they suggest that fuel operating temperature is the key variable that determines fission

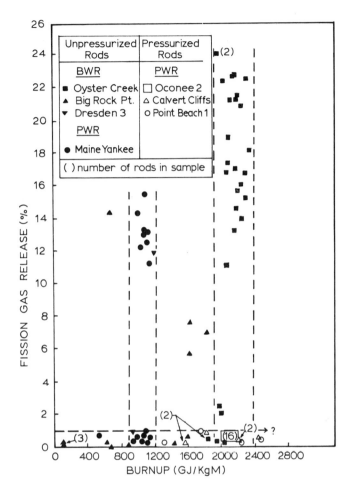

Fig. 2.24. Fission gas release vs. burnup data obtained from LWR fuel rods (adapted from Ref. 88).

gas release. The role of burnup appears to be secondary compared with that of fuel operating temperature, at least up to ~2600 GJ/kgM.

Figure 2-24 presents these recent results as a function of burnup. In contrast to the NRC and ANS model predictions, significant gas release (>1%) is not restricted to high burnup values; release values ranging from 5.7 to 15.3% were observed in 12 rods with burnups less than 1730 GJ/kgM. At higher burnups, from 1730 to 2590 GJ/kgM, gas release values ranging from 6.9 to 24.2% were observed in 24 rods. The factor common to these rods was that they were unpressurized. This appears to be a necessary, but not sufficient, condition for high gas release, since small releases have been reported for other unpressurized rods at burnups to 2080 GJ/kgM. Low fission gas release (<1%) was observed in all pressurized rods.

The results indicate that for these unpressurized rods a transition from low gas release values to high release values occurs over a very narrow burnup range. This transition region, or threshold for significant gas release, occurs at about 1040 GJ/kgM for the Maine Yankee fuel rods and at about 2080 GJ/kgM for the Oyster Creek fuel rods. Comparison of gas release data from sibling rods irradiated in the Big Rock Point Reactor[88] also supports the idea of a threshold value for significant gas release. Data for three pairs of rods showed that for modest differences in burnup of about 15%, the gas release values could differ by more than a factor of 10. Such threshold values are in conflict with the form of the burnup dependence of gas release proposed by NRC and the ANS committee.[84]

Phenomenological models of gas release that predict the general trends observed in the Oyster Creek, Maine Yankee, and Big Rock Point rods (i.e., nearly zero fission gas release up to a threshold burnup value, followed by a transition to high fission gas release values over a narrow burnup range) have been reported by Dollins and Nichols,[89] Hargreaves and Collins,[90] and by Tucker and White.[91] These models have a common view of the fission gas migration and release process, namely that fission gases, in the form of atoms or bubbles, migrate to a surface that can communicate with the free space of the fuel rod, that is, a grain boundary or crack surface. Such migration occurs by a thermally activated diffusion process that obeys an Arrhenius equation, thus providing the temperature dependence in the gas release process. As long as grain-boundary gas bubble concentrations are low enough that bubbles do not interlink, gas remains low; but, once the gas bubbles interlink on the grain boundaries to form tunnels, gas release becomes rapid. Tucker and White[91] illustrated the strong temperature dependence of this process by predicting a factor-of-2 reduction in the irradiation time before significant gas release occurs, for a temperature increase from 1473 to 1573 K.

Fuel performance code analyses[7] demonstrated the sensitivity of fuel temperature to linear heat rating, fuel–cladding gap size, and the thermal conductivity of the gap. The gap size, in turn, is strongly influenced by clad creepdown and fuel densification rates, as noted earlier. The data of Fig. 2-24 suggest that gap conductivity in unpressurized rods can be degraded to such an extent by the lower-thermal-conductivity fission gases that fuel temperatures will increase as a result of only a small release of gas, and this will then provide the driving force for further releases (i.e., an autocatalytic effect). Very few fuel performance codes can predict this feedback effect, but the COMETHE IIIJ code[7] was able to reproduce experimental results. Prepressurization with helium apparently dilutes any fission gas effects and maintains high values of thermal conductivity across the gap so that gas release is minimized ($\leq 1\%$), even at burnups approaching 2600 GJ/kgM.

Based on this more recent data, therefore, the high gas release values reported earlier[79–84] could merely be the consequence of fuel densification and/or lack of helium prepressurization, both of which will effectively reduce gap conductivity and lead to higher fuel temperatures than would be predicted for that power level (i.e., a so-called power/temperature mismatch). This possibility points to the need to carefully categorize the fission gas release data set with respect to initial fuel rod design parameters. It also suggests that, inasmuch as all fuel vendors now use densification-resistant fuel, and all PWR fuel is routinely prepressurized and new BWR fuel reloads will be prepressurized, fission gas release values reported in the future should be lower.

In this instance, therefore, an improved data base should be sufficient to maintain thermal margins. However, if an increase in thermal margin is desired, fuel pellet designs that minimize gas release should be evaluated (e.g., refer back to PCI remedies, Section 2.2.1.1c).

2.2.2.2. Heat Flux Limits

Between the extremes of LOCA and RIA, various off-normal power or cooling conditions may occur in the reactor core. These are generally termed power-cooling mismatch (PCM) accidents because of failure of the coolant to successfully remove the heat generated within the core.

Numerous credible single and coincident initiating events that may lead to PCM accidents can be postulated (hence the term "moderate frequency transients"). If departure from nucleate boiling (DNB) occurs during a PCM event, fuel rod damage and subsequent release of radioactivity into the primary coolant system may occur. Hence, DNB criteria for PWRs, or the equivalent boiling transition criteria for BWRs, see broad application for defining fuel damage in conditions other than LOCA. The specific heat flux limits are in terms of the departure from nucleate boiling ratio (DNBR) for PWRs and the minimum critical power ratio (MCPR) for BWRs.[65] Progress in our understanding of fuel rod behavior under DNB conditions is summarized in this section.

The thermal-hydraulic origin of these fuel damage limits makes them particularly sensitive to fuel rod or fuel assembly distortions. One such phenomenon that has only recently been resolved is the effect of fuel rod bow on DNB limits. A second subsection reviews the intense industry efforts to understand and mitigate this problem.

2.2.2.2a. Departure from Nucleate Boiling. DNB or boiling transition criteria address only one potential fuel rod failure mechanism, namely cladding failure resulting from overheating (with accompanying oxidation and embrittlement). Current licensing criteria dictate that if a rod is postulated to have exceeded DNBR or MCPR limits, the rod is assumed to have failed. However, recent experimental evidence[92,93] from fuel rods tested in the Power Burst Facility (PBF) indicates that Zircaloy cladding will not necessarily fail even after several minutes of operation in DNB.

Mehner *et al.*[92] reported that fuel rods can operate under DNB conditions and incur significant damage without failure, because, above 920 K, the cladding has sufficient ductility to accommodate the strains associated with cladding collapse. Failure due to oxygen embrittlement during "film boiling" does not occur until the cladding has nearly completely reacted to ZrO_2 and oxygen-stabilized α-Zr. Rod failure during quench from DNB operation or during posttest handling, due to oxygen embrittlement, is predictable from out-of-reactor isothermal steam tests (e.g., Fig. 2-22).

Fuel pellet damage occurring during these tests included grain boundary separation, molten fuel relocation, and fuel swelling. However, cladding melting did not occur when molten fuel contacted the cladding, and reaction between the two was insignificant. Likewise, fuel swelling did not result in fuel rod failure up to burnups of ~1450 GJ/kgM.

The primary rod failure mechanism in unirradiated and irradiated fuel rods was oxygen embrittlement of the cladding, and the maximum oxidation levels measured after these tests significantly exceeded the 17% oxidation criterion.

This and other data on fuel heat transfer effects therefore point to the inadequacy of the current DNBR/MCPR criteria as fuel failure indicators. An improvement in power capability would result from relaxation of the DNBR limits in PWRs because the plant could then operate at higher heat fluxes or higher core inlet temperatures. Consequently, some PWR vendors have moved in the direction of modifying heat flux limits. Analyses show that a 10% improvement in DNBR limit would result in a 5% increase in power level and a 1% increase in availability; this translates into $6–8 million saving per reactor per year,[94] which is a strong incentive for completing the necessary research to justify this move.

Should the DNBR limits be relaxed, the next limit to be encountered affecting power capability would be the LOCA limit on fuel peak linear heat generation rate; however, for PWRs presently operating, the balance of plant design would probably restrict power output before the LOCA limit was reached.

In the case of BWRs, either MCPR or the LOCA limit (MAPLHGR) may be most limiting with respect to operation, depending on core exposure, operating mode, control methods, etc. It is clear that relaxation of MCPR limits would improve operational flexibility because, in some instances, the margin has been reduced to zero, necessitating control rod pattern adjustments and consequential lost output during the fuel reconditioning period. (Note that operating restrictions can interact to magnify the lost output—the example here of MCPR limits interacting with fuel preconditioning requirements is a current concern for certain BWRs.)

2.2.2.2b. Fuel Rod Bow. Fuel rod bowing may result from radiation-induced relaxation of the internal stresses or from an interaction between rods and the spacer grids of the assembly structure[9] (refer back to Fig. 2-2 for assembly features). Excessive fuel rod bowing can lead to local coolant flow restriction, which may result in a decreasing departure from nucleate boiling heat transfer (i.e., cladding overheating), and under certain circumstances even in enhanced corrosion at this point. Rod bow has been observed to some extent in most fuel assembly designs, but has been greatest in PWR designs that have the largest span between grid spacers and the most slender fuel rods. Recent thermal-hydraulic testing has indicated larger than previously expected rod bow DNB penalties,[10] and therefore programs to eliminate or reduce rod bow to acceptable levels have been re-emphasized. Margins to thermal limits on core coolant temperature and pressure, which are designed to protect against fuel rod DNB, are generally less severe with respect to their potential impact on plant operations than the LOCA limits. However, fuel rod bowing has tended to compromise these margins.

A significant effort has been expended, particularly by fuel vendors, to understand and subsequently predict rod bow (or fuel assembly channel closure). Several fuel surveillance projects have provided rod bow (or channel closure) data from a spectrum of fuel designs. Using videotapes of peripheral rods, boroscopes, periscopes, and "Sulo" probe strain gauges, measurements have been made on 14 × 14, 15 × 15, and 17 × 17 PWR fuel designs.[3,9,37]

The dependence of gap closure on burnup and assembly design is shown in Fig. 2-25. The data points are verified by probability analysis and give the maximum (i.e., worst) gap closure in 99.7% of all rods. It can be concluded from this figure that gap closure, or the amount of bow, increases with increasing burnup, but depends on design and initial bowing. The orientation of the bowing is observed to be random and reverses at each spacer grid, and the maximum bowing of fuel rods is generally found

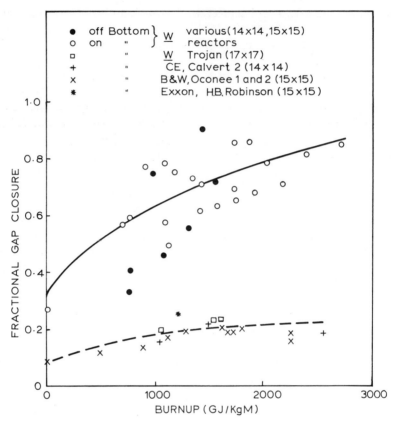

Fig. 2.25. Assembly channel closures due to rod bowing (maximum values in 99.7% of all rods) as a function of burnup for PWR fuel assemblies from different suppliers. Lines are plotted from the empirical equation (2-3) (adapted from Refs. 9, 21).

between the second and the third spacer from the bottom almost independently of the design at the bottom of the assembly (i.e., rods on or off-bottom).

An empirical relationship of the form

$$\Delta C/C_0 = [a + b(\mathrm{Bu})^{1/2}](1.2) \qquad (2\text{-}3)$$

where C_0 is the nominal pre-irradiation channel spacing, ΔC is the reduction of channel spacing with irradiation, a is the initial bow of the fuel rods due to fabrication tolerance, b is an empirical constant, and Bu is the burnup (region average or bundle average depending on the fuel designer), is used to describe the rod bow behavior. For licensing purposes a factor of 1.2 is included on the right-hand side of Eq. (2-3) to convert from the cold condition at which measurements are made to the hot operating conditions. As can be seen from Fig. 2-25, assemblies with different designs require different values of a and b for Eq. (2-3) to fit the data.

In comparing the various data plotted in Fig. 2-25, one must be aware of the nonuniformity in measurement technique and calculational procedures. Notwithstanding this, however, it is clear that certain fuel designs are more prone to bow than others. Bow of 25% or less, which saturates at early burnups, is exhibited by the B&W, Combustion Engineering, and Exxon Nuclear fuel assemblies and by the new 17 × 17 (8 grid) Westinghouse design. This degree of bow is similar to that recorded in the more rigid BWR fuel assembly designs.[21] The earlier Westinghouse 14 × 14 and 15 × 15 PWR fuel designs, however, exhibit rod bow greater than 50%, which could impact DNB margins.

Analysis of the individual and combined data sets leads to the conclusion that grid forces dominate the rod bow, with secondary contributions derived from differential axial growth of guide tubes and fuel rods, cladding creep strength, pellet perpendicularity in the tube, and tube wall eccentricity.[9,21] A complete mechanistic understanding of the phenomenon is still lacking,[9] but based on empirical models such as Eq. (2-3), appropriate fuel design changes have been made, where necessary, to minimize rod bow. The approach has been to balance grid forces against cladding creep strength, with the trend being toward an "all Zircaloy" assembly. Early results appear to support this approach and should lead to minimization of this concern in the near future.

2.2.2.3. Core Damage Assessment in Three Mile Island Accident

The accident at the Three Mile Island Unit 2 (TMI-2) Nuclear Plant[95] will ultimately provide a significant (though unexpected) statistical data base on the response of Zircaloy-clad fuel to overheating conditions. While it is clear that the core suffered extensive damage, the degree and nature of the damage must remain the subject of hypothesis until the vessel can be opened and the core examined and finally removed (1981–1982). Nevertheless, there is value in detailed prior analysis as a guide to what core damage to expect and how to handle it, etc. The extensive data base on Zircaloy behavior under overheating conditions described above is a major starting point for such work.† Various groups are involved in this activity[96–98]; the intent here is not to promote any one analysis but simply to report on the range of expectations that have

---

† It must be emphasized that this accident was not the hypothesized large-break LOCA for which most of the data were collected, but was in fact very similar to a small-break LOCA. The major differences between the two accident classes are rate of system depressurization and coolant loss, impact of fuel stored energy, and decay heat levels.

appeared in the public domain as examples of data interpretation and application.

"Reverse engineering" has been the principal method for assessing the degree of core damage since, at this point (1980), although some information relating to core damage is known (i.e., information on fission product release, hydrogen generation, and instrument readings), the overall level of detailed knowledge is insufficient for a definitive analysis, and the entire system response is open to different interpretations.

There appears to be a possibility that uncovering of the core occurred more than once during the early phases of the accident (<20 h), though there is a large uncertainty both on the absolute levels to which the coolant dropped and the rates of uncovering. Obviously, this aspect of the problem is critical to the core damage calculations because the amount and rate of coolant "boiling off" determines the cladding and fuel temperatures, the extent of steam oxidation of the cladding, and fission product release from the fuel.

Attention has been focused on the first uncovering period (6000–10,000 s into the accident) because decay heat was larger then, and because that period produced the radiation instrument signals in the containment, after 8500 s, indicating probable onset of major fuel damage. The calculations indicate that nearly all or all fuel rods failed. Burst conditions should start at temperatures greater than 1060 K, with "ballooning" occurring to some extent prior to failure. The possibility of mechanical damage, aggravated by overheating and thermal growth, cannot be ignored as a complicating factor that may have contributed to failure at lower burst thresholds. In any event, the mode of initial defecting versus continued progression of the accident may be irrelevant in light of the more subsequent damage.

According to isothermal kinetics data, Zircaloy oxidation will proceed at a fairly rapid rate above 1400 K.[71-73] Analyses have either assumed one-sided (O.D.) oxidation only—clad defects are small, and little or no steam enters the fuel rod—or two-sided (I.D./O.D.) oxidation—the cladding perforates and provides paths for steam entry. In the latter case, one should also take into account the possibility of steam interaction with the UO_2, which can result in hyperstoichimetric fuel. The Zircaloy–steam reaction is exothermic[4] and contributes significantly to the cladding heatup, but to offset this, heat transfer to steam (above the effective dryout level) is substantial. One can easily see, therefore, that from this point on predictions will vary depending on the choice of (a) coolant boil-off rate and resulting dryout level history (primary controlling factor), (b) oxidation route, and (c) heatup mode—decay heat plus exothermic reaction minus steam and radiative heat transfer.

Upper bounds on Zircaloy oxidation, consistent with the hydrogen measured in the plant, range from 40 to 65% of total Zircaloy inventory.

Calculations, using the oxidation kinetics data discussed earlier and varying assumptions noted above, indicate that most of this oxidation occurred during the initial period of uncovering (~6500–10,500 s) althouth conditions during portions of the ensuing period, to ~10 h elapsed time, probably were also severe enough to result in additional Zircaloy oxidation and further core damage. The region of severe oxidation was probably localized in the upper $\frac{1}{2}$ to $\frac{2}{3}$ levels of the core, and possibly did not include peripheral bundles.

The peak temperature attained in the fuel and cladding is likewise open to debate. In the hot upper central regions of the core, fuel temperatures probably reached or exceeded 2000 K, as indicated by the release of ≥50% of the total core inventory of noble gases. Gas release could have initiated at much lower fuel temperatures (1770 to 1870 K), however, if there had been significant stoichiometry changes in the UO_2 (e.g., O/U from 2.00 to 2.12) or if such temperatures persisted for much of the period through 10 h. If the cladding and UO_2 reached temperatures near the melting point of Zircaloy (2100–2270 K, depending on oxygen uptake) eutectic formation between ZrO_2, α-Zr and UO_2 would possibly occur and could result in formation of glasslike bonded structures. This is an endothermic reaction and should therefore tend to arrest temperatures at this point.

The oxidation reaction products, whether ZrO_2 or a eutectic (α-Zr + ZrO_2), and the UO_2 fuel pellets would react in a brittle manner and should fragment, upon quenching, into pieces ranging from millimeter size pieces to whole sections of rods. Much of this debris should have remained trapped in the upper core region because the upper end fittings have a grillage that would act as a screen. Furthermore, the compaction of fuel debris within the core region is limited because it is fabricated with a packing fraction of about 46% and the theoretical maximum packing fraction (for a bed of spherical particles) is only about 63%. However, it is possible that some core debris was distributed outside of the core region, possibly throughout the primary system as a result of thermal hydraulic events associated with quenching of the core.

In summary then, it is known that a high degree of Zircaloy oxidation did occur, with apparently concurrent high fractional release of core fission product inventory. Both of these conditions require sustained high temperatures, with maximum temperatures necessary to attain such damage and release strongly dependent on residence times that are well within the bounds of the first ≤20 h of the accident. It is possible that core damage progressed throughout much of this time span. Available data (in 1980) does indicate that: virtually all fuel rods failed; a great majority of the fuel assemblies were severely oxidized in the upper $\frac{1}{2}$ to $\frac{2}{3}$ of their length; regional temperatures in the core probably exceeded 1700 K for at least some portions of these ≤20 h; and, possibly, some local regions of the core reached Zircaloy melting temperatures or above. Beyond this, to be more specific with the current state of knowledge of physical evidence is academic and generally not technically supportable.

2.3. Plutonium Recycle Fuel Performance

If uranium and plutonium recovered by reprocessing of LWR fuel assemblies are used in power reactors, uranium needs and enrichment requirements can be reduced by more than 35%.[99] This incentive has motivated research and development of plutonium recycling for LWRs for many years, with the result that there is now sufficient reactor performance experience with experimental and demonstration fuel assembly irradiations to warrant large-scale commercial use of recycled plutonium fuels.[100] Unfortunately, the political controversy surrounding the reprocessing of spent LWR fuels and plutonium utilization is preventing the use of this valuable fuel resource in the U.S. (see Chapter 3, Section 3.6). Other nations, in Europe in particular, are, however, proceeding to adopt recycle fuels for LWR application.[99]

The materials properties and performance of the dilute (UO_2-3-6 wt.% PuO_2) MOX† fuels planned for recycle in commercial LWR cores are in many cases indistinguishable from the corresponding UO_2 fuels, and in all cases, the differences are small.[100] Inhomogeneity of fissile material in physically blended mixed-oxide fuel pellets could potentially cause a change in fuel performance, particularly in hypothetical accidents, but homogeneity can be controlled during fabrication. The nuclear property differences can be accommodated in most cases by using various rod placement and enrichment schemes that make it feasible to design fuel assemblies that are interchangeable with the spent UO_2 assemblies they replace.

It must be pointed out that plutonium fissioning in oxide fuels is not unique to the recycled plutonium fuels of current interest. A typical UO_2 core near the end of an equilibrium cycle, for example, at a core-average exposure of 1730 GJ/kgM, will derive approximately 50% of its power from fissioning of bred-in plutonium isotopes. Thus, in one sense, the use of plutonium as a fuel in LWRs does not represent a new situation.

Prototypical irradiations of MOX fuels have been undertaken or are underway in both BWRs and PWRs. Performance to date has been impressive, with peak burnups in excess of 2590 GJ/kgM being achieved

---

† The acronym MOX is widely used for LWR $(UPu)O_x$ mixed oxides in which x can vary from about 1.96 to 2.00, depending primarily on PuO_2 concentration.

without any apparent fuel rod deterioration. One such program in the Quad Cities BWR has the so-called "island" fuel assembly design comprised of MOX fuel rods containing both solid and annular pellets.[101] The latter is a better nuclear design for BWRs because of the resulting higher water-to-fuel ratios. This fuel form has so far behaved as designed up to about 2000 GJ/kgM burnup.

A recent study of the in-reactor densification behavior of MOX fuel pellets representing the range of microstructures and PuO_2 particle inhomogeneities expected in modern recycle fuels clearly demonstrated the similarity with UO_2 behavior (Fig. 2-26).[102] As for UO_2, the densification sintering is controlled by initial pore size and grain size, and PuO_2 particle size does not influence the sintering kinetics. MOX fuels stable to densification can be routinely fabricated.

Some recent PCI-related results with MOX fuels indicate that there may be some benefit with respect to cladding failure thresholds.[49] This is presumably due to the increased creep plasticity (a 20% increase in creep rate for 5 wt.% PuO_2) derived most probably from the combination of a slight shift toward hypostoichiometry and increased rates of diffusion of the plutonium atom over the uranium atom. More data are obviously needed to support this early result but, since no MOX rods have failed in power ramp experiments to date, utilization of recycle fuels as a PCI-resistant fuel design is an interesting future prospect.

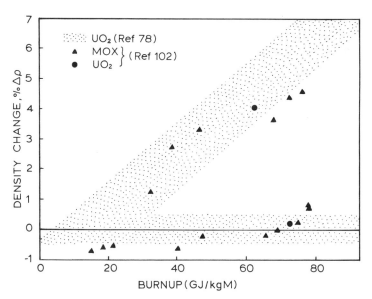

Fig. 2.26. Mixed oxide densification vs. burnup, compared to UO_2 behavior (Refs. 78 and 102).

2.4. Stainless Steel–UO$_2$ Fuel Experience

A few early, small commercial LWR cores utilize stainless steel, in particular Type 304, as cladding material. Stainless steel was an early choice for the cladding for LWRs, but lost favor during the early 1960s due to (a) the fuel cycle economic advantages of Zircaloy and (b) a rash of fuel failures in the BWRs (e.g., Dresden 1, Humboldt Bay, Big Rock Point).[103] Thus today, only one small BWR (LaCrosse) has a stainless steel core, while several PWRs (e.g., Haddam Neck,† San Onofre 1) still use stainless steel as cladding material.

A short discussion of stainless steel clad fuel behavior in LWRs is appropriate because the question of stainless steel versus Zircaloy has been reopened, not from a performance point of view, but from concerns about safety following the Three Mile Island accident. The large-scale adoption of Zircaloy cladding was made primarily on the basis of fuel cycle economics (i.e., improved neutron economy), and the safety implications of the Zircaloy–steam reaction were only realized much later during considerations of the hypothetical loss of coolant accident (LOCA). As has been discussed, these culminated in limits being established for both cladding temperatures and oxidation level (i.e., the 10CFR50 Appendix K criteria). Following the events at TMI-2, the primary concern with Zircaloy is the evolution and characteristics of the hydrogen bubble formed as a direct result of its reaction with steam, as well as the degree and extent of the core damage that results from its excessive oxidation and possible reaction with the UO$_2$. The question is whether stainless steel would behave better or worse in such an accident situation.

The data[104] and analyses[105] relevant to this question are quite limited. Taking the LOCA criteria specifically, stainless steel is both mechanically and chemically more stable up to about 1480 K. Above 1480 K, however, stainless steel is not always superior to Zircaloy in that the metal–water reaction is accelerated and the alloy melts and disintegrates ("froths") at about 1670 K. This latter point was the basis of the argument for maintaining a LOCA peak clad temperature of 1480 K, the same as for Zircaloy.[104]

Specific data comparisons are as follows: Austenitic stainless steel is less reactive than Zircaloy with steam up to 1480 K both in the rate and in the heat of reaction. Figure 2-27 compares the two parabolic rate plots. The heat release in the stainless steel–water reaction is a factor of 20 lower than the Zircaloy–water reaction, which would have a significant effect on the cladding heat-up rate (note Section 2.2.2.3 on TMI-2 core

---

† Also called Connecticut Yankee.

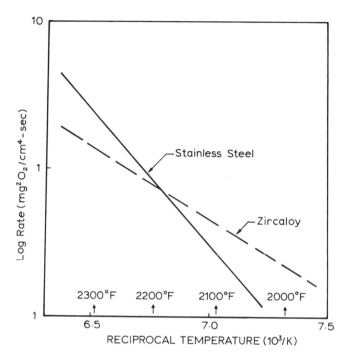

Fig. 2.27. Parabolic rates of metal–water reactions for Zircaloy and Type 304 stainless steel.

damage assessment). Also for the same weight of cladding, the hydrogen release is only two-thirds that for Zircaloy, which would reduce the possibility of hydrogen buildup in the core. Oxygen is relatively insoluble in stainless steel up to ~1600 K, thereby virtually eliminating the potential for oxygen embrittlement. Further, the mechanical properties of stainless steel are superior to those of Zircaloy with respect to both strength and ductility (Fig. 2-28), which would result in less cladding deformation and ballooning. However, the analyses discussed in Section 2.2.2.3 earlier indicate very little potential for ballooning (and coolant flow blockage) in Zircaloy cladding under the TMI-2 accident conditions.

In a U.K. study, Baker[105] utilized computer models to compare the behavior of the two fuel cladding materials, Zircaloy and Type 304 stainless steel, in a LOCA on a core-wide basis; that is, he computed the probability of fuel rod failure for all fuel rods in the reactor at various stages of fuel life. The aforementioned differences in properties are reflected in the results; in all cases examined (including different levels of prepressurization, burnup, and heat transfer coefficient) many more Zircaloy-clad fuel rods were predicted to fail and the associated extent of prior ballooning was about two times larger.

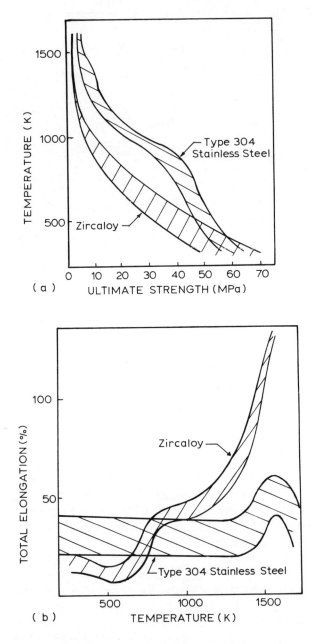

Fig. 2.28. Temperature dependence of (a) ultimate tensile strengths and (b) ductilities of Zircaloy and Type 304S stainless steel.

Clearly much more data on stainless steel behavior would have to be obtained to provide a statistically valid comparison with Zircaloy, but the data and analyses produced in the early 1970s at least justify the conclusion that a stainless steel core should be less susceptible to extensive damage under loss-of-cooling conditions up to 1480 K. If temperatures are allowed to exceed 1480 K, as was likely in the TMI-2 accident, the continued integrity of the stainless steel would be questionable due to its lower melting point.

Moving to the general area of fuel performance, the experience with LWR stainless steel cores is mixed. The BWRs were plagued with fuel failures during the early 1960s in the burnup range 520 to 1380 GJ/kgM,[103] with the most recent experience being that of the Type 348 stainless fuel in the LaCrosse BWR,[106] which culminated in a cladding specification change. On the other hand, the PWR stainless steel cores appear to have operated flawlessly until very recently when fuel failures were reported in the Haddam Neck reactor,[107] after 1730 GJ/kgM burnup. The cause of this latter occurrence is unknown at the time of writing, but on a macroscopic scale the cladding cracks resemble those observed earlier in BWR fuel. It is therefore worthwhile to briefly review the extensive R&D effort associated with elucidating the cause of failure in stainless-steel-clad fuel in BWRs.

A detailed postirradiation examination (PIE) program on both failed and incipiently defected fuel from several reactors revealed that the cladding cracks were intergranular and started at the fuel-rod outer surface and progressed inward through the cladding wall.[103] The cladding was not sensitized and so the problem is quite different from that currently being experienced in sensitized regions of Type 304 stainless piping in BWRs (see Chapter 4, Section 4.3.1.4.). There were significant differences in the orientation of the cracks in the cold-worked and annealed rods. The cracks in the former were longitudinal, whereas those in the latter were predominantly circumferential and usually occurred at pellet interfaces. All cracks occurred in the peak heat-flux region of the fuel rods, and there appeared to be a relationship between operating stress level of the cladding and time to failure.

The detailed laboratory study that followed, therefore, was based upon the premise that failure of the BWR fuel rods could be attributed to stress-assisted intergranular corrosion attack in the absence of grain boundary carbide sensitization.[103] Out-of-reactor corrosion testing of high-purity austenitic alloys with controlled impurity additions indicated that carbon, silicon, and phosphorus are of major importance in determining their intergranular corrosion resistance. Carbon additions affect the intergranular corrosion resistance but are dependent on heat treatment before testing. Manganese, sulfur, nitrogen, and oxygen had very little effect. Removal of the harmful Si and P impurities resulted in an alloy which was immune to various laboratory test media. In-reactor verification is currently underway with a new core for the LaCrosse BWR fabricated from Type 348 stainless steel with significantly lower Si and P levels. Thus, the demonstration of the remedy is in progress, but confirmatory data are still several years off. Interestingly, European specifications for stainless steel stipulate low P and Si levels, and there are no incidences of either fuel or control rod cladding failure with this steel.

Although the primary contributing feature was elucidated from the General Electric Programs[103] the basic mechanism(s) of the intergranular corrosion failure has eluded researchers. For instance, the reason why it has not (until recently?) appeared in PWRs was, originally, assigned to water chemistry differences (e.g., higher oxygen and/or presence of chloride in BWRs) and, more recently, in light of the Zircaloy PCI experience, thought to be associated with the higher stress levels experienced by BWR fuel. The latter suggestion warrants further examination because of its relevance to the recent Haddam Neck failure experience.

The BWR experience was that long incubation times were required before failure of the stainless cladding, and that higher cladding stresses (i.e., more severe duty) correlated with shorter times to failure.[103] In applying this informatin to the PWR situation, one could argue that, for normal PWR operations, the incubation time exceeded the fuel design lifetime. However, if situations arose in which the stress levels were increased, either through design or through operation, stress-assisted intergranular corrosion failures could in theory occur in PWRs. Specifically, in the Haddam Neck case, a unique duty cycle that involved preconditioning of fuel rods at reduced power for a long time period, followed by a rapid return to normal power late in life, did produce high clad stresses (at least according to fuel performance code predictions),[107] and the narrow fuel–clad gap and more stable fuel pellet might have combined to exaggerate these stresses. The particular duty cycle had not been experienced before at these high burnups with this particular fuel type.[107] In the absence of destructive-examination data it is not clear which of the two features, duty cycle or design, contributed most to the cladding stress levels, nor is it certain that the cladding cracks initiated at the outside surface. Nevertheless the assignment of these PWR fuel failures to a stress corrosion mechanism is attractive because it is difficult to explain how the observed multiple, "brittle" axial crack patterns could result from a purely mechanical fuel–cladding interaction and crack initiation at the inside surface.

From the foregoing it is evident that the "jury is still out" as far as stainless steel fuel performance is concerned. What is clear is that impure steels do not behave well in BWRs (note BWR control rod experience also, Section 2.5.1), but, clearly, this is not the entire answer or else PWR experience would also be poor. All indications are that, as for the Zircaloy PCI situation, there is a synergistic effect between stress, environment, and susceptible material, which will necessitate a significant effort to fully unravel. Easy design (less dense fuel, larger gap, etc.) or specification (lower P, Si) changes could provide the remedy, but, without the fundamental understanding or the data, the utility that takes this approach still runs the risk of fuel failures.

2.5. Control Materials

Neutron-absorbing materials are used either as burnable poisons or in control rods. Table 2-1 earlier identifies the materials in current use in the LWRs. This section will describe performance-related issues that have arisen in recent years, some of the corrective actions, and new alternative absorber materials currently under consideration for specific purposes. For background, the reader is referred to the excellent reviews by Murgatroyd and Kelly[108] of the physical, chemical, and metallurgical properties, and irradiation behavior of absorber materials in several reactor types, and by Meyer et al.[109] and Strasser et al.,[110] of specific problems in LWRs.

2.5.1. Control Rod Materials

Boron carbide (B_4C) is the primary control rod material. Even PWRs which have used the silver–indium–cadmium (80 wt.% Ag–15 wt.% In–5 wt.% Cd) alloy for many years are now switching, for economic reasons, to B_4C control rods. Typically, the B_4C powder or B_4C pellets are contained in austenitic stainless steel (BWR) or Inconel (PWR) cladding tubes.

Gas release (helium) and swelling are key phenomena that can determine the lifetime of B_4C control rods.[108] More detailed studies of these phenomena have been conducted for LMFBR applications, and this work will be described in Chapter 3. Only limited, somewhat incomplete data exist for thermal reactor conditions.

A summary of the gas release from boron carbide in a thermal flux is shown as a function of irradiation temperature in Fig. 2-29. It is readily apparent that there is considerable scatter in the data, with similar exposure conditions resulting in gas release fractions that vary by factors of 4 to 6 in some cases. Much of this inconsistency can be attributed to stress gradients (caused by self-shielding) that result in significant

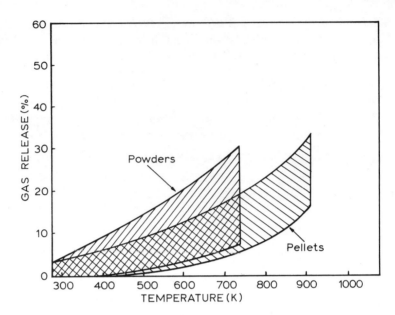

Fig. 2.29. Temperature dependence of gas release from B_4C control materials in thermal flux.

amounts of cracking, spalling, and complete fracture of the solids. This deterioration enhances gas release, and since its occurrence cannot be correlated with exposure conditions, the gas release data cannot be normalized to any baseline. In general, it appears that gas release fractions from powders are somewhat higher than those from hot-pressed shapes, but there is considerable overlap in the results. Other trends that have been observed are that gas release is inversely proportional to compact density and grain size and directly proportional to the B/C ratio. This latter effect may result from unreacted boron in the carbide.

These data, from which design correlations for the irradiation performance of B_4C have been made, were generated 15-20 years ago, and therefore their applicability to present-day absorber assembly designs is questionable. Although some B_4C control blades have operated successfully in BWRs for up to 16 years without replacement (expected life is 7 to 10 years), the trend currently is to replacement after 4 to 7 years.[110]

There are two main factors contributing to the reduced control blade lifetime and both relate to duty cycle. The theoretical control blade lifetime is established by the rate of ^{10}B burnout, or "nuclear" depletion rate; clearly core operations that call for blades to be inserted for long periods in high flux (central) positions will result in reduced lifetimes. In addition, the helium produced and the swelling of the B_4C, which, like the UO_2 fuel situation, is directly related to duty cycle, can stress the cladding and lead to tube failures. Subsequent leachout of the boron will result in accelerated ^{10}B depletion.

Cracking of the Type 304 stainless steel tubes has occurred in the control blades in several BWRs. An early scenario suggests that the B_4C sinters in-reactor, then swells and stresses the cladding tube, thus providing the stress necessary to initiate intergranular stress corrosion cracking from the outside surface.[110] All indications are that these failures are similar to those observed in BWR stainless-steel-clad fuel rods (Section 2.4 earlier) and attributed to high P and Si grain boundary impurities.[103] As noted then, the cleaner European steels do not seem to be prone to this form of failure and no European manufactured control rod tubes have failed to date.[110]

Unfortunately, a solution to the control rod tube cracking problem is only a (small) part of the answer to the current BWR control blade lifetime issue. An increasing trend in large, modern BWRs is toward core operations that minimize control-blade movements (to reduce fuel duty and increase plant capacity), which also means that certain control blades remain in the high-flux regions for complete cycles. In these circumstances the straight nuclear depletion of ^{10}B will occur rapidly and could result in control blade lifetimes as low as two to three years. The cost of new control blades is not as important as the potential for increased outage time to replace the blades and the associated replacement fuel costs.

Consequently, General Electric has recently embarked on a program to qualify an alternative absorber material, hafnium, for BWR control rod application. Hafnium can be used for extended irradiation without excessive reactivity loss;[108] the projected lifetime of a hafnium control blade is of the order of 40 years, or the plant lifetime. Hafnium control rods have operated successfully in the Shippingport PWR and submarine reactors, and therefore, the major uncertainty for BWR application appears to be the compatibility with the BWR coolant. General Electric's current research effort[111] will address this issue by measuring corrosion rates (and other key properties such as radiation growth) of several solid hafnium (alloyed with a small percent of zirconium) rods in standard BWR control blades in a commercial BWR.

The AgInCd alloy composition used in PWRs was chosen on the basis of neutronic considerations.[112] The major disadvantages of this alloy were its low creep strength and poor corrosion resistance in high-temperature, oxygenated water.[108] Consequently, the alloy is clad in Type 304 stainless steel or Inconel 627. Space is allowed inside the cladding for expansion and irradiation-induce swelling. However, experience is that dimensional stability is quite adequate—for example, volume increases of only 2% after neutron doses of 4×10^{24} n/m² ($E_n > 1$ MeV) at 590 K—and AgInCd control rods have performed as designed for 10 years.[110] Nevertheless, despite this good performance, the increasing cost of silver and indium is forcing a move for new PWRs to B_4C control rods. Given the problems being experienced in BWRs, in this writer's opinion it would be wise to move away from helium producers such as B_4C to control materials such as hafnium. Experience to date supports the use of hafnium in PWRs.

2.5.2. Burnable Poisons

Both of the LWR designs utilize burnable poisons, to prevent early-in-life power peaking and to assist in power shaping and optimum core burnup. Boron-based materials are again in use in PWRs, either as Al_2O_3–B_4C pellet compacts or borosilicate glass sections. In the BWRs, UO_2–Gd_2O_3 is the standard burnable poison.

The Al_2O_3–B_4C materials typically contain 3.5 to 6.9 vol.% B_4C and are used in relatively low density (<90% TD) pellet form, produced either by hot pressing or cold pressing and sintering powders. Gas release during irradiation is low. For instance, in one experiment[113] samples irradiated to 100% depletion of the ^{10}B were annealed at 660 K for 400 h with continuous collection of any gas evolved by the sample; the total volume of helium released was less than 2% of the theoretical volume generated. Irradiation-induced dimensional changes are also small under thermal reactor conditions, but some mechanical interaction with the Zircaloy cladding does occur. In the same study[113] hot-pressed (65 to 85% TD) Al_2O_3–3.5 to 6.9 vol.% B_4C pellets irradiated to full ^{10}B depletion exhibited diameter increases up to 1.2% and a length change of the order of 1%.

Therefore, with respect to the B_4C–Al_2O_3 materials used for burnable poisons in PWRs, the major concern appears to be the control of hydrogenous contaminants in the porous bodies to a very low level in order to avoid hydriding of the Zircaloy cladding. In general, burnable poison performance is adequate, but poison rod failures have occurred from hydriding in some Combustion Engineering fuel assemblies and, when cladding failures occur, the subsequent ingress of coolant leaches out the B_4C, thus redistributing the poison and causing a core power imbalance.[3, 109, 110] Procedures similar to those adopted for UO_2 fuel rod fabrication have been adopted to eliminate the potential for hydriding, and hot pressing is used by some vendors to maximize the closed porosity in the pellet.

The use of borosilicate glasses for burnable poisons eliminates the possibility of poison removal or migration through contact with water, but irradiation experience with borosilicate glass indicates that gas release approaches 100% and therefore fuel rod volume must be adjusted accordingly to avoid undue stressing of the cladding.[110] Little, if any, swelling occurs, however, and so mechanical interaction with the cladding is negligible. Borosilicate burnable poisons have operated successfully for several years in Westinghouse fuel designs.

With the trend to higher fuel burnups in PWRs, these burnable poisons will be kept in the core for a longer time, probably at least for an extra cycle (12 to 18 months). Additional irradiation testing will therefore be required to establish the performance capability of both types of boron-based burnable poisons for utilization in high-burnup PWR cores.

In addition to Al_2O_3–B_4C, several alternative burnable poison materials have been proposed for PWR application.[110, 114] These emphasize "nonhelium producers" and include UO_2–Gd_2O_3 (fuel-cycle cost advantages, no helium effects, and tritium disposal), boronated graphite (cheaper than Al_2O_3–B_4C) and Al_2O_3–Gd_2O_3 (no helium effects and no tritium disposal). All these materials are in various stages of development by PWR fuel vendors, with UO_2–Gd_2O_3 being furthest along. For instance, Exxon Nuclear now uses UO_2–Gd_2O_3 in PWR reloads, and KWU, in Germany, has already established a first-core design with UO_2–Gd_2O_3. In the KWU design,[115] every second fuel assembly is provided with eight UO_2–2 or 4 wt.% Gd_2O_3 rods. Calculations indicate that both the distribution and the effect on reactivity are similar to their current design with borosilicate glass.

Apart from a few test assemblies, however, the majority of performance data on UO_2–Gd_2O_3 is from BWRs, which have successfully utilized this ceramic for reactivity control for sometime.[109, 110] Since the UO_2–(1–5 wt.%) Gd_2O_3 burnable poisons in BWRs operate at lower powers and to lower burnups than the companion low-enrichment UO_2 fuel rods, their irradiation performance characteristics are bounded by the UO_2 fuel behavior. Moisture is controlled as for the UO_2 pellet rod to avoid hydriding. Likewise, although densification in-reactor is greater than for UO_2, the amount can be controlled by suitable processing techniques. Since the operating powers and exposures are low, fission gas release is correspondingly low (<0.1%) and PCI effects are absent. It has been generally concluded therefore that the in-reactor performance of homogeneous dispersions of Gd_2O_3 in UO_2 can be readily predicted with state-of-the-art data and analytical tools. Recently, however, some failures have occurred which point to possible inadequacies of the data base on UO_2–Gd_2O_3.[59, 110] One set of failures, General Electric rods in Vermont Yankee, can most probably be attributed to Zircaloy corrosion susceptibility rather than a UO_2–Gd_2O_3 phenomenon.[59] The other failure incidence, of Exxon Nuclear rods in Oyster Creek,[110] however, possesses all the characteristics of PCI effects. Examination is underway to determine whether the incidence of PCI is simply due to incorrect calculation of gadolinia burnout (that is, the fuel was in fact UO_2 at the time of failure) or whether UO_2–Gd_2O_3 can in fact induce PCI cladding strains (for example, the higher in-reactor densification could result in higher than expected temperatures, and hence more swelling and fission product release, as described for UO_2 in Section 2.2.2.1c).

With the need to increase Gd_2O_3 loadings for extended burnup BWR operation, and with the interest in this material for PWR application, it will be necessary to fully understand the origin(s) of failure and ensure that technology is in place to prevent their reoccurrence. Nonetheless, the magnitude of this problem to date has not given concern, and no move to alternative materials is anticipated for the BWR; similarly, a majority, though not all, of the PWR industry still believes that UO_2–Gd_2O_3 poisons represent the optimum configuration for their system.

2.6. Uranium Conservation Measures

One of the biggest issues in reactor technology today is the conservation of our diminishing uranium supplies, or, put another way, maximization of uranium utilization in LWRs. This issue has regained new momentum since the U.S. Government's decision to indefinitely defer the reprocessing and recycle of nuclear fuel because of its concern over the proliferation of nuclear bomb materials (see also Chapter 3). Without recycle, about 20 to 30% more uranium and separative work is required for each kilowatt-hour generated, and, with the current LWR fuel cycle practice, economic uranium supplies could well be exhausted by the year 2000.

There are two basic approaches to the uranium conservation (or utilization) question: on the one hand, alternative fuels can be introduced into the fuel cycle, which is exemplified, of course, by the recycling of plutonium through LWRs; on the other hand, the once-through (i.e., no recycle) fuel cycle can be reoptimized to minimize uranium and separative work requirements.

With the use of plutonium banished (temporarily?), the principal candidate alternative fuel is thorium, which on irradiation converts to the highly efficient fuel, U-233. A LWR thorium–U-233 fuel cycle would utilize the large deposits of thorium ore available, and has the potential for extending the life of our available uranium resources, initially comparable to plutonium recycle, but eventually 20% larger.[116] Thorium fuel experience will be discussed in more detail in Chapter 3, in connection with proliferation-resistant fuel cycles; suffice to say here that the limited thermal reactor experience with the mixed ThO_2–UO_2 fuels has been satisfactory and no major licensing or performance problems are anticipated. Needless to say, however, the data base is far too small to warrant near-term wide-scale introduction of thorium fuels into LWRs. A significant demonstration program in several modern LWRs will be required to develop the statistical performance data base, and this must be combined with safety-related studies in test reactors. However, the major impediment to thorium fuels is the lack of suitable fabrication and reprocessing facilities. The prospects for the implementation of the thorium fuel cycle would be greatly enhanced if dual-purpose thorium/uranium cycle fabrication and reprocessing facilities were available; the feasibility of dual purpose facilities should therefore be evaluated.

Reoptimization of the once-through fuel cycle in LWRs is the most straightforward approach to uranium conservation if the savings can be realized economically and without a long-term development effort. As shown in Table 2-5 provided by Lang,[117] several changes in LWR fuel and reactor design and operation can be identified which might be used to reduce uranium consumption by small percentages. Each change involves complex trade-offs between costs of materials, fabrication, operation, and capacity factor. In general, the small improvements are not simply additive. The potential for larger gains in uranium utilization by more basic design changes have also been under study by several industry

Table 2-5. Summary of LWR Improvements[a]

Improvement	Potential for uranium saving	Timeliness of introduction	Requirements for back fit
Increased burnup	10–20%	Near	None/fuel redesign
Lattice changes	~4%	Near/medium	Fuel redesign
Spectrum shift	Medium	Medium/long	Fuel and/or plant changes
Spatial variation of enrichment	Low/medium	Near/medium	None/fuel redesign
Full use of startup core	~1%	Near	None
Improved fuel management and control designs	Medium	Near/medium	None/fuel redesign
End of cycle stretchout	Low	Near	None
Reconstitution inversion of BWR fuel	~4%	Near/medium	None/fuel redesign
	Low: 0–2%	Near: by 1988	
	Medium: 2–10%	Medium: 1988–2000	
		Long: after 2000	

[a] From Ref. 117.

and government groups,[117] but these larger reductions in uranium requirements generally involve major changes in reactor design and licensing basis, extensive development efforts, substantial front-end costs, and marginal or noncompetitive economics under present conditions. Nevertheless, they are under continuing study for possible applications in the late 1980s or beyond.

Of the various options (Table 2-5), extended burnup of fuel has the most near-term prospect of practicality.[118, 119] If the fuel mechanical lifetime can be extended with sufficient reliability, there appears to be worthwhile net savings potentially attainable in uranium and in fuel storage requirements, and a possible small saving in separative work (even though higher enrichments are required for extended burnup fuel cycles). The increased burnup, if attained, can be exploited to provide longer runs between refuelings (18-month cycles) and, consequently, increasing plant average availability and capacity factors. However, as shown in Fig. 2-30, more savings in uranium and enrichment can be achieved by retaining annual refuelings, but for this fuel cycle there is essentially no gain in plant availability. The largest saving in uranium can come from even shorter refueling intervals, but the penalty from added unavailability for refueling makes this economically unattractive unless rapid turnaround can be routinely attained for refueling. Clearly, the optimum fuel cycle is one that optimizes uranium fuel savings and plant availability or capacity factor.

A preliminary analysis for PWRs[118] based on a five-batch, 12-month cycle, 5-year residence to 4320 GJ/kgM, shown in Fig. 2-30, indicates a savings of (a) 14 to 18% in uranium, (b) 2 to 4% in separative work units (SWU), and (c) 40% in storage and transportation costs. On a plant basis this could amount to a saving of $3 to $5 million per year (i.e., 0.5 to 0.8 mills/kWh) if mechanical reliability is attained. Burnups achievable in BWRs may be lower due to control limitations, but significant savings are still possible.

Several utilities are seeking to combine extended burnup fuel operations with an 18-month cycle; thus, two 18-month cycles will replace the three 12-month cycles, or four 18-month cycles replace six 12-month cycles. As expected (Fig. 2-30), the analysis indicates that uranium and separative work savings are not as high when using 18-month cycles (5–8% less U; 6–8% more SWU), but are still an improvement over lower-burnup (2590 GJ/kgM) 18-month cycles. For utilities with more than one plant to refuel, the increased flexibility in scheduling manpower and equipment, and the possibility of reduced total outage time over a 3-year period, become important added factors. One estimate of savings that might result from switching to 18-month cycles for a two-PWR plant site is as high as $6 million per year.[118]

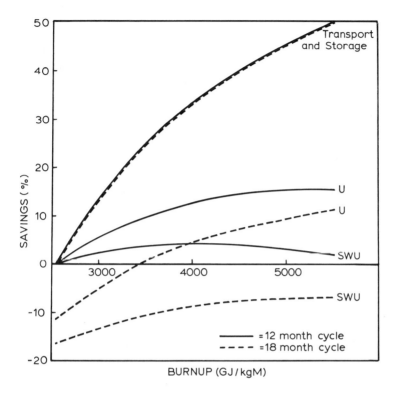

Fig. 2.30. Potential savings in uranium, enrichment, and transport and storage from attainment of extended burnup (Ref. 118).

Thus, the benefits of extending discharge-batch average burnup can include lower fuel cycle costs, facilitation of longer cycle length, reduction in spent fuel, and resource conservation. However, the extension of batch average burnup beyond 3555 GJ/kgM brings with it several uncertainties of a technical nature that must be answered before such fuel cycles can be widely applied. These uncertainties fall into three main categories.[118]

Fuel Performance. With extended burnups, peak fuel pellet burnups will now exceed 5180 GJ/kgM (up from 3455 GJ/kgM). At these exposures very little data exist on the statistical behavior of Zircaloy-clad fuel, particularly with respect to its dimensional stability, susceptibility to PCI distortions or defects, waterside corrosion resistance, and the effects of increased fuel rod pressure due to released fission gas. Integral fuel assembly data are even more limited. Since there is increasing emphasis by regulatory agencies on limiting coolant radioactivity and reducing occupational exposures, fuel reliability is clearly the main factor limiting the realization of savings from increased burnup.

The first objective then should be to solve the PCI problem. This in itself will provide a significant contribution to more efficient uranium utilization inasmuch as it will reduce the prospect of premature discharge of fuel, and, at the same time, allow utilities to regain the capacity lost due to operating restrictions. The fuel research described in section 2.2.1.1 provides a basis for modifying the UO_2 pellet to increase the resistance of the fuel rod to PCI. The emphasis during the next 3 to 5 years should be on minor modifications that fall within the capabilities of current fabrication technology (if not within current specifications) and do not raise new safety or licensing questions. In this category are the annular pellet and the large grain size (Nb_2O_3 doped) pellet. A single pellet combining both features could be in commercial production by 1982 following a successful short-cycle fuel-assembly demonstration program.[21] Combined with some equally straightforward modifications to the Zircaloy cladding, this revised fuel rod design could result in 0.5 to 1% reduction in plant capacity loss, which might well be sufficient for PWRs to operate as originally designed. However, further developments will be required to allow load-follow operation of BWRs. It is conceivable that the duplex fuel pellet (i.e., low-enrichment solid pellet located within a higher-enrichment annular pellet) or Spherepac fuel might fall in this category, but it is the opinion of this writer that a metal barrier will be necessary to protect the Zircaloy from possible chemical attack by fission products since it has a very high susceptibility to iodine SCC. A much longer demonstration program is necessary to justify wide-scale adoption of barrier fuel designs, and 1986 seems an optimistic target date.

The design remedies for PCI minimization are compatible with the goal of higher fuel burnups. For instance, the designs to maintain low fuel temperatures, minimize fission product release, and inhibit clad crack formation are all conducive to longer fuel rod lifetimes. Further, in theory at least, the potential for PCI should diminish with burnup since, in most fuel management plans, the fuel power level and hence duty cycle is reduced correspondingly. At the higher exposures, then, "wearout" phenomena must be considered, and at least two candidates should be of concern at burnups in excess of 3555 GJ/kgM. The first, and perhaps foremost, is the waterside corrosion of Zircaloy which will no doubt limit PWRs but could also impact BWR operations. The second lesser effect, at least in PWRs, is Zircaloy mechanical failure through fatigue. Crack initiation and slow propagation is a common mechanical failure problem in many components (see, in particular, Chapters 4 and 5). There is no reason to believe that Zircaloy will be immune to low-cycle fatigue, particularly in the corrosive environment that exists at both I.D. and O.D. surfaces (i.e., corrosion fatigue) and when temperatures are raised (perhaps into the creep-fatigue regime). Unfortunately little data exist on these phenomena in Zircaloy, and this omission may prove to be a problem later.

Core Performance. Achievement of the higher burnups will necessitate higher U-235 enrichments and hence higher initial loadings of burnable poison rods, which will stay in the core longer. Therefore, long-term behavior of burnable poisons needs to be evaluated. Alternative materials may prove to be more optimum (refer to Section 2.5.2). Also, fuel management schemes must be carefully studied to ensure that power-change duty and operation flexibility are not adversely affected while maintaining fuel rod integrity.

System Interactions. The switch to 18-month cycles will have significant effects on inspection schedules both for core and balance-of-plant components. Increased burnup and exposure time generally result in increased probability of fuel leakers in the core, which in turn, can impact such operations through increased radiation levels. The savings from extended burnup can be offset by unscheduled outages to remove failed fuel assemblies. The cost of one day of unscheduled outage ranges from $300,000 to $500,000 in replacement fuel costs. Thus, 10 days of unscheduled outage time can eliminate all the economic benefit, and some of the uranium and storage space benefits also.

U.S. DOE and industry[117, 118] have embarked on a three-phase, 10-year program to demonstrate a high-reliability, safe fuel design that can routinely achieve discharge burnups in excess of 3555 GJ/kgM. The first phase involves extending the irradiation of existing fuel to achieve assembly average burnups of 3555 to 4320 GJ/kgM. This work will provide statistical information on key aspects of the long-term behavior of Zircaloy-clad UO_2 fuel, such as fission gas release, clad corrosion, PCI, and assembly skeleton integrity. The second phase will involve irradiation to high burnups of lead test assemblies of modified fuel assembly designs; and the third phase will involve a full reload demonstration of the high burnup capability of one or more optimized fuel assembly designs. Such a phased program should enable fuel vendors to warranty progressively higher burnup fuel reloads over the next 10 years, while maintaining the necessary conservatism required by regulatory agencies.

References

1. *Proceedings of the Joint Topical Meeting on Commercial Nuclear Fuel Technology Today* (April 28–30, 1975, Toronto Canada, American Nuclear Society and Canadian Nuclear Association, CNS ISSN 0068-8517, 75CNA/ANS-100.
2. *Proceedings of the American Nuclear Society Topical Meeting on Water Reactor Fuel Performance,* May 9–11, 1977, St. Charles, Illinois.

3. *Proceedings of the American Nuclear Society Topical Meeting on Light Water Reactor Fuel Performance*, April 29 to May 3, 1979, Portland, Oregon.
4. MATPRO—Version 11: A Handbook of Materials Properties for Use in the Analysis of Light Water Reactor Fuel Rod Behavior, NUREG/CR-0497, TREE-1280 (February 1979).
5. R. H. Koppe and E. A. Olson, Nuclear and Large Fossil Unit Operating Experience, EPRI NP-1191 (September 1979).
6. M. Levenson and E. Zebroski, The Nuclear Fuel Cycle, *Annu. Rev. Energ.*, **1**, 645–674 (1976).
7. H. Freeburn, S. R. Pati, I. B. Fiero, and M. G. Andrews, LWR Fuel Rod Modeling Code Evaluation, EPRI NP-369, Project 397-1, Final Report (March 1977).
8. J. A. L. Robertson, Nuclear Fuel Failures, Their Causes and Remedies, in: *Proc. Joint Topical Meeting on Commercial Nuclear Fuel Technology Today*, April 28–30, 1975, Toronto, Canada, CNS ISSN 0068-8517, 75-CNA/ANS-100, pp. 2–2 to 2–14.
9. F. Garzarolli and R. von Jan-H. Stehle, The Main Causes of Fuel Element Failure in Water-Cooled Power Reactors, *At. Energ. Rev.*, **17**, 31–128 (1979).
10. A. A. Strasser and K. O. Lindquist, Reliability and Operating Margins of LWR Fuels, in: *Proceedings of the ANS Topical Meeting on Water Reactor Fuel Performance*, May 9–11, 1977, pp. 8–27.
11. D. Pomeroy and J. Waring, Methods for Determining the Cost of Fuel Failures in Nuclear Power Plants, EPRI NP-854 (August 1978).
12. C. L. Mohr, P. J. Pankaskie, J. C. Wood, and G. Hessler, PCI Fuel Failure Analysis: A report on a Cooperative Program Undertaken by Pacific Northwest Laboratory and Chalk River Nuclear Laboratories, PNL-2755, NUREG/CR-1163 (December 1979).
13. W. J. Baley, C. L. Wilson, L. J. MacGowan, and P. J. Pankaskie, State-of-the-Technology Review of Fuel–Cladding Interaction, Battelle Northwest Laboratory Report COO-4066-2, PNL-2488 VC-78 (December 1977).
14. N. Fuhrman, V. Pasuputhi, D. B. Scott, S. M. Temple, S. R. Pati, and T. E. Hollowell, Fuel Performance in Maine Yankee Core 1, EPRI NP-218 (November 1976).
15. Determination and Characterization of Incipient Defects in Power Reactor Fuel Rods, NP-812, Final Report on Project RP829 (July 1978), compiled by J. Perrin and V. Pasuputhi.
16. G. R. Thomas, The Studsvik InterRamp Project—An International Power Ramp Experimental Program, Meeting on Ramping and Load-Following Behavior of Reactors, November 30 to December 1, 1978, Petten, The Netherlands.
17. P. Knudsen, C. Bagger, and M. Fishler, Characterization of PWR Power Ramp Tests, in: *Proceedings of the American Nuclear Society Topical Meeting on Water Reactor Fuel Performance*, May 9–11, 1977, St. Charles, Illinois, pp. 243–252.
18. B. Cox and J. C. Wood, Iodine-Induced Cracking of Zircaloy Fuel Cladding, in: *Corrosion Problems in Energy Conversion and Generation*, C. S. Tediman, Jr., ed., Electrochemical Society, New York (1974), pp. 275–321.
19. D. Cubicciotti and R. L. Jones, EPRI-NASA Cooperative Project on Stress Corrosion Cracking of Zircaloys, EPRI NP-717, Research Project RP455-1 (March 1978).
20. J. T. A. Roberts, E. Smith, N. Fuhrman, and D. Cubicciotti, On the Pellet–Cladding Interaction Phenomenon, *Nucl. Technol.*, **35**, 131–144 (1977).
21. J. T. A. Roberts and F. E. Gelhaus, Zircaloy Performance in Light Water Reactors, in: *Proceedings of the 4th International Conference on Zirconium in the Nuclear Industry*, T. Papazoglou, ed., Am. Soc. Test. Mater. Spec. Tech. Publ. 681 (1979), pp. 19–39.
22. R. Holtzer, D. Knodler, and H. Stehle, Pellet–Clad Interaction: Experience, Testing and Evaluation. A KWU Review, in: Proceedings of the American Nuclear Society Topical Meeting on Water Reactor Fuel Performance, May 9–11, 1977, St. Charles, Illinois, pp. 207–218.

23. J. T. A. Roberts, R. L. Jones, E. Smith, D. Cubicciotti, A. K. Miller, H. F. Wachob, and F. L. Yaggee, An SCC Model for Pellet–Cladding Interaction Failures in LWR Fuel Rods, in: *Proceedings of the 4th International Conference on Zirconium in the Nuclear Industry*, T. Papazoglou, ed., Am. Soc. Test. Mater. Spec. Tech. Publ. 681 (1979), pp. 285–305.
24a. R. F. Mattas, F. L. Yaggee, and L. A. Neimark, Iodine Stress-Corrosion Cracking in Irradiated Zircaloy Cladding, in: *Proceedings of the American Nuclear Society Topical Meeting on Light Water Reactor Fuel Performance*, April 29–May 3, 1979, Portland, Oregon, pp. 128–140.
24b. Liv Lunde and Ketil Videm, The Influence of Environmental Variables and Irradiation on Iodine Stress Corrosion Crack Initiation and Growth in Zircaloy, paper presented at IAEA Specialists Meeting on Water Reactor Fuel Element Performance Computer Modeling, March 17–21, 1980, Blackpool, UK.
25. M. Peehs, H. Stehle, and E. Steinburg, Out-of-Pile Testing of Iodine Stress Corrosion Cracking in Zircaloy Tubing under the Aspects of the PCI Phenomenon, in: *Proceedings of the 4th International Conference on Zirconium in the Nuclear Industry*, T. Pagazoglou, ed., Am. Soc. Test. Mater. Spec. Tech. Publ. 681–(1979), pp. 244–260.
26. B. L. Syrett, D. Cubicciotti, and R. L. Jones, Some Observations on the Influence of Texture and of Second-Phase Particles on Iodine SCC of Unirradiated Zircaloy, in: *Proceedings of the American Nuclear Society Topical Meeting on Light Water Reactor Fuel Performance*, April 29 to May 3, 1979, Portland, Oregon, pp. 113–127.
27. E. Smith and A. K. Miller, Stress Corrosion Fracture of Zircaloy Cladding in Fuel Rods Subjected to Power Increases: A Model for Crack Propagation and Failure Threshold Stress, *J. Nucl. Mater.* **80**, 291–302 (1979).
28. A. K. Miller, K. D. Challenger, E. Smith, G. V. Ranjan, and R. C. Cipolla, Zircaloy Cladding Deformation and Fracture Analysis, EPRI NP-856, Projects 700/971, Interim Report (August 1978).
29. E. Rolstad and K. D. Knudsen, Studies of Fuel–Clad Mechanical Interaction and the Resulting Interaction Failure Mechanism, in: *Fourth United Nations International Conference on the Peaceful Uses of Atomic Energy*, September 6–16, 1971, Geneva, Switzerland
30. P. J. Fehrenbach, In-Reactor Measurement of Clad Strain, in: *Proceedings of the American Nuclear Society Topical Meeting on Light Water Reactor Fuel Performance*, April 29 to May 3, 1979, Portland, Oregon, pp. 255–263.
31. D. Lee and R. B. Adamson, Modelling of Localized Deformation in Neutron-Irradiated Zircaloy 2, in: *Proceedings of the 3rd International Conference on Zirconium in the Nuclear Industry*, August 10–12, 1976 Quebec City, Canada, Am. Soc. Test. Mater. Spec. Tech. Publ. 633 (1977), pp. 385–401.
32. A. K. Miller, Progress in Modelling of Zircaloy Nonelastic Deformation Using a Unified Phenomenological Model, in: *Proceedings of the 3rd International Conference on Zirconium in the Nuclear Industry*, August 10–12, 1976, Quebec City, Canada, Am. Soc. Test. Mater. Spec. Tech. Publ. 633 (1977), pp. 385–401.
33. A. K. Miller and O. D. Sherby, Development of the Materials Code, MATMOD (Constitutive Equation for Zircaloy), EPRI NP-567, Project 456-1, Final Report (December 1977).
34. R. R. Hosbons, C. E. Coleman, and R. A. Holt, Numerical Simulation of Tensile Behavior of Nuclear Fuel Cladding Materials, in: *3rd International Conference on Structural Mechanics in Reactor Technology* (1975), pp. 1–10.
35. S. Oldberg, Jr., A. K. Miller, and G. E. Lucas, Advances in Understanding and Predicting Inelastic Deformation in Zircaloy, in: *Proceedings of the 4th International Conference on Zirconium in the Nuclear Industry*, June 26–29, 1978, Stratford, U.K., T. Papazoglou, ed. Am. Soc. Test. Mater. Spec. Tech. Publ. 681 (1979), pp. 307–392.
36. J. H. Gittus, Theoretical Analysis of the Strains Produced in Nuclear Fuel Cladding

Tubes by the Expansion of Cracked Cylindrical Fuel Pellets, *Nucl. Eng. Des.*, **18**, 69–82 (1972).
37. J. T. A. Roberts, F. E. Gelhaus, H. Ocken, N. Hoppe, S. T. Oldberg, G. R. Thomas, and D. Franklin, LWR Fuel Performance Program: Progress in 1978, EPRI NP-1024-SR, Special Report (February 1979).
38. D. Cubicciotti, J. E. Sanecki, R. V. Strain, S. Greenburg, L. A. Neimark, and C. E. Johnson, The Nature of Fission-Product Deposits inside LWR Fuel Rods, in: *Proceedings of the American Nuclear Society Topical Meeting on Water Reactor Fuel Performance*, May 9–11, 1977, St. Charles, Illinois, pp. 282–294.
39a. D. Cubicciotti, State-of-the-Art Model for Release of Volatile Fission Products from UO_2 Fuel, EPRI Project RP355-11, Final Report (December 1978).
39b. D. Cubicciotti, A Quantitative Model for Volatile Fission Product Release from UO_2 Fuel, EPRI Project RP355-11, Final Report (June 1979).
40. J. C. Wood, D. G. Hardy, and A. S. Bain, Mechanistic Studies of Power Ramping Defects and Remedies, in: *Proceedings of the American Nuclear Society Topical Meeting on Light Water Reactor Fuel Performance*, April 29 to May 3, 1979, Portland, Oregon, pp. 169–178.
41. B. Tomkins and J. H. Gittus, Stress Corrosion Crack Growth for Short Cracks with Particular Reference to the Zircaloy 2/Iodine System, in: *Fracture 1977*, Vol 4 ICF 4, Canada (June 1977).
42. A. Garlick, A Quantitative Treatment of Stress Corrosion Failure in Irradiated Zircaloy Cladding, private communication, August 1978.
43. A. K. Miller, Bundle-of-Sticks SCC Model, EPRI Internal Progress Report, on RP700-3 (1979).
44. W. J. Penn, R. K. Lo, and J. C. Wood, CANDU Fuel-Power Ramp Performance Criteria, *Nucl. Technol.*, **34**, 249–268 (1977).
45. D. Pomeroy, S. Sorenson, and E. Rolstad, Nuclear Performance Evaluation, EPRI RP509-1, Final Report (April 1977).
46. R. L. Jones, E. Smith, and A. K. Miller, Observations on Damage Accumulation during Iodine-Induced Stress Corrosion Cracking of Zircaloy Cladding, *Nucl. Technol.*, **42**, 233–236 (1977).
47a. R. Christenson, SPEAR Fuel Reliability Code System: General Description, EPRI NP-1378, Project 971, 700-3 Interim Report (March 1980).
47b. S. T. Oldberg, Fuel Failure Modeling, *Electr. Power Res. Inst. J.*, **5**, 57, (April 1980).
48. F. E. Gelhaus, D. L. Pomeroy, M. T. Pitek, Impact of Fuel Damage on Nuclear Power Plant Operation, invited paper, American Nuclear Society Winter Meeting, Washington, D.C. (November 12–16, 1978).
49. D. H. Locke, Pellet–Cladding Interaction for Water Reactors: Experience Testing and Evaluation, At. Energ. Rev., **15**, 779–792 (1977).
50. F. W. Buckman, C. E. Crouthamel, and M. D. Freshley, FCI Remedy Development, in: *Proceedings of the American Nuclear Society Topical Meeting on Light Water Reactor Fuel Performance*, April 29 to May 3, 1977, Portland, Oregon, pp. 158–168.
51. R. F. Mattas and F. L. Yaggee, The Effect of Zirconium Oxide on the Stress Corrosion Susceptibility of Irradiated Zircaloy Cladding, paper presented at Fifth International Conference on Zirconium in the Nuclear Industry, August 4–7, 1980, Boston, Massachusetts.
52. D. G. Hardy, J. C. Wood, and A. S. Bain, CANDU Fuel Performance and Development, paper presented at the 18th Annual International Conference of the Canadian Nuclear Association, Nuclear Fuels and Recycle Session, Ottawa, June 14, 1978.
53. J. T. A. Roberts, Ceramic Utilization in the Nuclear Industry; Current Status and Future Trends, *Powder Metall. Int.*, **11**, Part I (February 1979), pp. 24–29; Part II (May 1979), pp. 72–80.

54a. D. S. Tomalin, R. B. Adamson, R. P. Gangloff, The Performance of Irradiated Copper and Zirconium Barrier-Modified Zircaloy Cladding under Simulated PCI Conditions, in: *Proceedings of the 4th International Conference on Zirconium in the Nuclear Industry*, Am. Soc. Test. Mater. Spec. Tech. Publ. 681 (1979), pp. 122–144.

54b. J. H. Davies, E. L. Esch, L. D. Noble, R. L. Schreck, J. S. Armijo, H. S. Rosenbaun, J. M. Isaacson, G. R. Parkos, and R. A. Proebstle, Power Ramp Tests of Potential PCI Remedies, in: *Proceedings of the American Nuclear Society Topical Meeting on Light Water Reactor Fuel Performance*, April 29 to May 3, 1979, Portland, Oregon, pp. 275–283.

55. Private communication, Tokyo Electric Power Company (March 1979).

56. F. W. Trowse, R. Sumerling, and A. Garlick, Nodular Corrosion of Zircaloy 2 and Some Other Zirconium Alloys in Steam-Generating Heavy Water Reactors and Related Environments, in: *Proceedings of the 3rd International Conference on Zirconium in the Nuclear Industry*, Am. Soc. Test. Mater. Spec. Tech. Publ. 633 (1977), pp. 236–267.

57. A. B. Johnson, Jr., and R. M. Horton, Nodular Corrosion of the Zircaloys, in: *Proceedings of the 3rd International Conference on Zirconium in the Nuclear Industry*, Am. Soc. Test. Mater. Spec. Tech. Publ. 633 (1977), pp. 295–311.

58. I. K. Dickson, H. E. Evans, and K. W. Jones, A Comparison between the Uniform and Nodular Forms of Zircaloy Corrosion in Water Reactors, *J. Nucl. Mater.*, **80**, 223–231 (1979).

59. R. Grube, Yankee Atomic Company, private communication (July 1979).

60. M. Zukor, Oyster Creek Fuel Channel Surveillance Program, paper presented to Utility Conference on Fuel Performance Experience, Gatlinburg, Tennessee (October 1977).

61. A. W. Urquhart, D. A. Vermilyea, and W. A. Rocco, A Mechanism for the Effect of Heat Treatment on the Accelerated Corrosion of Zircaloy 4 in High-Temperature, High-Pressure Steam, *J. Electrochem. Soc.*, **125**, 199–204 (1978).

62. L. Lunde and K. Videm, Effect of Surface Treatment on the Irradiation Enhancement of Corrosion of Zircaloy 2 in HBWR, in: *Zirconium in Nuclear Application*, Am. Soc. Test. Mater. Spec. Tech. Publ 551 (1974), pp. 514–526.

63. S. B. Dalgaard, Long-Term Corrosion and Hydriding of Zircaloy 4 Fuel Clad in Commercial Pressurized Water Reactors with Forced Convective Heat Transfer, in: Extended Abstracts of Proceedings of the Electrochemical Society, Washington, D.C. (May 2–7, 1976).

64. F. Garzarolli, D. Jorde, R. Manzel, G. W. Parry, and P. G. Smerd, Review of PWR Fuel Rod Waterside Corrosion Behavior, C-E, NPSD 79, Combustion Engineering, Windsor, Connecticut (June 1979).

65. M. Tokar, Development of Improved LWR Fuel Damage Limits for Reactor Licensing, invited paper, American Nuclear Society Winter Meeting, Washington, D.C. (November 1978).

66. R. O. Meyer, The Analysis of Fuel Densification, Report of U.S. Nuclear Regulatory Commission, NUREG-0085 (July 1976).

67. W. V. Johnston, Zircaloy—Three Years after the Hearings, in: *Proceedings of the 3rd International Conference on Zirconium in the Nuclear Industry*, Am. Soc. Test. Mater. Spec. Tech. Publ. 633 (1977), pp. 5–23.

68. Water Reactor Safety Research Program, Publication of U.S. Nuclear Regulatory Commission, NUREG-0006 (February 1979).

69. *Proceedings of the 3rd International Conference on Zirconium in the Nuclear Industry*, Am. Soc. Test. Mater. Spec. Tech. Publ. 633 (1977), Session on High-Temperature Transient Behavior (eds. A. L. Lowe Jr. and G. W. Parry), pp. 5–208.

70. *Proceedings of the 4th International Conference on Zirconium in the Nuclear Industry*, T. Papazoglou, ed., Am. Soc. Test. Mater. Spec. Tech. Publ. 681 (1979), pp. 479–627.

71. H. Ocken, R. Biederman, C. Hann, and R. Westerman, Evaluation Models of Zircaloy

Oxidation in the Light of Recent Experiments, in: *Proceedings of the 4th International Conference on Zirconium in the Nuclear Industry*, T. Papazoglou, ed., Am. Soc. Test. Mater. Spec. Tech. Publ. 681 (1979), pp. 514–536.
72. R. R. Biederman, R. D. Sission, J. K. Jones, and W. G. Dolison, A study of Zircaloy 4 Steam Oxidation Reaction Kinetics: Part II, EPRI NP-734, Final Report on Project 249 (March 1978).
73. R. E. Westerman and R. M. Hesson, Zircaloy Cladding I.D./O.D. Oxidation Studies, EPRI NP-525, Final Report on Project 251 (November 1977).
74. H. M. Chung, A. M. Garde, and T. F. Kassner, Development of an Oxygen Embrittlement Criterion for Zircaloy Cladding Applicable to LOCA conditions in Light Water Reactors, in: *Proceedings of the 4th International Conference on Zirconium in the Nuclear Industry*, T. Papazoglou, ed., Am. Soc. Test. Mater. Spec. Tech. Publ. 681 (1979), pp. 600–627.
75a. C. R. Hann, C. L. Mohr, K. M. Busness, N. J. Olson, F. R. Reich, and K. B. Stewart, Transient Deformation Properties of Zircaloy for LOCA Simulation, EPRI NP-526, Vols 1–5, Final Report on Project 251 (March–December 1978).
75b. L. R. Bunnell, G. B. Mellinger, J. L. Bates, and C. R. Hann, High-Temperature Properties of Zircaloy–Oxygen Alloys, EPRI NP-524, Final Report on Project 251 (March 1977).
76. D. A. Powers and R. O. Meyer, Evaluation of Simulated LOCA Tests That Produced Large Fuel Cladding Ballooning, NUREG-0536, March 1979, Office of Nuclear Reactor Regulations, USNRC, Washington, D.C.
77. D. W. Brite, J. L. Daniel, N. C. Davis, M. D. Freshley, P. E. Hart, and R. K. Marshall, EEI/EPRI Fuel Densification Project, EPRI RP-131, Final Report (March 1975).
78. M. D. Freshley, D. W. Brite, J. L. Daniel, and P. E. Hart, Irradiation-Induced Densification of UO_2 Pellet Fuel, *J. Nucl. Mater.*, **62**, 138–166 (1977).
79. The Role of Fission Gas Release in Reactor Licensing, Report of Core Performance Branch, U.S. Nuclear Regulatory Commission, NUREG-75/077 (November 1975).
80. R. O. Meyer, C. E. Beyer, and J. C. Voglewede, Fission Gas Release from Fuel at High Burnup, Report of U.S. Nuclear Regulatory Commission, NUREG-0418 (March 1978); also published in *Nucl. Safety*, **19**, 699–708 (1978).
81. W. L. Baldewicz, The State-of-the-Art of Fission Gas Release from LWR Fuels, Report prepared for Electric Power Research Institute, UCLA-ENG-7740 (May 1977).
82. C. Vitanza, Fission Gas Release from UO_2 Fuel, paper presented at European Nuclear Conference, May 6–11, 1979 Hamburg, West Germany; abstract published in *Trans. Am. Nucl. Soc.*, **31**, 163–166 (1979).
83. F. Garzarolli and R. Manzel, High-Burnup UO_2 Performance in LWRs, paper presented at European Nuclear Conference, May 6–11, 1979, Hamburg, West Germany; abstract published in *Trans. Am. Nucl. Soc.*, **31**, 162–163 (1979).
84. L. D. Noble, ANS-5.4 Fission Gas Release Model I. Noble Gases at High Temperature, *Proceedings of the ANS Topical Meeting on Light Water Reactor Fuel Performance*, 1979, Portland, Oregon, pp. 321–327.
85. R. O. Meyer, Fission Gas Release from High-Burnup Fuel, presentation to ACRS Reactor Fuel Subcommittee, Washington, D.C., May 20, 1977.
86. S. R. Pati, Gas Release and Microstructural Evaluation of One- and Two- Cycle Fuel Rods from Calvert Cliffs-1, Report to EPRI on Project 586-1 Task A, NPSD-75 (March 1979).
87. H. Ocken and J. T. A. Roberts, Comments on "Fission Gas Release from Fuel at High Burnup," *Nucl. Safety*, **20**(4) 417–420, 1979.
88. B. Zolotar, private communication, Interim Report on EPRI Project 306 (March 1979).
89. C. C. Dollins and F. A. Nichols, Swelling and Gas Release in UO_2 at Low and Intermediate Temperatures, *J. Nucl. Mater.*, **66**, 143–157 (1977).

90. R. Hargreaves and D. A. Collins, A Quantitative Model for Fission Gas Release and Swelling in Irradiated Uranium Dioxide, *J. Br. Nucl. Energ. Soc.*, **15**, 311–318 (1976).
91. M. O. Tucker and R. J. White, The Release of Unstable Fission Products from UO_2 during Irradiation, *J. Nucl. Mater.*, **87**, 1–10 (1979).
92. A. S. Mehner, R. R. Hobbins, S. L. Seiffert, P. E. MacDonald, and R. K. McCardell, Damage and Failure of Unirradiated and Irradiated Fuel Rods Tested under Film Boiling Conditions, in: *Proceedings of the ANS Topical Meeting on Light Water Reactor Fuel Performance*, April 28 to May 3, 1979, Portland, Oregon, pp. 207–217.
93. R. R. Hobbins, S. L. Seiffert, S. A. Plager, A. S. Mehner, and P. E. MacDonald, Embrittlement of Zircaloy-Clad Fuel Rods Irradiated under Film Boiling Conditions, in: *Fourth International Conference on Zirconium in the Nuclear Industry*, T. Papazoglou, ed., Am. Soc. Test. Mater. Spec. Tech. Publ. 681 (1979), pp. 586–599.
94. P. Bailey, private communication, EPRI (May 1979).
95. Kemeny Commission Report on Three Mile Island-Unit 2 Accident (October 1979).
96. Analysis of Three Mile Island-Unit 2 Accident, Report of Nuclear Safety Analysis Center, NSAC-1 (July 1979) and NSAC-1 Supplement (October 1979).
97. Rogovin Report on Three Mile Island-Unit 2 Accident, Vol. II (January 1980).
98. Three Mile Island-Unit 2 Core Damage Assessment Report, prepared by Nuclear Safety Analysis Center, NSAC (March 1980).
99. H. J. Schenk, K. L. Huppert, and W. Stoll, Experience with the Use of Recycle Plutonium in Mixed-Oxide Fuel in Light Water Reactors in the Federal Republic of Germany, *Nucl. Technol.*, **43**, 174–185 (1979).
100. Final Generic Environmental Statement on the Use of Recycle Plutonium in Mixed Oxide Fuel in Light Water Reactors, NUREG-0002, Vols. 3 and 4 (August 1976).
101. Electric Power Research Institute Research Project 497. Project Manager, B. Zolotar, private communication (May 1979).
102. M. D. Freshley, D. W. Brite, J. L. Daniel, P. E. Hart, and R. K. Marshall, Plutonia Fuel Study, Electric Power Research Institute, Research Project 396, Final Report NP 637 (1978).
103. R. Duncan, Stainless Steel Failure Investigation Program, Final Summary Report, General Electric Report GEAP-5530 (February 1968).
104. F. D. Coffman, Evaluation of Stainless Steel Cladding for Densification and LOCA Conditions, USNRC Core Performance Branch, Technical Note No. 74-1 (January 1974).
105. J. N. Baker, Fuel Element Integrity and Behavior in a Loss of Coolant Accident, paper presented at OECD Committee on Reactor Safety Technology (CREST) Specialist Meeting on Emergency Core Cooling for Light Water Reactors, October 18–20, 1972, Munich, West Germany.
106. G. F. Rieger, Post-Irradiation Evaluation of Fuel Rods from the LaCrosse Boiling Water Reactor, General Electric Report NEDC-12633 (June 1976).
107. M. Pitek, Northeast Utilities, Connecticut Yankee Batch 8 Fuel Failure Analysis, private communication (July 1979).
108. R. A. Murgatroyd and B. T. Kelly, Technology and Assessment of Neutron-Absorbing Materials, *At. Energ. Rev.*, **15**, 3–74 (1977).
109. R. O. Meyer, M. D. Houston, M. Tokar, and F. E. Panisko, Control Materials Behavior in Commercial Reactors, paper presented at the 23rd Annual Meeting of The American Nuclear Society, June 1977, New York.
110. A Strasser and W. Yario, Control Rod Materials and Burnable Poisons: State-of-the-Art. Report to Electric Power Research Institute (October 1979).
111. R. O. Brugge, Proposed Peach Bottom Atomic Power Station Unit 3 Alternate Absorber Control Blade Test Program, NEDO-24213 Class 1, August 1979, General Electric Co.

112. I. Cohen, Silver and Silver Base-Alloys, in: *Neutron Absorber Materials for Reactor Control,* W. K. Anderson and J. S. Theilacker, eds., USAEC (1962), Chapter 5, pp. 405–530.
113. W. R. Jacoby, P. R. Mertens, and J. S. Theilacker, Boron Compounds Dispersed in Ceramic Systems, in: *Neutron Absorbers for Reactor Control,* W. K. Anderson and J. S. Theilacker, eds., USAEC (1962), Chapter 4, pp. 214–223.
114. J. T. A. Roberts and D. Franklin, Ceramic Control Materials in Light Water Reactors: An Overview, paper presented at 81st Annual Meeting of American Ceramic Society, April 28 to May 2, 1979, Cincinatti, Ohio; abstract published in *Am. Ceram. Soc. Bull.* (March 1979).
115. H. Roth-Seefrid and J. G. Fite, First Core Design of a 1300 MW(e) PWR with Gadolinium, paper presented at European Nuclear Conference, May 6–11, 1979, Hamburg, West Germany; abstract published in *Trans. Am. Nucl. Soc.,* **31,** 233–234 (1979).
116. Assessment of Thorium Fuel Cycles in PWRs, Electric Power Research Institute, EPRI NP-359, Final Report Project 515-1 (February 1977).
117. P. Lang, private communication, Department of Energy (July 1979).
118. J. T. A. Roberts, R. F. Williams, and E. L. Zebroski, Incentives for Extended Burnup Nuclear Fuel Operation, Electric Power Research Institute Journal (December 1978), pp. 47–48.
119. S. E. Turner, W. J. Elgin, and R. P. Hancock, Historical Survey of Nuclear Fuel Utilization in U.S. LWR Power Plants, Final Report DOE/ER/10020-Ti (August 1979).

3

LMFBR Core Materials

The extensive experience with the thermal reactor core materials (uranium dioxide, boron carbide, and stainless steel) has provided the basis for LMFBR core materials development. However, as noted in Chapter 1, the LMFBR fuel system must survive a much more hostile environment than the LWR fuel system with respect to temperature and fast neutron fluence. Typical fuel rod powers of 40 to 90 kW/m (compared to a maximum of ~40 kW/m in LWRs) yield fast neutron fluences of ~2 × 10^{27} n/m² ($E > 0.1$ MeV), or ~90 dpa, with maximum cladding temperatures of the order of 920 K (Fig. 3-1). These more extreme conditions are manifested in more profound chemical and microstructural changes in the fuel and cladding; for instance, neither fission gas release approaching 100% nor void swelling of cladding are observed in LWR fuel rods, but they are normal occurrences in some LMFBR fuel systems and must therefore be accounted for in design. Similar allowances, although of a lesser nature, have had to be made in the LMFBR control rod design.

The current choice for a fuel system to contend with this environment is a uranium oxide, 25 ± 5 wt.% plutonium mixed oxide fuel, and austenitic stainless steel cladding and duct materials (refer to Table 2-1). Both pellet (solid and annular) and packed particle (Vipac, Spherepac,† etc.) fuel types have been seriously considered for the mixed oxide, and two austenitic steels, Types 304 and 316, in both annealed and cold-worked conditions, have been evaluated for cladding and duct materials. However, first-generation cores are comprised of the solid pellet fuel contained in 20% cold-worked Type 316 stainless steel cladding and ducts, which makes maximum use of LWR technology. The initial fuel and core designs discussed in Section 3.1 are very conservative, emphasizing high

† Names for particle fuel processes.

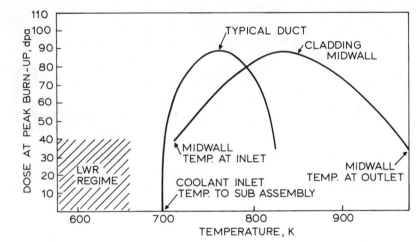

Fig. 3-1. Typical radiation dose–temperature profiles for LMFBR cladding and duct at 8640 GJ/kgM, compared to LWR cladding conditions (adapted from Ref. 43). 1 dpa $\approx 2 \times 10^{25}$ n/m^2 ($E > 0.1$ MeV).

reliability at the expense of breeding ratio and doubling time.† Some of the essential fuel rod design features are commonly accepted worldwide (Table 2-2 for example), while a lot of uncertainty is still to be resolved regarding the details.

The successful high-burnup irradiation behavior (10,000 GJ/kgM burnup) in fast flux reactors of mixed oxide fuel rods with stainless steel cladding has been demonstrated by the French in Rapsodie,‡ the U.K. in DFR,‡ the Russians in BOR-60,‡ and the U.S. in EBR-II,‡ as well as others. These development programs, covering the period 1965 to the present, have generated substantial information about the physical phenomena which must be considered in fuel rod design, and have provided a basis for improved performance designs. Important findings of this research effort will be described in Section 3.2, on "Current" mixed oxide fuel behavior.

Improved mixed oxide fuel designs utilizing larger rod diameters, reduced cladding thickness-to-diameter ratios, higher fuel smeared densities, bottom plenums, modified spacers, slotted ducts, and alternative

---

† Doubling time, in simple terms, is a measure of the time required for a breeder reactor to breed an amount of surplus plutonium equal to the original inventory. In actuality, it is a complex function of a great number of core parameters, including breeding ratio, that are considered outside the scope of this book. Breeding ratio is defined as the ratio of the number of fissile atoms produced from fertile material per fission event; a value >1 denotes "breeding" and the higher the value the better the "breeder."

‡ All are experimental fast breeder reactors.

cladding and structural materials are now being developed in support of potential improvements in doubling time, plutonium inventory requirements, or power density.[1, 2] Such changes are necessary, since to reduce plutonium doubling time to less than 15 years with mixed oxide fuels requires that the fraction of the core volume occupied by fuel be increased to 40 to 45% (compared to 30 to 35% in current designs) with corresponding decreases in the amount of steel and sodium in the core. The progression in design improvements to accomplish this are shown in Fig. 3-2.[1]

High burnup capability, >8600 GJ/kgM is also desirable to minimize fuel cycle costs. Success here requires development of "advanced" cladding and core structural materials that exhibit reduced irradiation swelling and creep, because these phenomena are currently limiting fuel assembly lifetime. Recent progress in advanced oxide fuel, cladding, and duct development will be reported in Section 3.3.

Any further improvements in breeding gain and doubling time in LMFBRs, however, can only be achieved by switching to an alternative fuel type, for example, the carbide, nitride, or metal. The marked advantages of carbide and metal fueled cores over reference oxide cores were demonstrated in a recent 1000 MWe LMFBR design study.[3] In this study, breeding ratios and compound system doubling times were computed for both the conventional homogeneous and alternative heterogeneous core configurations (refer to Fig. 3-5). Results, summarized in Fig. 3-3, indicate that both carbide and metal cores can achieve breeding ratios in the range of 1.5 to 1.6 versus 1.45 for oxide (Fig. 3-3a) and, of more importance, the carbide and metal cores can achieve 10-year doubling times (Fig. 3-3b). These differences translate into millions of dollars sav-

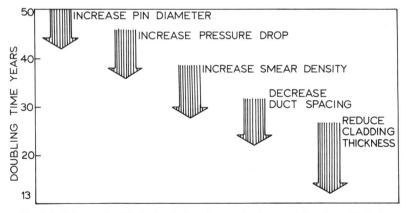

Fig. 3-2. Advanced oxide fuels: design changes for low doubling times (Ref. 1).

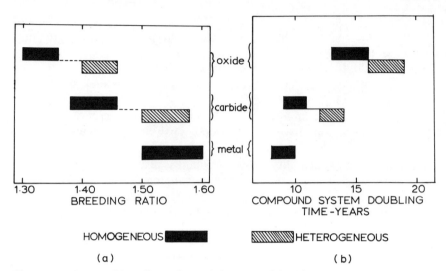

Fig. 3-3. (a) Range of breeding ratios and (b) range of doubling times for reference fuel assembly designs using oxide, carbide, and metal fuels (Ref. 3).

ings in fuel cycle costs and hence make LMFBRs more economically attractive. As will be discussed in Section 3.4, carbide fuel development is therefore being continued at a modest level to provide a backup to the mixed oxide design. Work on nitride fuels has been largely terminated since no significant advantages over carbides have been seen. Metal fuel studies have reappeared in the context of proliferation resistance and will therefore be discussed in Section 3.6.

As for the LWR, control material reliability will be extremely important to LMFBR operations. Based on development efforts and favorable experience in thermal reactors, boron carbide (B_4C) was specified for the LMFBR. In Section 3.5 the LMFBR development efforts on B_4C, and the backup control rod material, europia (Eu_2O_3), will be explored.

A final section, 3.6, is included for completeness to summarize the leading candidate alternative fuel cycles that might be more proliferation resistant than the U–Pu fuel cycle.

3.1. Fuel and Core Designs

The evolution of the reference LMFBR mixed oxide fuel rod design from the LWR fuel designs can be seen from Tables 2-1 and 2-2. The choice of an oxide pellet for the first-generation LMFBRs was based on the extensive experience with UO_2 fuels in thermal reactors. Inasmuch as neutron economy is of less importance in fast reactors, stainless steel components can be used in the core region. Type 304 stainless steel was an original candidate, again based on thermal reactor experience, but has now been superseded by Type 316 stainless steel which possesses better resistance to swelling and high-temperature mechanical properties.

The basic fuel rod concept is the same for LWR and LMFBR, that is, pellets stacked in a sealed metal-alloy tube. However, specifics of the LMFBR design differ from its LWR counterpart: (a) the fuel rod diameter and clad wall thickness are smaller (thickness-to-diameter ratios are not too different, however); (b) the active fuel length is shorter; (c) the plenum-to-fuel ratio is considerably higher; and (d) since the LMFBR is designed to "breed" nuclear fuel, axial fertile "breeding" blanket regions of depleted UO_2 (U-238) pellets are included above and below the active

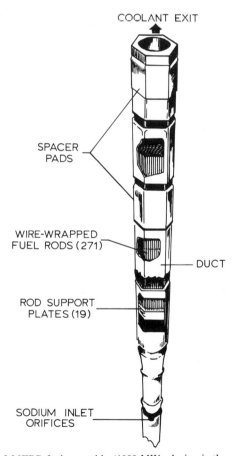

Fig. 3-4. LMFBR fuel assembly (1000 MWe design is the example).

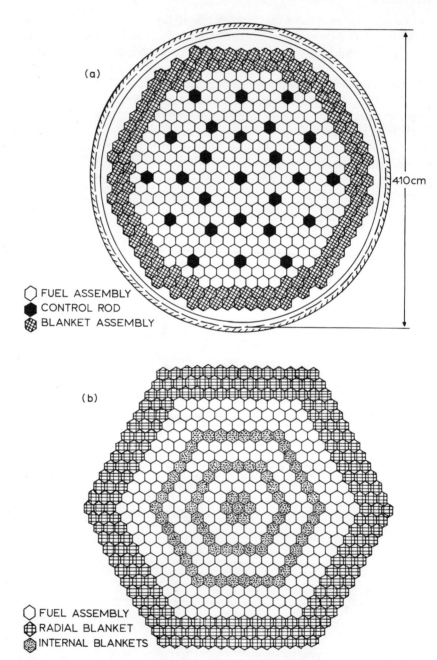

Fig. 3-5. Cross section of LMFBR core showing (a) arrangement of assemblies and control rods in homogeneous design, and (b) heterogeneous core design.

fuel column. It is important to note that the fuel rod design parameters for the mixed oxide are not necessarily optimum for the alternative fuels; for example, for carbide fuels larger diameter rods would be utilized to take advantage of the higher fuel thermal conductivity, and smaller plenums would be acceptable because of the lower fission gas release.

However, regardless of whether oxide, carbide, or metal fuels are used in LMFBR core designs, the fuel and radial "breeder" blanket assemblies are designed basically as hexagonally shaped components. Again, using the reference mixed oxide fuel as an example, the fuel assembly is comprised of a bundle of over 200 fuel rods, containing fuel and blanket pellets, a surrounding "hex-can," and upper and lower hardware, as shown in Fig. 3-4. Rods are positioned and spaced by helically wrapped wire ("wire-wrap"), of 1.4 mm diameter, wrapped in a clockwise helical spiral with 300 mm pitch. Blanket assemblies are similar in overall dimension to the fuel assemblies, but the radial blanket rods are larger in diameter because of their lower power densities.

The core of an LMFBR is essentially comprised of driver fuel assemblies, fertile "blanket" assemblies, and control rods. Burnable poisons are unnecessary in LMFBRs. Driver fuel and blanket assemblies can be arranged in different configurations, but two leading concepts are shown in Fig. 3-5. In the homogeneous core, Fig. 3-5a, which is the conventional approach adopted by first-generation systems, the driver fuel central core is completely surrounded by the blanket assemblies, and, typically, two fuel-enrichment zones are employed to flatten the radial power profile. In contrast, in the heterogeneous core, Fig. 3-5b, internal blanket assemblies are utilized in the core region to increase the breeding ratio by being exposed to the higher neutron flux (refer to Fig. 3-3). This concept is accordingly referred to as the bull's-eye core design. In addition to the radial blanket assemblies, a reflector zone is likely to be used in either design to protect the vessel wall against high neutron irradiation and to achieve greater neutron economy. Also, two independent control rod systems, utilizing B_4C pellets clad in Type 316 stainless steel, are likely to ensure reliability.

3.2. Performance of Current Mixed Oxide Fuel Designs

The result of some 15 years of worldwide irradiation test programs[4,5] is an impressive body of data on the behavior of the UO_2, 20 wt.% PuO_2–20% cold-worked (CW), Type 316 stainless steel fuel rod system in high flux, flowing sodium environments. In the U.S. alone, approximately 1500 experimental fuel rods clad with 20% CW stainless steel (out of a total of 2800 rods) have been irradiated in EBR-II, about

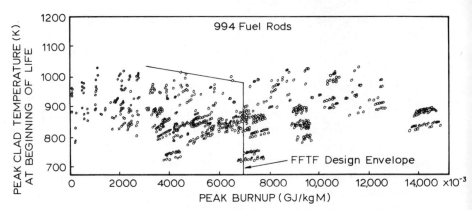

Fig. 3-6. Conditions of clad temparature and burnup for U.S. reference mixed oxide fuel rods (994) tested in EBR-II compared to fast flux test facility (FFTF) fuel design envelope (courtesy C. Cox, HEDL).

1000 of which are prototypic of the U.S. reference design. Of the prototypic fuel rods, about 400 have exceeded the target burnup and about 150 fuel rods have been irradiated to the design peak fluence of 1.2×10^{27} n/m^2 ($E > 0.1$ MeV). Fuel rods were irradiated over a wide range of powers and temperatures.[1] Figure 3-6 illustrates the breadth of the data field with respect to temperature. Similar numbers of rods have been irradiated in the U.K., French, and German programs.

The fuel rod assembly design that evolved from this work is conservative. Selection of fuel smear densities† of the order of 85%, fuel pellet stoichiometries (O/M) of 1.94 to 1.97, plenum-to-fuel length ratio of 1.17, and cladding and duct wall thicknesses of 0.38 mm and 3 mm, respectively, are intended to (a) minimize fission product attack of the cladding inside surface; (b) eliminate or minimize fuel pellet–cladding mechanical interaction; (c) allow for 100% fission gas release; and (d) account for significant irradiation creep-swelling interactions in both cladding and duct. The latter, primarily through duct distortion, turns out to be the performance-limiting factor; thus peak burnups are limited to about 6910 GJ/kgM for current designs in order to avoid deleterious duct bulging and potential duct-to-duct interference (refer to Fig. 3-6).

The high reliability of this fuel assembly design has already been demonstrated on a statistical basis in the French development program. At the beginning of July 1978, after 12 irradiation cycles, more than 58,000 fuel rods had been irradiated in the Phenix reactor, without a single failure, to burnups up to 7344 GJ/kgM.[6] Since there is some level of cooperation in the exchange of information between countries with breeder reactor

---

† Net density of fuel in the rod when the fuel–cladding gap is taken into account.

programs, there is an overall benefit accruing from this mounting experience base.

As a result of an extensive worldwide research effort, both the data base and understanding of potential performance-limiting phenomena (e.g., fuel swelling and gas release, fission product attack of the cladding, cladding mechanical properties, and external cladding corrosion) are at an advanced stage, in some areas better understood than the LWR UO_2–Zircaloy system. Much of this data is contained in several volumes of the *Nuclear Systems and Materials Handbook*[7] prepared by Handford Engineering Development Laboratory under the auspices of USDOE. Also, highly sophisticated computer models have been developed that are able to predict with reasonable accuracy the time evolution of the physical and chemical changes in the pellet, the microstructural changes in the cladding, and hence the overall dimensional changes in the cladding. One such code, the LIFE code, has been adopted as the U.S. National LMFBR Fuel Rod Performance Code and is continuously updated with data and submodels by all contributors to the U.S. Program. Similar computer codes are being developed in other countries.[8]

The book by D. Olander, *Fundamental Aspects of Nuclear Reactor Fuel Elements*,[9] provides a detailed (analytical) account of the mechanisms, submodels, and computer code developments for the mixed oxide–stainless steel system. Therefore, as for the LWR situation, to avoid unnecessary duplication the following subsections will simply highlight phenomena in relation to their potential contribution to current fuel assembly design limits, and as a basis for the fuel and cladding improvements to be discussed in later sections.

3.2.1. Mixed Oxide Fuel Behavior

Because of the higher power ratings (40 to 60 kW/m) and corresponding higher fuel centerline temperatures, and higher burnups ($\gtrsim 8600$ GJ/kgM) planned for fast-breeder fuels compared to LWR fuels, more attention has been focused on temperature gradient effects, such as development of microstructural and compositional gradients, and their influence on fuel temperature, fuel swelling, and fission product release. The reader is referred to a review by Nichols[10] on analytical treatments of transport phenomena in nuclear fuels under severe temperature gradients.

Figure 3-7 illustrates the effects of the more extreme LMFBR thermal conditions. Exaggerated grain growth results from the dragging of grain boundaries as the as-fabricated pores move up the temperature gradient by a vaporization/condensation mechanism.[9, 10] This effect leads to the formation of a zone of columnar grains. Subsequently, the fission gas

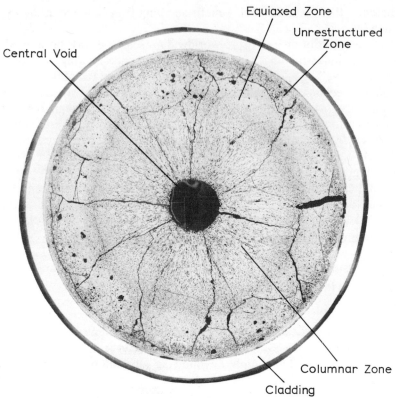

Fig. 3-7. Cracking and structural changes in LMFBR mixed oxide pellet, centerline temperature 2070 K (courtesy L. A. Neimark, ANL).

produced in the fuel pellet migrates to the grain boundaries, diffuses through the columnar grain boundaries, and, eventually, is released to the central void that has formed by the earlier porosity migration.

The extent to which this fuel restructuring occurs is, of course, determined by the fuel rod power level and, hence, the fuel temperature and temperature gradient. In at least 20% of the core, the initially uniform, solid mixed oxide fuel pellet transforms to what is essentially a structure consisting of three zones, i.e., outer unrestructured; equiaxed, higher density; and fully dense, columnar zones, and an annulus.

This thermal effect is also manifested in oxygen diffusion and measurable segregation of plutonium and uranium atoms in these fuels.[9, 10] With respect to the former, oxygen diffusion can either be down or up the thermal gradient depending on whether the fuel is, respectively, hypo- or hyperstoichiometric; regardless, an O/M gradient is established in the fuel pellet. Plutonium segregation typically results in a local concentration increase of 35 to 50% at the central-void edge, and no segregation is observed when a central void is not formed. This effect is important because the maximum allowable power level is constrained by the linear heat rating to incipient melting (typically "power-to-melt" is ~60 kW/m at startup) and therefore segregation of fissile material toward the fuel centerline will lower the allowable power. Working counter to this, however, is the closure of the initial fuel-to-cladding gap and restructuring that occurs early in life, both of which improve power-to-melt by ~20%.[11] Power-to-melt at high burnup therefore appears to be similar to fresh fuel.

The data base on fuel swelling and fission product release is extensive.[11] As noted earlier in Chapter 2 (Section 2.2.2.1c) fission gas release from LMFBR fuels basically follows the same kinetics as for thermal reactor UO_2 when appropriate corrections are made for different diffusion coefficients and fuel temperatures. Mechanistic models for gas release in UO_2 and mixed oxide fuels are generally identical and provide reasonable predictions of the observed behavior. High gas release in LMFBR fuel (100% at 6910 GJ/kgM) is accommodated in the large plenum provided in the fuel rod (refer back to Table 2-2).

The correlation between fuel swelling and available initial fuel porosity is also sufficiently well developed so that suitable fuel densities (of the order of 89 to 91% TD) can be specified and fabricated to accommodate fission-product swelling without gross distortion of the cladding to design burnups.

Two types of fuel-cladding chemical reactions have been identified; the first is uniform oxidation (matrix attack) of the stainless steel, which results in a general decrease in the thickness of the cladding: the second involves the intergranular penetration by oxygen and fission products along the grain boundaries in the cladding. Matrix attack rarely exceeds a depth of 50 μm; however, intergranular attack, which is nonuniform in depth and sporadic in occurrence, may penetrate the entire thickness of the cladding. Nevertheless, it must be noted that no cladding failures have been attributed to this type of attack.

Fission products play important roles in fuel-cladding chemical interactions. Cesium is universally observed in the attack region and is usually associated with chromium from the cladding. Iodine is observed almost as frequently and is generally associated with cesium. Tellurium is observed less frequently, and numerous cladding attack regions have been observed in which tellurium has not been detected. The role of fission products in cladding attack has been extensively investigated in out-of-pile experiments, which show that cladding attack readily occurs in the presence of one or more of the following: cesium (in the presence of oxygen), Cs_2O, CsOH (liq.), I_2, and Te. If oxygen is not present, cesium is unreactive toward stainless steel. The fuel and the remaining fission products do not promote intergranular attack separately or together.

Although the parameters that affect cladding attack have not yet been completely defined, the research to date has concluded that the cesium attack of stainless steel is controlled by the oxygen partial pressure in the fuel rod and the cladding temperature[12a] and, most importantly, the local temperature difference between the outer surface of the fuel and the

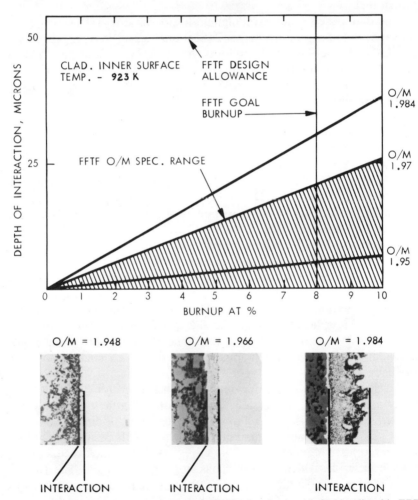

Fig. 3-8. Fuel–cladding chemical interaction (FCCI) in mixed oxide fuel irradiated in EBR-II (Ref. 11).

inner surface of the cladding.[12b] Fuel O/M values near 2.0 and cladding temperatures above ~810 K favor the formation of cesium chromate; therefore, current practice is to utilize mixed oxide fuels with O/M in the range 1.94 to 1.97, which results in a fuel surface O/M of less than 2.0 and keeps the fuel-cladding attack to below ~20 μm at end of life (Fig. 3-8).[11]

In the U.S. at least the long-term development thrust in this area is to either utilize lower O/M fuels (e.g., <1.94) or oxygen buffer/getter materials in the fuel rod. One study in EBR-II[13] evaluated fuel having an O/M of 1.90 and four different buffer/getter materials (V, Nb, Cr, Ti), incorporated either as coatings on the pellet surface, as layers at pellet–pellet interfaces, or as coatings on the cladding inside surfaces. Both approaches proved effective in reducing attack. The low O/M fuel exhibited no attack after 4320 GJ/kgM burnup, and two fuel rods with Nb and Cr coatings have successfully reached a peak burnup of 8640 GJ/kgM.

One potential problem with very low O/M fuel is the increased mobility of cesium within the fuel pellet column. Axial migration of cesium to the colder ends of the fuel column can be as detrimental to fuel rod performance as is radial migration to the fuel–cladding gap, primarily because cesium will react with the UO_2 insulator pellets, forming the larger-volume Cs_2UO_4, which locally strains the cladding.[14] Additionally, the insulator pellets might fragment, and the central void may become restricted. Therefore, ideally one should choose a fuel O/M that would avoid both excessive radial and axial cesium migration–in other words, "tie up" the cesium in the fuel pellets. To assist in this goal, cesium buffers are also being tested in the U.S. program along with the aforementioned oxygen buffers.[15] TiO_2, for example, when added to fuel reacts with cesium and avoids the formation of Cs_2UO_4. Also, it is possible that the niobium, used as an oxygen getter in the EBR-II experiments noted earlier, also acts as an efficient cesium getter once it is oxidized to the +5 valence state.[15]

Despite the extra fuel swelling expected from fission gas bubble accumulation, the mechanical interaction between fuel pellet and cladding was considered to be less important than the chemical interactions because of two counteracting phenomena: first, stainless steel cladding is known to swell (increase specific volume and, hence, diameter) with increasing fast fluence; second, the plasticity (creep rate) of the mixed oxide is increased over UO_2 by the deviation from stoichiometry that results from the addition of plutonium (20 wt.% PuO_2 results in stoichiometries of 1.94–1.97), the plutonium content itself, and the higher fuel temperatures. As more data have been collected, however, at the higher burnups it has been shown that fuel swelling can overtake cladding swelling and that significant cladding hoop stresses can be generated.[11, 16, 17] Also, fission product accumulation in the fuel–clad gap can contribute significantly to cladding strain.[14] Preliminary evidence that reintroduces this issue into LMFBR fuel performance considerations will be discussed later.

3.2.2. Cladding and Duct Mechanical Behavior

As outlined in Chapter 1, production of defects in metals by fast neutron irradiation may lead to enhanced creep (from enhanced diffusion), grain matrix hardening (from the combined "barrier" action of defect clusters and voids), grain boundary embrittlement (from helium bubbles), and, swelling (again from voids). The effects of these defects on properties were illustrated in Figs. 1-18 to 1-23. It was therefore considered that fuel assembly lifetime might be limited either by the strength and ductility of the fuel-rod cladding or by swelling and creep of the ducts. Experience now shows that the latter phenomena control assembly lifetimes in most of the current designs. Nevertheless, research emphasis has historically been concentrated on providing a quantitative description of both cladding strength and ductility degradation and duct creep-swelling interaction to end-of-life fluences of the order of 1.2×10^{27} n/m^2 ($E > 0.1$ MeV).

The procedures used, as one might expect, have paralleled those used in Zircaloy mechanical property research, namely, the extensive use of "post-irradiation" mechanical property characteristics and the development of equation-of-state concepts that allow the results from these simple tests to be applied to engineering situations.[18–21]

In-reactor, the strength of 20% CW Type 316 stainless steel is governed by the interplay between irradiation hardening and thermal recovery. At temperatures below about 770 K the hardening component dominates, but recovery effects above 770 K lead to a reduction in yield strength (Fig. 3-9).[18]

The principle cause of ductility loss in stainless steel at high temperatures (>820 K) (see Fig. 1-18) is helium embrittlement. To reiterate, this is the formation of helium from n,α reactions and subsequent migration to grain boundaries and growth of bubbles under applied stress. An increase in grain-boundary helium concentration is manifested in a reduction in uniform and total creep and tensile elongation, and as a reduction in stress rupture life.[22, 23] In addition, irradiation also introduces the possibility of crack propagation through locally heavily deformed regions in the matrix. This is the so-called "channel fracture" mechanism discussed in Chapter 2 earlier in relation to Zircaloy stress corrosion

LMFBR Core Materials 151

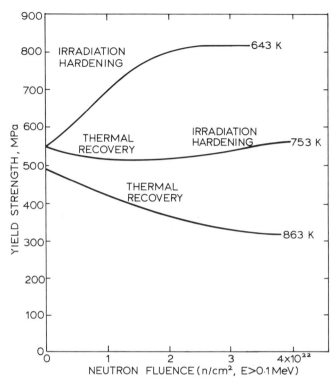

Fig. 3-9. Effect of temperature and fluence on the yield strength of 20% cold-worked type 316 stainless steel (Ref. 18).

cracking (Section 2.2.1.1.). In Type 316 stainless steel, it is likely to occur at fluences above about 2×10^{26} n/m^2 ($E > 0.1$ MeV) in the temperature range from 570 to 750 K.[23] The fracture map in Fig. 3-10 shows the various fracture fields, and it can be seen that the intergranular fracture field covers a significant portion of the fuel-cladding operating regime. Channel fracture would be expected only at high stresses, but such locally high stress concentrations could develop in a PCI situation, as hypothesized for LWR fuel (Chapter 2, Section 2.2.1.1.).

Important for fuel rod design strain limits is the fact that, despite the synergism between grain-boundary weakness and severe helium embrittlement, ductility does not approach zero but saturates at a low (1–2% strain) but manageable value for normal operating conditions. Only in cases where specimens are fractured at significantly higher temperatures than at which they were irradiated—to simulate an off-normal temperature excursion, for example—did fracture strains fall to dangerously low

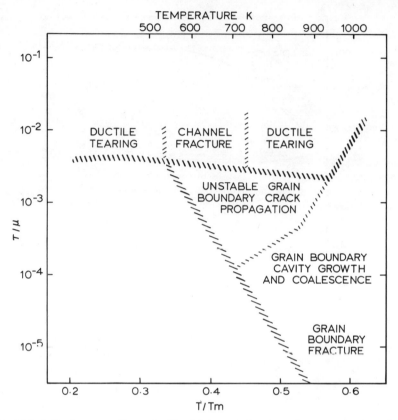

Fig. 3-10. A fracture map for austenitic stainless steel: schematic representation of the fracture behavior after irradiation to high fast neutron fluences as a function of normalized stress and temperature (Ref. 23).

levels of ≤0.1%. The low ductility found in these tests has not resulted in a reduction in fuel rod life, however, since load-bearing capacity remains sufficiently high to meet goal lifetime requirements.[23]

For fuel rod design purposes, various empirical correlations have been developed to describe the history-dependent cladding strength and ductility changes. Straalsund et al.[19] successfully used a Larson–Miller parameter to correlate tensile, burst, and simulated transient and stress rupture data for failure predictions under transient (accident) loading conditions. However, approaches with perhaps potentially broader application have recently been developed separately by Miller[21] and by Johnson et al.[20] The MATMOD approach developed by Miller has already been described in some detail in earlier chapters. The latter workers used a modification of Hart's equation-of-state method (refer to Chapter 1, Section 1.3.3.1.) for stress rupture correlation. This correlation produces a linear relationship between the temperature-compensated rupture time and a function of stress, with complete separation of stress and temperature dependence. The advantage of this approach, like MATMOD, is that major parameters are traceable to basic material properties and thus extrapolations outside the data fields are valid. Extension of this method to include irradiation effects has not been completed yet, but Hart and Li[24] have suggested a modification to the plastic equation-of-state theory which could account for both flux and fluence effects.

In the LMFBR core environment irradiation creep is essentially the only deformation process of interest for ducts, whereas both irradiation creep and thermal creep need to be considered in fuel cladding applications. This behavior was depicted in Fig. 1-23. As noted at the outset, the irradiation creep–swelling interaction in the duct now proves to be life limiting for the entire fuel assembly.[25] Combined "bulging" (Fig. 3-11a) and "bowing" (Fig. 3-11b) of the ducts can be sufficiently severe to cause interference between neighboring ducts. Bulging (or dilation) of ~3% and bowing up to 25 mm has been measured in the Phenix CW Type 316 steel ducts.[25]

Bulging distortion is a direct consequence of irradiation creep and is concentrated in the lower portions of the duct where sodium pressure is higher than average to maintain the required sodium flow rate through the fuel rod bundle.

The bowing phenomenon is much more complex inasmuch as it is the result of an interaction between swelling and creep. In general there will be temperature and neutron flux gradients through the core so that the temperature and flux dependence of irradiation-induced swelling will create a bowing tendency in the ducts. The core restraint mechanism will tend to keep the ducts straight so that a restraining force between ducts can be generated as a result of swelling. The diagram in Fig. 3-11b, taken from a paper by Gilbert *et al.*,[26] attempts to illustrate how opposite duct-bowing configurations could be achieved depending on the relative magnitudes of swelling and creep. Case B would result if there was more swelling on the inside duct surface than on the outside surface and little relaxation of stresses by creep. On the other hand, if a very large amount of irradiation creep occurs so that the swelling-induced stresses are completely relaxed during irradiation, upon cooling, the core could take on the shape shown in Case A, which represents distortions due to the thermal expansion changes occurring during cooling. It is, therefore, apparent that, for duct bowing, the relative amounts of the two phenomena are more important than the absolute magnitude of each.

A major effort, both experimental and analytical, has therefore been expended to quantify the irradiation creep-swelling behavior of the ref-

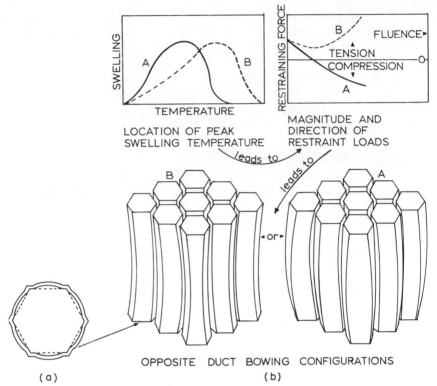

Fig. 3-11. Schematic representations of (a) duct bulging and (b) core restraint loads from differential swelling of ducts (adapted from Ref. 26).

erence stainless steel materials. A priority objective has been the collection of high-fluence creep data. Several measurement methods have been used, including pressurized tubes, springs, uniaxial tension/compression tests, beams, and slit-tube stress relaxation. Gilbert et al.[26] have reviewed progress on the data base. Comparison of in-reactor and thermal creep behavior is presented in Fig. 3-12, with the former derived from pressurized tubes in EBR-II. In general, for a given stress, the creep rate is dominated by irradiation creep processes at low temperatures and is eventually swamped out by normal thermal creep at high temperatures. (Also refer back to Fig. 1-23.) Total irradiation creep strain is therefore assumed to be the sum of irradiation-induced creep and thermal creep strains, i.e.,

$$\epsilon_{\text{in-reactor}} = \epsilon_{\text{irrad}} + \epsilon_{\text{thermal}} \qquad (3\text{-}1)$$

Additional results are contained in several papers published in 1977 in the *Proceedings of the International Conference on Radiation Effects*

LMFBR Core Materials

in *Breeder Reactor Structural Materials*,[27] and in 1979 in the *Proceedings of the International Meeting on Irradiation Behavior of Metallic Materials for Fast Reactor Core Components*.[28] A collective summary of key observations is as follows.

1. Irradiation creep rates can be correlated with swelling rates; that is, creep rate increases with fluence in a manner similar to, but not exactly equal to, that of swelling rate, and the temperature dependence is similar to that of swelling below the temperature for maximum swelling (Fig. 3-13).

2. Over a broad range of stress, neutron flux, and temperature conditions, the creep strain rate varies linearly with stress and flux. However, new data presented by Bergmann *et al.*[29] and shown in Fig. 3-14 indicate that, at the highest levels of stress, flux, and temperature, nonlinear behavior occurs. A similar observation was also reported by Paxton *et al.*[30] for the ferritic and solid-solution-strengthened stainless steels (refer to Section 3.3.2).

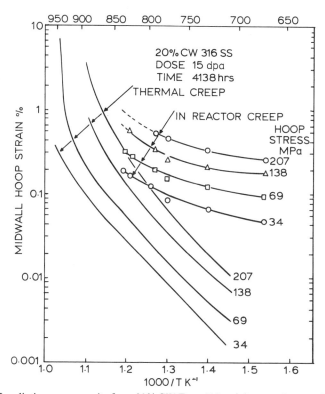

Fig. 3-12. Irradiation creep results from 20% CW Type 316 stainless steel pressurized tubes, in EBR-II (Ref. 26).

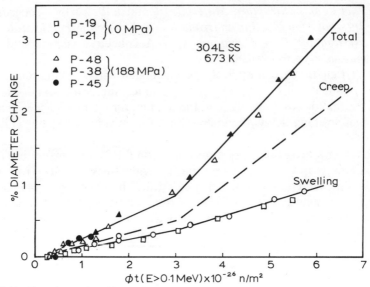

Fig. 3-13. Experimental evidence for a relationship between irradiation creep rate and swelling rate (Ref. 26).

3. Multiaxial creep is described by effective strain and stress, and is independent of stress history, at least up to fluences of 4×10^{26} n/m² ($E > 0.1$ MeV).

The theory of radiation-induced creep and swelling is in an advanced state, as evidenced by several papers on the topic in the aforementioned conference proceedings.[27, 28] Basically, creep results from the stress-induced preferential absorption of irradiation-produced interstitial defects. This process leads to a creep rate which is linearly proportional to the applied stress and the irradiating flux (note that the recent nonlinearity data are not explained by current models). Swelling results from dislocation–interstitial interaction and the accompanying void growth, and is influenced by both the applied stress and the helium pressure developing in the voids.

The complex microstructural theories can be condensed into more practical constitutive equations that relate creep rate to the swelling rate, and have the general form

$$\dot{\bar{\epsilon}}_{irrad} = B_0 \bar{\sigma} \phi + D \dot{S} \bar{\sigma} \tag{3-2}$$

and

$$\dot{S} = \dot{S}_0 (1 + P\sigma_{hyd}) \tag{3-3}$$

where $\dot{\bar{\epsilon}}$ is the deviatoric strain rate, B_0 is the zero-swelling creep rate coefficient, $\bar{\sigma}$ is the deviatoric stress, ϕ is the neutron flux, D is the swelling-enhanced creep coefficient, $\dot{S}$ is the swelling rate, $\dot{S}_0$ is the zero-stress swelling rate, P is the stress-enhanced swelling coefficient, and σ_{hyd} is the hydrostatic stress. The possibility of a linear stress effect on swelling is allowed for in Eq. (3-3).

These two equations, although approximations, have been used with some success in independent predictions of creep–swelling behavior.[26, 30] What is found is that creep rates generally tend to increase with increasing

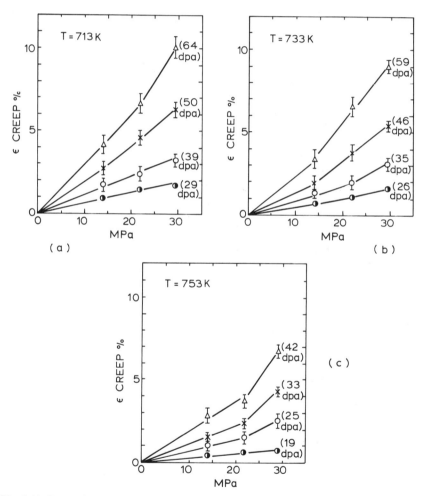

Fig. 3-14. Stress, flux, and temperature dependence of irradiation creep strains from internally pressurized 14% CW stainless steel (German equivalent to Type 316, W-Nr 14981) (Ref. 29).

fluence and, at high swelling and swelling rates, the in-reactor creep rate coefficient $B(\phi t) \sim D\dot{S}$. However, at fluences near or below the incubation fluence, $B(\phi t) > D\dot{S}$.

In summary, the overall emphasis at this juncture is to complete the 20% CW Type 316 stainless steel data base to end-of-life fluences and to develop improved design relationships that will cover the effects of neutron fluence, flux, stress, time, temperature, and history. The recently observed nonlinear creep behavior at the high end of the spectrum (Fig. 3-14) requires corroboration since, if proven a valid phenomenon, current models would underestimate creep in that operating regime. On the other hand, the use of post-irradiation data appears to overestimate the situation, predicting degradations in strength and ductility that are not observed in the few *in situ* experiments run so far. More in-pile data must be obtained to verify these early results, but there is a possible basis for less conservative designs which would allow fuel-assembly burnups to be increased.

3.2.3. Fuel–Cladding Interactions†

As was discussed earlier in Section 3.2.1, deleterious chemical interactions (FCCI) between fuel and cladding were observed early on in the LMFBR fuel rod development program and, as a result, the effects of fission product corrosion of stainless steel cladding were the subject of detailed studies.[4, 5, 12] As shown in Fig. 3-8, significant cladding wastage is possible in current-design LMFBR fuel rods. This corrosion of austenitic stainless steel cladding by fission products degrades the mechanical properties and contributes to the reduction in life of cladding alloys. The strength reduction can be quantitatively explained by assuming that the effect of grain-boundary penetration is equivalent to the reduction in the tube-wall thickness of the specimen (i.e., the inner diameter a, for specimens exposed to cesium oxides, becomes equal to $a + 2x$, where x is the maximum depth of grain-boundary penetration[31]). As noted earlier, however, cladding ID corrosion can be minimized (a) with hypostoichiometric fuel, and (b) by maintaining cladding temperatures below ~810 K; the conservative design limit established for the first-generation fuels should not lead to any performance problems.

Only recently has the potential for significant mechanical interactions (FCMI) emerged, as a result of more detailed analyses of fuel rod dimension changes during irradiation.[1, 11, 16, 17] The masking effect, of course, is the cladding swelling phenomenon which causes dilation of the cladding in the absence of stress. In fact, in the reference design considerations it was estimated that a combination of cladding swelling and low

---

† LMFBR researchers prefer the acronyms FCI, FCMI, FCCI to PCI, etc.

fuel smeared densities would preclude any significant FCMI. However, with the trend to higher smeared densities and low-swelling cladding to improve burnup and breeding gain, the concerns about FCMI phenomena have reemerged.

Boltax and Biancheria recently reviewed the U.S. LMFBR data base for information that relates to FCMI effects.[16] Of the data analyses reported, two are worth mentioning. In the first, the effect of fuel smear density was evaluated:[16] Fig. 3-15 indicates that the influence of fuel smear density (i.e., fuel pellet density and fuel–cladding gap size) on cladding strain is observed well before the start of significant cladding swelling, i.e., below 9500 GJ/kgM burnup and fluence levels below 6×10^{26} n/m^2 ($E > 0.1$ MeV). In the second study, Hilbert et al.[32] presented an analysis of mechanical strain data on a wide variety of mixed oxide fuel rods irradiated in fast test reactors. The results of this study are shown in Fig. 3-16. The data are indicative of FCMI effects since fission gas pressure effects alone would only account for relatively small strains (0.1 to 0.3% at 8640 GJ/kgM burnup). In addition, the following points should be noted:

(a) All of the data were obtained in fast test reactors where the

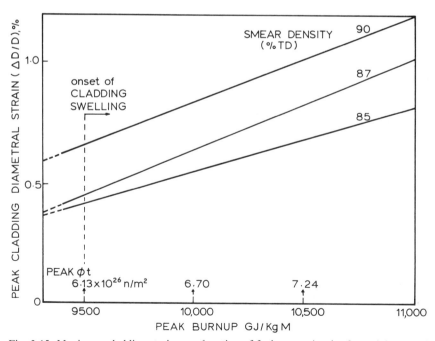

Fig. 3-15. Maximum cladding strain as a function of fuel smear density for stainless steel fuel rods irradiated in EBR-II (Ref. 16).

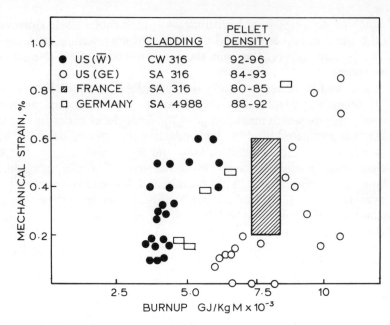

Fig. 3-16. A compilation of data on cladding mechanical strain for mixed oxide fuel rods as a function of burnup (Ref. 32).

fluence-to-burnup ratio was approximately one-half of that expected in prototype LMFBRs.
(b) The data indicate, also, that the high-density fuel exhibits greater mechanical strain than low-density fuel.
(c) Mechanical strains of up to 1% have been achieved without cladding failure.

Boltax and Biancheria[16] used the LIFE-III code to compute cladding hoop stresses of 69 to 138 MPa to account for these FCMI effects. These calculations neglected locally high cladding stresses that might be caused by accumulation of cesium at the fuel–cladding interface.

Although it is apparent that steady-state FCMI effects have not given rise to cladding failures, the FCMI stress level may have significant effects on the margin to failure in subsequent overpower events. Boltax and Biancheria[16] reported on both experimental and analytical studies of FCMI in power increase situations. In the experiments, fuel rod failures were actually generated by power increases in the range of 20 to 100%. Calculated cladding deformations for the failed rods were 0.5 to 0.9%, in contrast to 0.08% for the unfailed rod. In the analytical studies, the LIFE-III code was again used to assess cladding loads during a slow overpower event (a 10% power increase over a one-hour period). Figure 3-17 summarizes the results, which indicate significant cladding stresses and strains.

From the foregoing discussion, one can develop a picture not unlike the situation that exists in LWR fuel, that is, there is potential for synergism between locally stressed cladding regions and fission-product chemicals to yield cladding cracks. Unirradiated 20% CW Type 316 stainless steel has been shown to crack at low stresses and strains (602 MPa stress and a calculated strain of 0.04) in iodine vapor at 823 K.[33] The probability of such events occuring in-reactor will, no doubt, increase with the severity of the duty cycle, i.e., power, magnitude of the power increase, and burnup. Future requirements for LMFBRs to increase reactor ramp rates and to operate in a load-follow mode will therefore exacerbate the problem. The trend to higher fuel smear densities and low-

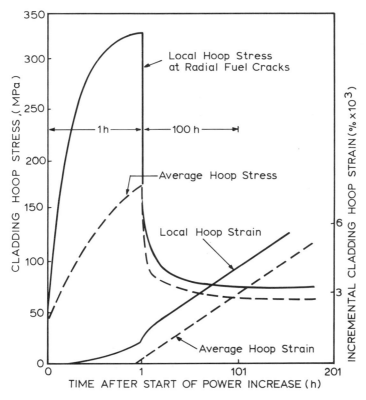

Fig. 3-17. Cladding stress and strain history following a 10% increase in power for a mixed oxide fuel rod: LIFE code calculation (Ref. 16).

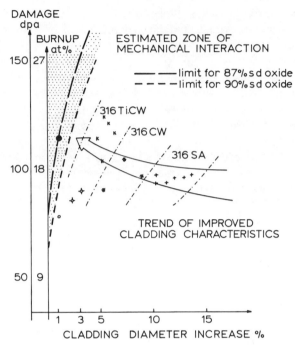

Fig. 3-18. Increasing potential for mechanical interaction between oxide fuel and cladding in LMFBR fuel rods as cladding swelling is reduced and fuel smear density is increased (adapted from Ref. 6).

swelling cladding alloys will accelerate the onset of fuel–cladding interaction. Figure 3-18, adapted from a recent French paper,[6] recognizes the "collision course" set by increasing smeared densities and swelling-resistant alloys. Hopefully, the LMFBR fuel developers will learn from their LWR counterparts on this particular issue. It is clear from work to date that some of the design remedies for PCI in Zircaloy, such as annular, large grain size, or duplex fuel pellets, and cladding barriers are equally applicable to FCI in stainless steel.

3.2.4. External Cladding Corrosion

Sodium influences the structure and composition of the stainless steels in two important ways. First, sodium preferentially leaches out the chromium and nickel from the steels, with the latter being responsible for conversion of the original austenitic phase to ferrite. A 10- to 20-μm-thick ferrite layer forms at the outside cladding surface. Second, the transfer of interstitial elements, in particular carbon, in the liquid sodium/

stainless steel system can result in either carburization or decarburization of the steel, depending on the temperature and carbon activity (i.e., concentration) of the sodium.[9]

In contrast to fission product attack which is predominantly intergranular, external corrosion due to sodium is uniform. The corrosion behavior of austenitic stainless steels in a liquid sodium environment is well documented over the temperature range from 720 to 1030 K, oxygen content of 1 to 30 ppm in the sodium, and sodium velocities between 0.1 and 12 m/s.[34-36] In general, the corrosion rate for Types 304 and 316 stainless steel reaches a steady-state value after an initial period of rapid metal loss. The steady-state corrosion rate increases exponentially with temperature and linearly with oxygen content and sodium velocity up to ~3 m/s. The corrosion rate becomes independent of velocity at the higher values. The corrosion rate correlations for Type 316 stainless steel that are used for design calculations of fuel rod wall thinning in the LMFBR are shown in Fig. 3-19. These empirical relations are derived from a compilation of data from investigations in the U.K., Netherlands, France, Japan, and the U.S.,[34] and they refer to steady-state corrosion rates determined at locations of maximum corrosion, i.e., maximum upstream position. The downstream or positional effect, metallurgical condition of the material (annealed or cold-worked), and minor variations in the nickel, chromium, niobium, or titanium contents of the stainless steels have relatively little influence on the corrosion rates in comparison with the temperature, oxygen content, and velocity of the sodium.

Interestingly, actual in-reactor experience indicates that corrosion is about a factor of 10 lower than indicated by Fig. 3-19. Measurement of the extent of sodium–cladding chemical interaction on high-temperature, 948 to 973 K, outside cladding surfaces showed a decreasing corrosion rate with time.[37] A thickness loss of about 2.5 μm is approached after 6000 to 7000 h, and remains at this value to about 11,000 h, which is the longest time for which data are available.

The carburization/decarburization and the sodium leach-out phenomena can both influence the mechanical properties of stainless steel, but all indications are that these effects are of minor significance to cladding integrity. For instance, the ferrite layer is typically included in the overall cladding wastage allowance (i.e., it is treated simply as a reduction in load-bearing capacity), and the effects of carburization/decarburization on important cladding creep and strength parameters are small ($\leq 10\%$)[35] compared to the effects of irradiation, and are included implicitly in the "safety factor" applied to the design curves. These phenomena become much more important in considerations of LMFBR pressure boundary (Chapter 4) and heat exchanger (Chapter 6) integrity and, therefore, further discussion will be deferred until later. Overall, therefore, coolant

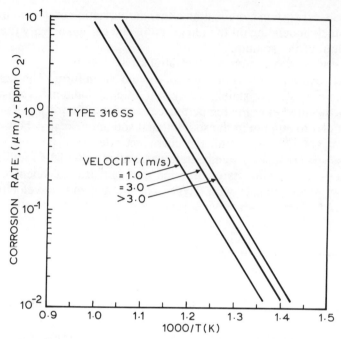

Fig. 3-19. Effect of temperature on the corrosion rate of Type 316 stainless steel in flowing sodium (Ref. 34).

corrosion is not considered a limiting factor in the performance of LMFBR fuel, although it must be continually monitored for new effects such as are periodically seen with Zircaloy and water in LWRs (Section 2.2.1.2).

3.3. Advanced Oxide Fuel Development

As summarized in Fig. 3-2, several design changes will be necessary to reduce the doubling time to less than 15 years. Of these, decreasing the duct spacing and reducing cladding thickness can only be achieved through the development of cladding alloys with reduced irradiation creep and swelling characteristics. This goal is the subject of a large, international R&D effort. Fuel pellet changes are less profound, being primarily limited to diameter and density increases, that is, larger-diameter fuel rods with higher smeared densities. Also, alternative fuel forms such as annular pellets and particle fuels are being evaluated by the U.K. Because remote fabrication methods are facilitated by the use of liquids rather than solids, particle fuels derived from solutions (e.g., sol–gel fuels) will no doubt gain in interest if proliferation resistance remains an issue (see Section 3.6).

3.3.1. Optimization of Mixed Oxide Fuels

As noted earlier, fuel pellet–cladding mechanical interaction will be a life-limiting consideration in advanced oxide systems. Detailed evaluation of experimental data, such as that shown in Figs. 3-15 and 3-16, utilizing fuel code calculations, shows that a combination of smear density and gap size controls the extent of FCMI, rather than smear density alone.[1]

Thus, the most optimum pellet design, from the point of view of improving mechanical performance under steady power operation, would be to use the highest practical fuel density, say 95% TD, with the gap size adjusted to yield the required lifetime. With a 0.18 mm diametral gap, for example, the smear density could be increased to 88% TD while retaining the same level of mechanical performance as the FFTF reference rod design.[1] Model predictions indicated that the burnup could be increased to 8640 GJ/kgM by the use of a 0.12 mm diametral gap and a smear density of 90% TD. The U.S. P-40 experiment recently demonstrated this improvement.[38] Fuel rods with 0.25-mm-thick, 20% CW 316 stainless steel cladding, and 90% smear density fuel reached a peak burnup of 11,200 GJ/kgM before failures were detected. Analysis of earlier EBR-II experiments also supports use of 88 to 90% smear density fuel rods to burnups in excess of 8640 GJ/kgM. Practical core designs using this design can achieve a doubling time of about 13 years, which is comparable with preliminary carbide core values[3] (refer to Fig. 3-3).

Alternative approaches to increasing breeding gain include the use of annular pellets (which can operate at higher powers without melting, and exhibit reduced FCMI) and particle fuels (which can achieve high smear densities without FCMI). By comparison to pellet fuel, the work on particle mixed oxide fuel (Spherepac, Vipac, etc.) is limited, but, nevertheless, significant progress has been made in understanding the key physical phenomena that control performance in-reactor. Several assemblies of vibro-compacted particle fuels, UO_2–15 to 20 wt.% PuO_2 (73% TD), have been irradiated in the U.K. program in DFR,[39] and irradiation tests were conducted on Spherepac sol gel fuel as part of the U.S. LMFBR fuels program in EBR-II.[40, 41]

Some of the information of principal interest for LMFBR application of these particle fuels is shown in Fig. 3-20, which is a transverse view of a Spherepac 20% PuO_2 fuel rod irradiated in EBR-II to 4320 GJ/kgM at 46 kW/m.[40] A general observation (in contrast to Fig. 3-7, for example) is an annulus of unsintered fuel at the periphery of the fuel column, about 0.5 mm thick. This annulus of unsintered fuel apparently reduced both the stresses on the clad (clad strains ranged from only 0.4 to 0.75% $\Delta D/D$) and fission-product transport to the cladding (no fission-product attack of the cladding is observed in Fig. 3-20 even though the temperature at the cladding inside surface is about the threshold temperature where chemical reactions should proceed very quickly).[40]

Irradiation of this fuel was continued to about 8640 GJ/kgM at approximately the same power levels, 33-50 kW/m, without failure. No further examinations have been conducted, however, due to a deemphasis of the fuel type in the U.S. LMFBR program. In contrast, particle fuel

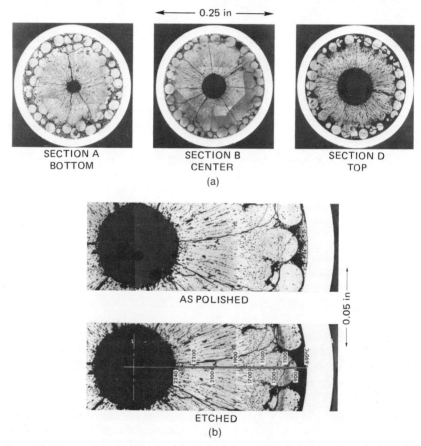

Fig. 3-20. Transverse views of mixed oxide Spherepac fuel rod irradiated in EBR-II to 4320 GJ/kgM at 46 KW/m (Ref. 40). (a) Transverse metallographic sections from PIN S-1-E, (b) peak heat rate, Section G, PIN S-1-E showing microstructure and calculated temperatures.

is of major interest to the U.K., and, in the DFR program, fuel rods operated at ~40kW/m achieved burnups as high as 14,500 to 18,000 GJ/kgM before failure occurred. The cause of failure at these very high burnups was considered to be a genuine endurance limit.

3.3.2. Optimization of Cladding and Duct Materials

The search for advanced cladding and duct alloys has concentrated on minimizing void swelling and irradiation creep strains, while still maintaining an appropriate balance of other desirable properties (e.g., ease of fabrication and welding, adequate mechanical strength and ductility, compatibility with coolant, etc.). This is an area where cooperation between the various countries is limited, with a consequence that much of the data have not been released to the open literature.

Initially, the search for low-swelling alloys followed the marked trend of decreasing void densities with increasing nickel content (Fig. 3-21). Johnston et al.[42] showed, through heavy ion irradiation simulation studies, that increasing the nickel content from 15 to 35% reduced void swelling by a factor of 100. Unfortunately, Type 316 steel was located in the region of maximum swelling. This type of study motivated the U.K., for example, to develop the high-nickel alloy, PE-16, for use in the core of PFR. Experience with this alloy shows a low in-service swelling rate combined with improved resistance to creep.[43] Further work is continuing to optimize high-temperature mechanical properties through boron concentration variations (18–107 ppm by weight) and/or heat treatments. With respect to the latter, beneficial effects result from the introduction of a 1173 K carbide precipitation treatment after solution annealing and prior to aging.[43]

High-nickel alloys suffer several disadvantages, however, that will prevent their wide-scale use in advanced cores. These include the production of high amounts of activated corrosion products, higher neutron absorption and hence lower breeding ratios, and possibly intensified helium embrittlement, particularly during off-normal temperature transients. The concerns about increased sodium corrosion rates has largely been dispelled by recent loop corrosion tests at 973 K, which have shown that corrosion rates are not directly proportional to nickel content but can be quite low even at nickel levels of 40% or more.[44]

Disadvantages of high-nickel alloys prompted the return, first to optimization of lower-nickel, austenitic alloys and, more recently, to the zero-nickel, higher-chromium, ferritic stainless class of steels. Both the U.K.[43] and France[6,28] have adopted a Ti and/or Si modified, cold-worked Type 316 steel for the next generation of fuel assemblies (note Fig. 3-18 earlier), and the U.S. will likely follow suit with their D9 alloy

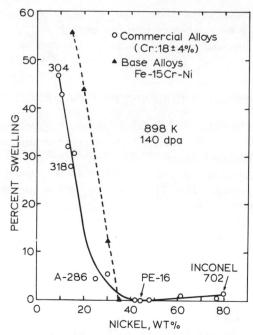

Fig. 3-21. Swelling versus Ni content of pure Fe–15Cr–Ni alloys and commercial alloys containing 18 ± 4 wt.% Cr (Ref. 42).

(see Table 3-1). All three countries now favor the ferritic stainless class of steels for subassembly ducts.

Extensive in-reactor creep/swelling studies have been undertaken by the French in order to qualify their titanium-modified Type 316 stainless steel for use in Super Phenix.[45] Most of the data have been obtained on two alloys: 47-2, which contains 0.3% Ti and 0.83 Si, and 109-2, which contains 0.44% Ti and 0.47% Si. In the 20% cold-worked condition, both alloys exhibit a slightly lower creep rate coefficient† than the reference Type 316 steel (Fig. 3-22). The creep modulus of both modifications is independent of temperature from 670 to 770 K and shows little variation with dose up to 80 dpa.[45] Inasmuch as creep and swelling are related, swelling strains are also correspondingly lower in the Ti-modified alloys. Therefore, when applied to duct and fuel rod lifetime predictions, the use of Ti-modified 316 stainless in Super Phenix is expected to double the dose limits, from 80 to 160 dpa (5360 to 10,700 GJ/kgM).[28]

A systematic survey of the void-swelling response of a wide range of ferritic materials after fast reactor irradiation has been completed by

---

† Defined as $\epsilon/\sigma\phi t$, i.e., B_0 in Eq. (3-2).

Table 3-1. Advanced LMFBR Cladding and Duct Development Alloys

Alloy class	Potential application	Commercially available alloys	New developmental alloys	Expected characteristics
Ferritics	Duct only	HT-9[a] (12 Cr–Bal Fe)	D57[c] (10 Cr–6 Mo–Bal Fe)	Very low swelling; strength too low for cladding
Solid-solution-strengthened austenitic	Duct and cladding	310 CW 330 CW (20,35 Ni–25,16 Cr)	D9[a,c] D11 (14,20 Ni–14,7 Cr)	Moderate swelling; strength marginal for cladding
Precipitation-strengthened austenitic	Cladding and duct	A286 M813 Inconel 706 PE-16[b] (25–43 Ni–15–18 Cr)	D21[a] D25 D32 D66, D68[a] (25–45 Ni–9–12 Cr)	Low to very low swelling; high strength acceptable for cladding
High-nickel austenitic	Control assembly	Inconel 718 (52 Ni–19 Cr)	D42 (60 Ni–15 Cr)	Very low swelling and high strength; nickel too high for fuel assembly

[a] Selected for advanced stage of U.S. program.
[b] In U.K. PFR.
[c] Similar to alloy chosen for Super Phenix.

the U.K. group.[46] Irradiation temperatures in DFR ranged from 653 to 888 K, up to doses of 30 dpa. The magnitude of swelling in all the chromium-based commercial alloy steels, e.g., $2\frac{1}{4}$Cr–Mo, and 12Cr-type martensitic steels, were below the detection limit of 0.1% for all irradiation conditions examined. Furthermore, voids were totally absent from the 12% Cr steels, indicating the operation of strong void-swelling suppression mechanisms in this class of materials.

Ferritic steels have also been irradiated in Rapsodie and in Phenix test reactors as part of the French program to study their creep-swelling characteristics.[47] Two groups of steels have been evaluated, containing 12% Cr and 17% Cr, and swelling was essentially absent for doses up to 150 dpa (equivalent to 10,000 GJ/kgM burnup), which are much higher than achieved in the U.K. experiments. In addition, in pressurized tube tests the 17% Cr ferritic steel did not exhibit irradiation creep (i.e., the creep modulus was essentially zero) in the temperature range 673 to 773 K (Fig. 3-22).

The status of the U.S. program to study the in-reactor creep characteristics of alternative cladding and duct alloys was reported by Paxton et al.[30] Only data on commercial alloys were available, but the trends were similar to those observed by the French, summarized in Fig. 3-22. The commercial HT-9 ferritic stainless steel alloy exhibited lower creep

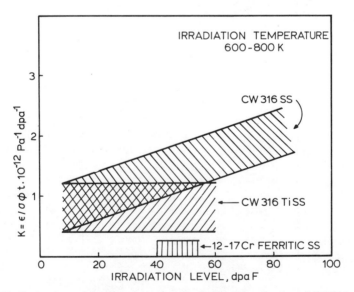

Fig. 3-22. Comparison of the French data on irradiation-induced creep of CW 316 stainless steel, Ti-modified CW 316 stainless steel, and 12–17Cr ferritic stainless steel as a function of irradiation dose (adapted from Refs. 45 and 46).

strains than either the Ti-modified Type 316 (factor of 2 lower) or the CW Type 316 stainless steels (factor of 4 lower), after irradiation to 2 and 4 × 10^{26} n/m² ($E > 0.1$ MeV) at 813 K, under a 175 MPa hoop stress. Interestingly, under the same conditions, the high-strength precipitation-hardened alloys (Inconel 706 and 718 and nimonic PE16) displayed the greatest resistance to in-reactor creep, an order of magnitude lower than HT-9.

Although the creep-swelling behavior of the ferritic stainless class of steels is favorable for LMFBR core application, their near-term use will likely be limited to temperatures below 723 K because of the strong decrease in uniform elongation and in rupture strain that occurs at this point. This is not a helium embrittlement effect (in fact, this class of materials does not appear to be subject to helium effects), but, rather, the effect of coarsening of the $M_{23}C_6$ or M_6C carbide precipitates. In any event, they are not considered acceptable for fuel rod cladding with current core operating parameters, but they should provide a definite improvement over the austenitic stainless steels for fuel assembly ducts. Developing alloys in this class, such as the Combustion Engineering 9Cr–1MoNb stabilized steel, discussed in Chapter 6 (refer to Fig. 6-16) might enable temperature limits to be raised.

The prospect of dissimilar alloys for cladding and duct, and hence the potential for carbon transfer between the two and subsequent property degradation, has motivated the development of dispersion-strengthened ferritic stainless steels, in which the high-temperature strength is improved through additions of either titanium oxide or yttrium oxide particles. These steels have performed well in preliminary irradiation tests.[48] A burnup of 10,370GJ/kgM was achieved in the thermal reactor, BR2, before a failure occurred despite a high smear density (90%), a high clad temperature (943 K), and a high linear power (56 kW/m). The diametral increase of the failed rod was 3.3% $\Delta D/D$. Post-test mechanical property measurements indicated that the dispersion strengthening effect was not reduced by irradiation and that helium embrittlement effects were absent, the latter being in agreement with data on the nondispersion strengthened ferritics. Some of the samples in this study were also examined by transmission electron microscopy. Interestingly, for the TiO_2 dispersion alloy, no voids were found in either the ferritic matrix material or the chi-phase (the chi-phase has an α-Mn structure with the approximate formula $Fe_{36}Cr_{12}Mo_3Ti_7$), but some voids were observed in the TiO_2 particles. No voids were found in the yttria particles of the yttria dispersed alloys, however. These results were sufficiently encouraging for the decision to be made to irradiate three rods in the Rhapsodie reactor under more prototypical conditions.

Theoretical understanding of the role(s) played by solutes, impurities, or precipitates in reducing the irradiation creep and swelling behavior of these alloys is lagging the experimental work. While it is clear that the additives are acting as effective defect traps, and thereby increasing the probability of vacancy–interstitial recombination and annihilation, the precise nature of the trapping process is unknown and the relative trapping ability of various solutes or particles for interstitials and vacancies has yet to be quantified. For instance, no satisfactory explanation of the composition dependence of swelling in the higher-nickel alloys has been given, although microscopic evidence ties the swelling reduction to the γ and γ' precipitates.[42] Likewise it is not fully understood why the zero-nickel ferritic stainless steels also exhibit very low creep and swelling rates, although Little,[49] in his review of this topic, points to the effectiveness of three possible mechanisms: (a) point defect/solute atom trapping to enhance mutual point-defect recombination, or dislocation/solute interactions that (b) reduce the dislocation bias for preferential self-interstitial capture and/or (c) inhibit the climb rate of dislocations. Fortunately, this lack of basic understanding of the mechanism has not hindered the alloy optimization studies in any obvious way.

To summarize, research worldwide is concentrated on three principal alloy classes: (1) austenitic stainless steels, (2) iron–nickel base, precipitation-strengthened alloys, and (3) high-strength ferritic steels. Some of the more promising alloys that have been under consideration in the U.S. program are shown in Table 3-1.[50] Note the conservatism inherent in the choice of the six prime candidates for further development. While the ferritic stainless steels, either HT-9 or the developmental alloy D57, are firm candidates for duct materials, the dispersion-strengthened ferritics are not considered "far enough along" for use as cladding alloys. Therefore, both solution-strengthened and precipitation-strengthened austenitics are being promoted, with no clear decision yet as to which will be the primary cladding alloy, although, as noted earlier, a Ti-modified Type 316 alloy (D9) is a leading candidate. This philosophy is also evident in the U.K.[43] and French[51] development programs. What might produce a change in emphasis, however, is the move toward lower sodium outlet temperatures in recent commercial LMFBR designs.[52] The operating regime is now compatible with the use of ferritic stainless steels as both duct and cladding materials.

All the advanced alloys, regardless of class, exhibit significantly improved swelling and creep resistance over that of the reference alloy, Type 316, based on early irradiation experience. The projected end-of-life swelling and creep behavior of the advanced alloys is compared to the properties of the reference alloy Type 316 in Fig. 3-23,[44] and the indications are that the dual goals of <5 vol.% swelling and <1% creep strain at fluences of 2×10^{27} n/m^2 ($E > 0.1$ MeV) are achievable.[28]

When factored into core design studies, these values yield substantial increases in fuel system burnup capabilities (13,000 GJ/kgM) and extended core component lifetimes. Further, if combined with the use of less metal in the core, i.e., reduced cladding and duct wall thickness, significantly lower doubling times are predicted.

Since almost all these alloys possess adequately improved "steady-state" properties, it follows that the extent to which they will be used in future core designs will depend on (a) properties under off-normal or transient conditions; (b) compatibility with other primary-system materials in sodium; and (c) fabricability and cost. Some interesting trends are apparent from results to date. With respect to the ferritic steels, for instance, both HT-9 and D-57 alloys behave very well in transients, exhibiting neither helium embrittlement effects nor significant dilation or "ballooning." Concerns about fabricability, particularly welding technology, have been largely solved. The key remaining issue is compatibility in sodium in the presence of austenitic steels (Type 316 and 304) in the primary system. Successful operation in the planned EBR-II and loop tests[44] could result in selection of ferritic steels for both duct and fuel cladding materials in U.S. commercial plants.

Notwithstanding the foregoing, there is a strong incentive to continue the development of austenitic steels because the ferritics are an unknown quantity in nuclear systems. The extensive experience with Types 304 and 316 stainless steel justifies the continuation of basic studies and evaluation irradiations that seek the optimum minor element and/or thermomechanical treatment for retention of creep-swelling resistance to higher fluences. The Ti and/or Si additions to cold-worked Type 316 stainless steel (e.g., D9 alloy) is a good example of such work, and there will no doubt be more alloy modifications of this type for cladding use.

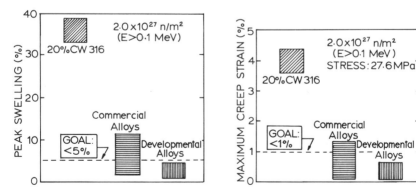

Fig. 3-23. Expected high-fluence swelling and creep behavior of advanced alloys, relative to the reference material, as extrapolated from current data base (Ref. 44).

The concern of this writer is that such subtle modifications to microstructure seem inherently unstable and, therefore, subject to modification during the long-term complex thermalmechanical-irradiation histories to which core components are subjected. This is not the case for ferritic stainless steels.

With respect to the high-nickel alloys, the future is somewhat clearer because rather obvious disadvantages exist. If high-nickel alloys are to be used, their quantity must be limited to, for example, control assembly components (Table 3-1). Both neutronic and safety conditions will prevent the use of high-nickel austenitic steels as fuel cladding in commercial cores.

3.4. Advanced Fuel Development

In principle, much better breeding gains and improved utilization of fuel resources can be achieved by increasing both the fissile atom content and thermal conductivity of the fuel. This means replacing the oxides by carbide, nitride, or even the metal fuel forms[3] (for example, refer to Fig. 3-3). By comparison with oxides, however, little work has been done to qualify these fuels for large commercial LMFBR application, and hence they are termed "advanced fuels," possibly destined for later cores.

The mixed (U–Pu) carbides and nitrides, in both pellet and particle form, have been actively considered as alternative fuels to the oxides. In the U.S., pellet carbide fuel has been emphasized as the follow-on to the mixed oxide, while particle carbide fuels are under evaluation in Europe. Nitride fuel work has largely terminated since the nitrides do not appear to offer any major performance advantages over the carbide, and yet fabrication costs are significantly higher due to the need for enrichment with ^{15}N. Consideration of metal fuels is limited to a small group within the U.S., and has only sparked renewed interest in the context of a proliferation-resistant fuel cycle; recent developments in this area will, therefore, be described in Section 3.6.

Thus, this section will review carbide and nitride fuel performance experience gained over the past ten years. It will become apparent that, whereas the design limits for the oxides are well established, insufficient work has been done on either the carbide or the nitrides to firmly establish their design limits.

3.4.1. Carbide Fuel Development

The U.S. program, perhaps more than any other, has persevered with the development of a pellet carbide fuel for the LMFBR. Therefore, results and trends from this program will be used as the example of progress in pellet carbide fuel development worldwide (U.S., U.K., Japan, France, and Germany).

Two types of pellet fueled rods have been irradiated in the U.S. Program.[53-55] The first type is similar to the mixed oxide fuel rods, that is, they contain helium as the gap heat-transfer medium (so-called "helium-bonded carbides").[53, 54] The second type uses sodium as a more effective heat-transfer medium ("sodium-bonded carbides"), thus allowing higher power levels to be attained without fuel melting.[53, 55] Early irradiation tests examined a wide number of material and design variables, with the consequence that their effect on failure rate can only be established as trends.

With respect to the helium-bonded mixed carbides, the U.S. tests indicated that failure rate was not appreciable until burnups in excess of 6910 GJ/kgM, and several fuel rods achieved burnups in the range 7770 to 10,370 GJ/kgM without failure. Fuel rods containing high-density (95% TD) annular pellets appeared to perform better than those with solid pellets of like density, but, overall, the low- to moderate-density (77–91% TD) fuel had lower cladding failure rates. These fuels released more fission gas than the high-density fuels, but the effect of increased gas pressure on clad strain was overshadowed by the better fuel swelling accommodation of the low-density pellets. Gas release in the most recent helium-bonded carbide fuel rods was of the order of 15 to 20% at 7000 to 8000 GJ/kgM burnup.

The characteristic appearance of a cladding failure in a carbide fuel rod is shown in Fig. 3-24. Cladding ridges were observed just as in LWR fuel rods, but much higher strains were measured.[53] Also, extensive bonding of the fuel to the cladding in these fuel rods may have contributed to the cladding strains. Interestingly, despite a macroscopic appearance that is quite similar to the stress corrosion cracks in Zircaloy LWR fuel rods (e.g., compare Fig. 2-7 to Fig. 3-24), Barner et al.[53] conclude that the primary failure mode in carbide fuel rods is purely mechanical; in other words, strong fuel–cladding mechanical interaction (FCMI) leads to cladding ductility exhaustion.

The bonding of the fuel and cladding appeared to be greater in single-phase (U, Pu)C fuel than in fuel containing the sesquicarbide (U, Pu)$_2$C$_3$; this is a result of increased fission-product migration out of the fuel, the kinetics of which are controlled by the carbon activity. Therefore, although the presence of sesquicarbide leads to increased cladding carburization, since no failures could be attributed to excessive carburization, carbide fuel with 5–10% (U, Pu)$_2$C$_3$ was selected for the next phase of the U.S. program to avoid the bonding effect.[53]

A major design innovation in the sodium-bonded carbide fuel rods is the introduction of a "shroud tube," which is intended to maintain the cracked carbide pellets in their approximate original geometry and thus

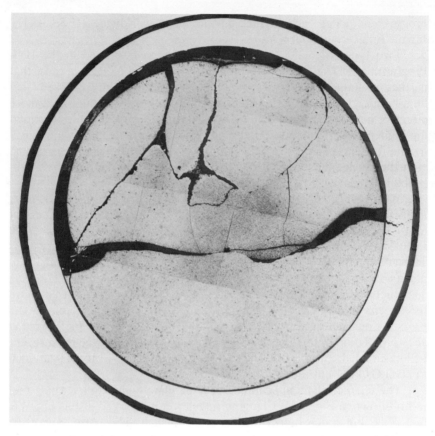

Fig. 3-24. Transverse section of a helium-bonded carbide fuel rod showing a cladding crack located at the end of a large fuel crack (Ref. 53). Note that the apparently large fuel–cladding gap is the result of the section being taken at a pellet–pellet interface, in the proximity of the chamfered edge of a pellet.

prevent fuel pieces penetrating the sodium bond to mechanically interact with the cladding.[53, 55] This rearrangement of fuel-pellet fragments is highly undesirable since it not only promotes FCMI failures but also alters the thermal performance of the fuel. Several shroud tube rods have achieved over 8640 GJ/kgM burnup to date without failure.[55] However, it must be pointed out that "unshrouded" carbide fuel rods have performed equally well.

In sodium-bonded fuel rods, no significant behavioral differences have been observed using fuel pellets that range in density from 88–99% TD. Therefore, the density of the fuel should be as high as can be fabricated economically; current development work specifies fuel >98% TD.

Gas release in these high-density sodium-bonded fuel rods is about half that of the helium-bonded fuel at comparable burnups, due to the lower fuel temperatures.

In localized regions of high temperature, increased swelling of the fuel was observed, which was dependent on the grain size. Thus, as is the trend in UO_2 fuels, a mixed carbide fuel with a relatively large and uniform grain size would be desirable.

Carbon transport from the fuel to the cladding is greater in sodium-bonded rods than in helium-bonded rods. Depth of cladding carburization in sodium-bonded carbide fuel rods clad with Type 316 stainless steel (and Alloy 800) was two to three times greater than in comparable helium-bonded rods. The amount of $(U, Pu)_2C_3$ in the fuel appeared to have a significant effect on the depth of cladding carburization (Fig. 3-25). Although carburization of the cladding leads to a lessening of the ductility of the cladding, the overall effect of various degrees of carburization on the mechanical properties of the cladding has not been determined because of the complicating effects of other irradiation factors such as void formation, transmutation products, and irradiation strengthening. As noted earlier, however, no fuel failures have thus far been attributed to cladding carburization effects.[53]

In summary, the overall performance of carbide pellet fuels to date indicates that defense against mechanical overstraining of the cladding

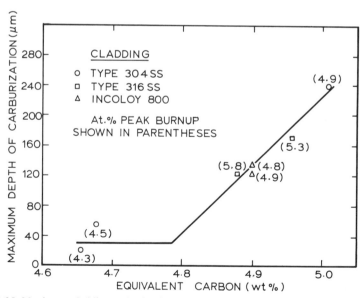

Fig. 3-25. Maximum cladding carburization of sodium-bonded carbide fuel elements versus initial equivalent carbon content of the fuel (Ref. 55).

is the key issue. Fission gas release levels are substantially lower than those commonly encountered in oxide fuel rods, reflecting the lower in-reactor fuel temperatures that prevail in carbides. However, fuel swelling is inexorable and attempts to restrain it fully call for extreme measures. Migration of fission products out of the fuel can be controlled, if not eliminated, by using the sesquicarbide form. In any event, no evidence of fission product attack of the cladding has been reported, and the only chemical interaction observed, carburization, does not appear to exert significant influence over the failure potential.

In the helium-bonded fuel rod, fuel densities of 81–87% TD and smear densities of 75–81% TD are recommended to provide adequate space for swelling accommodation.[54] In the sodium-bonded fuel rod the current emphasis is on fuel densities >98% TD, in conjunction with a perforated shroud tube (0.08 mm thick), to prevent the rearrangement of fuel pellet fragments within the fuel rod. Both helium- and sodium-bonded fuel rods emphasize hyperstoichiometric (U, Pu)C containing 10 vol.% $(U, Pu)_2C_3$.

First post-irradiation examinations of shrouded, sodium-bonded carbide fuel rods indicated that this design has superior irradiation characteristics with respect to fuel–cladding mechanical interaction and behavior during power increases. Although the fabrication of this type of rod is somewhat more complex than helium-bonded or unshrouded sodium-bonded fuel rods, its potential for better irradiation performance and higher burnup capability may overshadow any economical penalty associated with its fabrication. One major drawback might be the metallurgical interaction between the fuel pellet and the shroud over the long run. If this occurs, and it can be expected, the restraining feature of the shroud would be jeopardized. Finally, it should be noted that future irradiation evaluations of carbide pellet fuels, in the U.S. at least, will likely utilize the advanced claddings in order to assess overall performance levels.[3]

With the importance of FCMI effects on carbide fuel performance demonstrated, it should not be surprising that attention is being given to the use of microspheres or sphere-pac fuel. It was mentioned earlier that sphere-pac oxides exhibited improved mechanical performance in both LWR and LMFBR environments. Most of the work on the development of the (U, Pu)C shpere-pac fuel since 1971 has been undertaken by the Swiss.[56] A series of screening tests using fuel of two size fractions, 600/800 μm and 40/60 μm (smear density 76–77% theoretical; Pu content about 15%) revealed the following trends:

- Performance of sphere-pac mixed carbide fuel with sesquicarbide levels below 5% is satisfactory at ratings up to 100 kW/m to at least 5200 GJ/kgM. Peak clad temperatures were in this case 923 K.

- With fuel containing >5% $(U, Pu)_2C_3$, however, carburization occurs sometime after 860 GJ/kgM at temperatures ~813 K. Carburization becomes heavy above 2600 GJ/kgM and 80 kW/m, and under these conditions cladding failures occurred.
- In the intact rods total diametral strains of the order of 0.15% $\Delta D/D$ per 860 GJ/kgM have been recorded. Gas release levels, although increasing rapidly after 2600 GJ/kgM, are in line with values obtained in pellet carbide studies.

In the same context, the U.K. program has a "vibro fuel" design consisting of a low-density (70–75% TD) vibro-compacted carbide particle fuel (gel precipitated or a mixture of coarse and fine granules), which they believe can be exploited early as a substitute for the mixed oxide, at moderate heat ratings (about 100 kW/m).[57]

3.4.2. Nitride Fuel Development

Nitride fuel development has also seen considerable emphasis in the U.S.[58, 59] As for carbide fuels, sodium bonding is considered necessary to assure good heat transfer between nitride fuel and cladding and also to accommodate fuel swelling. Therefore, in a first series of tests in EBR-II, both sodium-bonded and helium-bonded nitride fuel rods were irradiated. Some of the sodium-bonded rods contained the shroud tube to minimize fuel relocation, and two out of the four helium-bonded fuel rods contained annular fuel pellets.[58] In many respects, the performance of the two fuel rod designs compared well with that of the carbide fuel. No failures were observed in either the helium-bonded or the shrouded, sodium-bonded fuel rods. However, failures were observed in the unshrouded fuel rods.

For the sodium-bonded fuel rods, fission gas release ranged from 0.3 to 3.7%, with a tendency to increase with increasing burnup (up to 6050 GJ/kgM). Fuel swelling data indicate a rate of about 1.5 vol.% per 1000 GJ/kgM burnup, compared to a value of 2.5 vol.% for carbide fuels. In the shrouded rods the shroud appears to provide an effective means of preventing FCMI until the fuel–cladding gap is consumed. No evidence of mechanical interaction (nor chemical interaction) was observed in these test rods, presumably because the cladding began swelling at a rate greater than the fuel before the gap was consumed. In the unshrouded fuel rods, pellet-fragment relocation occurred and FCMI was clearly evident as the cause of the observed fuel rod failures (similar to Fig. 3-24).

In the helium-bonded fuels, the effect of both fuel temperature and fuel density on fission gas release was evident. Fuel rods containing annular pellets exhibited the lowest central temperatures and also the lowest gas release percentage. The low-density fuel exhibited the highest gas release, in agreement with the oxide and carbide data. Also, fission gas release increased with increasing burnup to ~8000 GJ/kgM. Interestingly, the annular pellet did not appear to accommodate fuel creep and/or swelling and, thereby, reduce cladding strains.

A second series of test irradiations that involved only helium-bonded mixed-nitride fuel rods[59] produced failures after only about 4300 GJ/kgM burnup. Detailed post irradiation examinations revealed small axial cracks in a number of fuel rods, containing both high- (87% TD) and low- (81% TD) density fuel pellets. The only common feature was the small fuel–cladding gap (nominal 0.127 mm). Therefore, although the diametral expansions of the fuel rods were found to be related to smear density, clearly the gap played a more dominant role than the pellet density. Further, failure is apparently not related either to maximum average cladding strain (the rods that failed exhibited the smallest $\Delta D/D$ in their group) or maximum local strain (failures did not occur at the ridges).

Although the final conclusion on the most likely failure mechanism has yet to be reported, the occurrence of these early fuel rod failures, when combined with the large fabrication penalty associated with ^{15}N enrichment requirement, was sufficient cause for termination of the U.S. nitride fuel development program.

3.5. Control Rod Material Development

Based on development efforts and favorable experience in thermal reactors, boron carbide (B_4C) was selected for fast-reactor control-rod applications. As for the LWR control rod, performance criteria are based on gas release (helium) and swelling characteristics during exposure. Since fast-reactor operating temperatures, neutron flux levels, and energy distributions are significantly higher than those in thermal reactors, a considerable amount of development effort has been required to qualify boron carbide for fast-reactor service. The results of the B_4C development program have been widely reported, specifically in the review by Murgatroyd and Kelley[60] and in the reports by Mahagin and Dahl.[61, 62]

Experience with B_4C pellets irradiated in both EBR-II and DFR has been reported.[63–65] The instrumented tests in EBR-II[64] permitted almost continuous plenum pressure measurements so that a large amount of helium release data could be determined over a spectrum of burnup levels. In Fig. 3-26, helium releases from 92% TD B_4C pellets (20 to 92% ^{10}B enriched) irradiated at three different temperatures are compared. At low burnup levels ($<10 \times 10^{26}$ captures/m^3), the B_4C exhibited a high helium release rate, i.e., the so-called primary release region. At higher

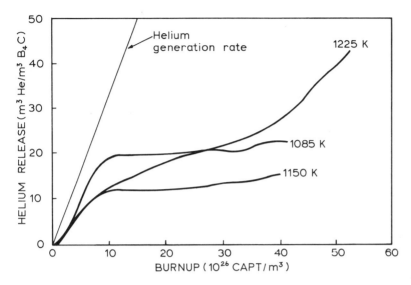

Fig. 3-26. Helium release from boron carbide at three temperatures (Ref. 64).

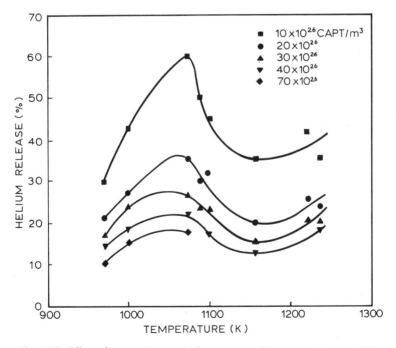

Fig. 3-27. Effect of temperature on helium release of boron carbide (Ref. 64).

burnups, the release rate declined to an almost steady-state value, which is temperature dependent, i.e., the secondary-release region.

The temperature dependence of helium in fast-reactor irradiations is more complex than that observed in thermal reactor irradiations (refer to Fig. 2-29). Fractional helium release, at various burnup levels, is plotted against temperature in Fig. 3-27. A complex relationship results at low burnup levels, involving a maximum at ~1030 K and a minimum at ~1150 K. At higher burnup levels the influence of the primary region is reduced and hence the magnitude of both the maximum and minimum is reduced. The helium release fraction itself decreases to only 10 to 20% at high ^{10}B burnups. As will be discussed below, this decrease in helium release is attributed to helium bubble formation within the B_4C grains (refer to Fig. 3-28).

Contrary to experience in thermal reactors, boron carbide swelling in fast reactors exhibits a definite temperature dependence. In the range 773 to 1143 K, dimensional instability is greatest at low exposure temperatures and decreases as the irradiation temperature is raised. In this case, a boron burnup dependence is also noted, with the swelling increasing monotonically up to exposures of about 18×10^{26} captures/m^3. For design purposes,[60, 61] the fast-reactor swelling dependence (in terms of pellet $\Delta D/D$) on both temperature and burnup is described by an empirical equation:

$$\Delta D/D = C\left(-2 \times 10^{-8} + \frac{3 \times 10^{-4}}{9T - 2297}\right) \qquad (3\text{-}5)$$

where $\Delta D/D$ is given in percent, T is irradiation temperature (K), and C is boron burnup (10^{26} captures/m^3).

Also, in contrast to thermal reactor experience, gas release and swelling rates of boron carbide in fast-reactor spectra are relatively insensitive to material variables such as pellet density, boron enrichment, and grain size after a burnup level of about 10^{27} captures/m^3.

Unlike the situation for fission gases in fuels, theories specifically developed for helium bubble formation in B_4C are not yet available. Hollenberg et al.[66] are attempting to develop more physically based correlations for B_4C swelling and gas release by systematically measuring the density, shape, and size of helium bubbles irradiated over a wide temperature range (773 to 2123 K). These workers reported significant differences in bubble characteristics at low ($T < 1173$ K) and high (>2123 K) temperatures. In the former regime, bubbles are small (0.01 μm), lenticular shaped, and of high density, and are nucleated homogeneously. Note the strain fields around the bubble shown in Fig. 3-28. In the latter regime, there is a low density of large bubbles (1.0 μm), which are equiaxed and faceted and either nucleated heterogeneously or the result

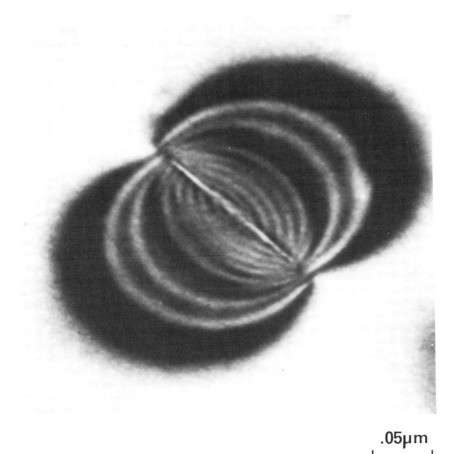

.05μm

Fig. 3-28. Strain fields surrounding helium bubbles in low-burnup boron carbide (11×10^{26} captures/m^3) (Ref. 66).

of coalescence of smaller bubbles. This temperature effect on helium bubbles in B_4C is consistent with the theoretical expectations derived from nuclear fuel-fission gas bubble models (refer to Chapter 2, Section 2.2.2.1), and hence provided the basis for a preliminary swelling and gas release model for B_4C. Comparison of model predictions to data such as those in Fig. 3-26 was good at temperatures below about 1250 K, but model and data deviated considerably at higher temperatures.[67]

In addition to gas-induced swelling, significant pellet volume increases can result from the formation of a large density of microcracks.[68, 69] Formation of microcracks is related to both high strains generated by helium bubbles (Fig. 3-28)[68] and to free graphite, which is generally observed in the pore regions of the B_4C pellet.[69] Thus, care must be taken to avoid the presence of unreacted carbon in the pellet, or additional clad strain will result.

The chemical interaction between B_4C pellets and stainless steel cladding can be controlled. Results of two studies[65] indicate that no interaction is observed below 773 K and it becomes just detectable at 973 K with the formation of a layer of approximately 10 μm of Fe_2B. The compatibility limit is in fact between 1373 and 1473 K, well above normal operating temperatures. Compounds with excess boron can cause more rapid attack of the steel, but B_4C density is not important.

In summary, the period of service for B_4C control rods in LMFBRs appears to be limited largely by the ability of the cladding to withstand stresses imposed by swelling (gas bubble plus microcracking) of the B_4C. Clad failure has occurred at strains in the range 2 to 5%, and, therefore, as noted earlier in Table 3-1, development of cladding of adequate ductility has been initiated to improve fast-reactor control-rod lifetimes. The development of both vented and sealed control rods, the latter with a plenum to contain released helium, is proceeding in the U.K., France, the U.S., and Germany.[60, 64]

Backup control rod materials, primarily europium compounds such as europia (Eu_2O_3) and europium hexaboride (EuB_6), are being developed for the LMFBR. Evaluations of Eu_2O_3 have been terminated in the U.S., but are proceeding in Great Britain, Germany, Russia, and France. A small EuB_6 qualification program continues in the U.S.

In the U.S., investigations on Eu_2O_3 date from 1972, and the program has included studies on synthesis, fabrication, physical and chemical attributes, compatibility with stainless steel cladding and sodium coolant, and irradiation behavior. Progress reports were published by ORNL in 1977[70, 71] at the termination of the program. Key results of this work will be summarized below as an illustration of progress in qualification of this ceramic material for reactor application.

Current control rod designs require the Eu_2O_3 to be in the form of cylindrical pellets (15–17 mm diameter, 28–40 mm long) with as high a density as practical, in excess of 90% TD. Hence, both sintering and hot-pressing techniques have been investigated for pellet fabrication. More than 200 pellets clad in Type 316 stainless steel tubes were utilized in the fast reactor irradiation evaluation in EBR-II, and these were subjected to extensive pre- and post-test characterization.

The irradiation program to date has produced no experimental evidence to indicate that monoclinic Eu_2O_3 would be unsuitable for fast reactor application as a control rod material. Pellets crack during irradiation, as expected from the thermal stress calculations and the thermal performance tests, but this cracking does not appear to be detrimental to control assembly performance. Eu_2O_3 is chemically compatible with the cladding at anticipated operating temperatures, both inside and outside the core. Radiation swelling appears to be a function of temperature, flux, and/or fluence in a manner which is not presently well understood. It is in this area that critical future studies are required. This is especially true given the difference between the neutronic environment of the test pellets in EBR-II and those of anticipated demonstration and commercial LMFBRs. Determination of other physical properties of irradiated pellets, such as elastic moduli, thermal conductivity and expansion, etc., would provide data very useful in absorber assembly design.

The cubic-phase Eu_2O_3 has a slightly different set of thermal and mechanical properties, with the result that cracking of the pellets is not observed during irradiation. However, chipping or spalling of pellet edges was observed, and this effect may lead to serious problems of mechanical interaction with the cladding if the small pieces become trapped in the pellet–cladding gap. Further, high-fluence irradiation caused shrinkage of cubic-phase material, an effect which would increase the pellet temperatures by causing an increase in the pellet–cladding gap. This could conceivably accelerate the process of temperature increase, until the phase transformation temperature is surpassed, at which point a large volume contraction (>9%) would occur. The overall effect would be significantly higher temperatures in the rods than allowed in the original design. Such a sequence of events could result in significant problems concerning chemical compatibility with the cladding. Thus, the irradiation behavior of cubic europia would require significantly more study if this is to be serious candidate material.

Overall, the ORNL irradiation tests[70, 71] showed that Eu_2O_3 has desirable properties as a solid neutron absorber material up to a fluence of 6×10^{26} neutrons/m² ($E < 0.1$ MeV) at an average pellet temperature near 973 K. However, further data are required at higher flux levels to determine if the good performance characteristics observed at flux levels typical of EBR-II are also observed at neutron fluxes more typical of anticipated commercial LMFBRs.

3.6. Proliferation-Resistant Fuel Cycles

So far, discussion has assumed a U–Pu fuel cycle that involves both the recycle of Pu through LWRs (described in Section 2.3, Chapter 2) and use of Pu in LMFBRs. In recent years, however, public concern has grown over the possibility that the planned extraction of Pu from spent nuclear fuel would provide an opportunity for theft of material from which a nuclear bomb could be made by terrorists, international outlaws, or nations suddenly adopting a militaristic policy.

In response to this concern, the Carter Administration deferred indefinitely the reprocessing of LWR fuel and the introduction of LMFBRs in the U.S. Other countries that are more dependent on nuclear power, such as France, Germany, and Japan, have expressed deep concern over the U.S. decision and, in consequence, emphasis in the nuclear fuel cycle area over the past few years has been on identification of "(more) proliferation-resistant fuel cycles," either as modified forms of the U–Pu fuel cycle or alternative fuel cycles, which would provide the necessary extension of uranium supplies. Examples of approaches considered by the Nonproliferation Alternate System Assessments Program (NASAP) and the International Nuclear Fuel Cycle Evaluation (INFCE) Groups and progress made will be discussed in this section. At this stage, however, it is apparent that there is no near-term replacement for the U–Pu fuel cycle. Although thorium-based fuels and metal fuels should be explored over the long run because they do offer advantages (albeit differant) in extending fission fuel supplies, neither fuel cycle appears to be any more proliferation-resistant than that based on U–Pu mixed oxides or carbides.

3.6.1. Modified Mixed Oxide Fuel Cycles

Whether there is any difference in proliferation risks between the U–Pu breeders and alternative fuel cycles is still hotly debated. Nevertheless there are ways to make the U–Pu fuel cycle "more" proliferation-resistant if it is deemed necessary for public acceptance of the LMFBR. The Civex fuel cycle [72] that was proposed recently is a good example since it would still enable us to capitalize on the advanced stage of development of the mixed-oxide fuel. The basic objective of the Civex process is to reduce the possibility for Pu diversion by not separating Pu from U in a reprocessing plant. In contrast to the Purex flow sheet used for military separations, which produces three product streams—pure plutonium, pure uranium, and all the waste products—the Civex process adopts a "coreprocessing" approach that separates out the bulk of the wastes, leaving a small percentage of the wastes with the U and Pu in such dilute concentrations that they are useless for weapons purposes, but at the same time unapproachable (sometimes called "spiking" of the fuel). The mixed oxide powder is "coprecipitated" from this solution. Thus, the novelty of the Civex process is that it does not separate out Pu, makes it impossible to do so, and maintains a high level of radioactivity deliberately so that it may be handled safely only by remotely operated tools. Civex also departs from presently planned practice for nuclear fuel processing by combining or "colocating" in a single facility the spent fuel reprocessing and the refabrication, from recovered material, of new fuel, thereby eliminating the transportation of plutonium from reprocessing plant to refabrication plant. One can see that it combines the frequently proposed individual concepts of coreprocessing, coprecipitation, spiking, and colocation into one physical facility.

While in principle the Civex process can be designed either for the LWR or the LMFBR fuel cycle, the concept is better suited for the LMFBR cycle because the presence of fission products in the fuel does not influence the physics of the core. Refabricated fuel could be of pellet or particle geometry, but the necessity for remote fabrication puts emphasis on the use of solutions rather than powders, and hence makes sol–gel coprecipitated particle fuel (Spherepac) more attractive.

3.6.2. Alternative Fuel Cycles

Two alternative fuel cycles have received most attention recently, although both have been around for years, ever since the beginnings of nuclear power. The ceramic-based fuel cycle that would utilize UO_2 technology to a large degree is based on thorium. The metal-fuel-based fuel cycle is derived principally from EBR-II driver fuel experience.

3.6.2.1. Thorium-Based Ceramic Fuels

As noted earlier in Chapter 2, the principal candidate for a fuel cycle based on alternative fuels is thorium–U-233.[73] Several thorium core designs have been proposed for the LMFBR, including thorium either in the radial blanket, in both radial and axial blankets, or in the active core. These would operate in combination with U-233 fueled LWRs.[38] Early core designs, in comparison to the U–Pu oxide cores, indicate several trends that will have to be followed to utilize thorium in commercial LMFBRs.[38] First, due to the lower breeding potential of thorium [as either $(U,Th)O_2$ or $(Pu,Th)O_2$ mixed oxides] advanced cladding and duct materials will be essential. Second, the carbide forms would be more advantageous since they yield higher powers, and hence better breeding ratios, but experience on these materials is limited. Third, the use of remote fabrication will necessitate adoption of particle fuel technology, either as packed particle fuels or as pellets fabricated by pressing and sintering compacted particles.

It is of interest to briefly review the data on which decisions on the use of thorium-based fuels must be made. Past research on thorium-based fuels focused on ThO_2–UO_2 with UO_2 contents in the range of 2 to 10%.

Most of the existing data were measured during the late fifties and through the sixties when interest in these ceramics as fuels in the Light Water Breeder Reactor (LWBR) and converter reactor concepts was strong. Little new data have been obtained since about 1970, other than from coated-particle HTGR fuels.

$(U,Th)O_2$ fuels that have undergone extensive irradiation evaluation have been fabricated with the U-235 isotope. These fuels have been fabricated into sintered pellets, microspheres (i.e., Spherepac), and coated-particle fuel forms. One program of interest fabricated a complete core of stainless steel clad pellet fuel for the Indian Point 1 commercial PWR, utilizing 93% U-235 and U-to-(Th + U) ratios that ranged from 0.031 to 0.1, and conventional cold pressing, sintering, and centerless grinding pellet processing steps.[74]

For fast-breeder reactor application, however, it is desirable to use instead of U-235 the U-233 isotope as the fissile material because of its higher fission density in the fast flux. U-233 inevitably contains U-232 as an impurity, which, through radioactive decay, produces gamma-active daughters that could pose a radiation hazard to the material-handling personnel. Therefore, fabrication with U-233 must be accomplished either remotely or very rapidly after chemical separation of the U-232. Since there is very little experience with remote fabrication of large core loadings, it is likely that fabrication procedures will have to be modified. Certainly, chemical processes such as those used in sol-gel particle manufacture look promising in this respect.

Detailed irradiation performance data on ThO_2–UO_2 fuels are limited, but all indications are that the fuel performs as well as UO_2 or the mixed (U,Pu) oxide. Perhaps the most significant comprehensive project to date has been the ThO_2–UO_2 fuel irradiation in Indian Point and the subsequent destructive examination of 13 of the rods.[74] During this operating period, the fuel achieved peak burnups as high as 2850 GJ/kgM and the borated type 304 stainless steel cladding received peak fast neutron fluences ($E > 0.1$ MeV) of about 3×10^{25} neutron/m^2.

Examinations showed that all 13 fuel rods had performed very satisfactorily. Nothing observed during the examination indicated that these core components could not operate to a considerably higher exposure. In-reactor exposure caused no major microstructural changes in the fuel. Lack of microstructural changes indicated that the centerline fuel temperature was probably less than 1873 K and the amount of fission gases released—1 to 2%—is consistent with results from fuel temperatures in the 1873 K range.

ThO_2–UO_2 fuel operated to exposures of 2850 GJ/kgM without introducing localized or general strains in the cladding. No ridging at the pellet interfaces was observed, even in the rod of highest exposure. The lack of pellet distortion and cladding strains and the fact that entire pellets could be removed from the cladding indicated that irradiation-induced fuel swelling was minimal. In contrast, UO_2 fuels operated under similar conditions would show some increase in volume and possibly some ridging at pellet interfaces.

This irradiation study and other limited ones,[73] while indicating that thorium-based pellet fuels should operate satisfactorily in LWRs, do not provide sufficient guidance as to their performance in LMFBRs. A factor of 3 or more in burnup and over 400 K increase in fuel temperatures are required for efficient LMFBR operations, and this regime is where the deleterious phenomena (swelling, gas release, corrosion, etc.) begin to limit useful life.

In an analytical study using the LIFE code and available property data on thorium fuels, Weber et al.[3] compared the expected behavior of these fuels with the reference plutonium-based fuels in an advanced alloy, 1000 MWe core. The $(Th,Pu)O_2$ fuels and ThO_2 blanket generally predicted larger cladding damage than either the reference oxide fuel and blankets or the sodium-bonded (U,Pu)C fuel and helium-bonded UC blankets for both steady-state and transient conditions. Thorium fuels swell more and creep less than the reference oxide because of their higher thermal conductivity, which effects result in high FCMI stresses. There was insufficient data on the (Th,U)C fuel and ThC blanket materials to allow any predictions to be made.

Although somewhat ambitious overall, this analytical study[3] was worthwhile in pointing out the areas where property and performance data are required. An extensive program appears necessary to bring the thorium-based fuels up to the current status of the plutonium-based fuels. Such a broad-based program has not yet been implemented, but a small number of scoping experiments are underway in EBR-II.[38] These involve irradiation testing of mixed Th–Pu oxides [Pu/(Pu + Th) ratio ranging from 0.19 to 0.33] and carbides [Pu/(Pu + Th) = 0.55] and denatured uranium ($^{233, 235}U$) fuels to check on key features of thermal and mechanical performance for both driver and blanket application, with near-term emphasis being on the latter.

Irradiation experience with other forms of thorium-based fuels is equally limited, but at the same time promising. For instance, in the German program at Julich Spherepac fuels of ThO_2–UO_2 have been irradiated in the FRJ2 reactor. After irradiation to a maximum burnup of 4320 GJ/kgM at maximum linear powers of 60 kW/m, the fuels gave results in excellent agreement with those reported on Spherepac UO_2 by the Dutch.[75]

The final point on the thorium cycle was made earlier in Chapter 2 but is worth restating here; that is, the major impediment to the deployment of the thorium fuel cycle is not data, but the lack of suitable fabrication and reprocessing facilities. On one hand, thorium utilization clearly is not viable unless the fuel cycle is closed, while on the other hand, private capital for such facilities is unlikely to be risked unless there is a demonstrated market. Prospects for the implementation of the thorium fuel cycle would be greatly enhanced if dual-purpose thorium/uranium cycle fabrication and reprocessing facilities were feasible; the feasibility of dual-purpose facilities should therefore be evaluated.

3.6.2.2. Metal Alloy Fuels

The first LMFBRs (EBR-I, EBR-II, DFR, and Fermi) all used uranium alloy metal fuels but, with the developing LMFBR objectives of high fuel burnups, on the order of 8600 GJ/kgM, and high sodium outlet temperatures, on the order of 870 to 920 K, metal fuel was de-emphasized in relation to ceramic fuels, due to the latter's improved stability at high temperatures.

This trend has been reversed somewhat recently, for, in addition to improved potential for proliferation resistance, renewed interest in metal fuels for LMFBRs has occurred because design studies of large commercial LMFBRs during the past few years have specified decreasing requirements for sodium outlet temperature.[52] Metal fuels have been developed that appear to have adequate compatibility with cladding materials at current temperatures of interest. Also, a preferred metal fuel rod design concept has been developed which should achieve fuel burnups of at least 8600 GJ/kgM with high reliability.[76, 77] Finally, as shown in Fig. 3-3 earlier, recently completed system studies of metal-fueled LMFBRs have shown that compound system doubling times in the range of 9 to 10 years are achievable in conservative reactor designs.[3, 78]

In-reactor performance of metal fuels is controlled by two important phenomena, i.e., fuel swelling and fuel–cladding metallurgical interaction. Recent developments in fuel rod design, reactor operating conditions, and fuel composition have enabled both the burnup and thermal performance of metal fuels to be significantly increased.[76, 77] A solid-metal fuel, sodium-bonded to stainless steel cladding, has proved to be the most feasible design for metal fuel rods. Analyses of fuel rod behavior and initial tests of prototype rods indicated that metal fuel rods with 75% fuel smear density, which allows 33 vol.% swelling, should be capable of achieving burnups of at least 8600 GJ/kgM, independently of the fuel composition.

A metal fuel rod design concept for LMFBR use has been developed based on the experimental findings and analyses. In-reactor performance of the design concept is now being demonstrated as the Mark II fuel rod in EBR-II. A Mark II assembly contains 3.30-mm-diameter U–Fs alloy† fuel rods, 343 mm long, sodium-bonded to 4.42-mm-diameter Type 316 stainless steel cladding. The sodium annulus thickness is 0.25 mm, while the cladding thickness is 0.31 mm, and the fuel smear density is 75% of theoretical.

Approximately 27,000 Mark II driver fuel rods have been irradiated to date in EBR-II. Fuel rods achieved burnups in excess of 8640 GJ/kgM prior to failure.[77] To ensure a conservative margin to failure during reactor operation, the burnup limit for the Mark II fuel has, therefore, been set at 6910 GJ/kgM. Approximately 800 rods have reached burnups of ~8640 GJ/kgM and some have successfully achieved burnups as high as 13,800 GJ/kgM.

Ideally, the metal fuel composition for an LMFBR would be a binary fertile-fissile alloy such as U–Pu or Th–Pu. These alloys, however, are not compatible metallurgically with conventional iron- or nickel-base cladding alloys at normal reactor operating temperatures. Metallurgical compatibility is essential to avoid excessive fuel penetration into the cladding at normal operating temperatures, and eutectic formation under transient over-temperature conditions.

Extensive development work has shown that zirconium additions to metal fuel alloys are highly effective in minimizing fuel/cladding penetration rates and in raising both the fuel/cladding eutectic temperature and the solidus temperature of the fuel alloy.[76, 77, 79] U–15 wt.% Pu–10 wt.% Zr alloy, for example, has been found in recent analyses of cladding cumulative damage fractions to be acceptable for use with Type 316 stainless steel at cladding temperatures up to 941 K, at 8640 GJ/kgM burnup.[78] Higher-strength cladding alloys should be able to accommodate temperatures above 973 K without undue loss of strength from fuel/cladding interaction.

Despite the advantages of metal fuels and encouraging performance in both experimental and analytical studies, their future for large LMFBR application is still uncertain. An irradiation test has been designed for insertion in FFTF,[38] and two fuels, U–15 wt.% Pu–10 wt.% Zr and Th–15 wt.% Pu–10 wt.% Zr, were specified for the test. However, no commitment has yet been made to implement the program. A new LMFBR concept, called the fast-mixed spectrum reactor (FMSR),[80] may serve to promote the interests of metal fuels in the near future. This reactor concept purportedly would use only one-fifth of the uranium used by an LWR, but would be proliferation resistant because it would not

† Fissium (Fs) is an equilibrium concentration of fission product elements left by the pyrometallurgical reprocessing cycle used for EBR-II. The simulated U–Fs alloy used as the driver fuel in EBR-II contains 2.4% Mo, 1.9% Ru, 0.3% Rh, 0.2% Pd, 0.1% Zr, and 0.01% Nb.

require reprocessing for plutonium. The fuel would remain in the reactor a very long time (~17 years) and the burnup of heavy metal achieved in this period would exceed 10,000 GJ/kgM. A metal fuel would be the most efficient fuel form if cladding and other components could be found that would be compatible for this period.

References

1. C. M. Cox, R. F. Hilbert, and A. Biancheria, U.S. Experience in Irradiation Testing of Advanced Oxide Fuels, in: *Proceedings of the American Nuclear Society Topical Meeting on Advanced LMFBR Fuels*, Tucson, Arizona (1977), pp. 136–148.
2. S. Kaplan, J. D. Stephen, R. M. Vijuk, and R. J. Jackson, Advanced Concepts for Fuel Assembly Design, in: *Proceedings of the International Conference on Fast Breeder Reactor Fuel Performance*, Monterey, California, March 5–8, 1979, Pub. No. ISBN: 0-89448-105-3, pp. 758–776.
3. C. W. Weber, V. W. Lowery, A. Biancheria, and R. P. Omberg, The Performance of FBR Core Designs with Alternative Fuel, in: *Proceedings of the International Conference on Fast Breeder Reactor Fuel Performance*, Monterey, California, March 5–8, 1979, Rep. No. ISBN: 0-89448-105-3, pp. 879–896.
4. *Proceedings of the Conference on Fast Reactor Fuel Element Technology*, New Orleans, Louisiana, April 13–15, 1971, R. Farimaker, ed., American Nuclear Society.
5. *Proceedings of the International Conference on Fast Breeder Reactor Fuel Performance*, Monterey, California, March 5–8, 1979, Pub. No. ISBN: 0-89448-105-3, Papers in Section IA, pp. 2–78.
6. H. Mikailoff, J. M. Dupouy, J. Ravier, M. Savineau, J. P. Pages, and J. M. Chaumont, Some Lessons Drawn from Operation of Phenix and Rhapsodie Cores, paper presented at European Nuclear Conference, Hamburg, Germany, March 6–11, 1979; abstract in *Trans. Am. Nucl. Soc.*, **31**, 179–180 (1979).
7. *Nuclear Systems and Materials Handbook*, Westinghouse Hanford Co., Richland, Washington, TID-26666, Vols. 1 and 2 [distribution limited to U.S. DOE approved recipients].
8. *Proceedings of the International Conference on Fast Breeder Reactor Fuel Performance*, Monterey, California, March 5–8, 1979, Rep. No. ISBN: 0-89448-105-3, Section III, Fuel Design and Modeling, pp. 619–757.
9. D. Olander, Fundamental Aspects of Nuclear Reactor Fuel Elements, Technical Information Center, ERDA, TID-26711-P1 (1976).
10. F. A. Nichols, Transport Phenomena in Nuclear Fuels under Severe Temperature Gradients, *J. Nucl. Mater.*, **84**, 1–25 (1979).
11. R. D. Leggett, E. N. Heck, P. J. Levine, and R. F. Hillbert, Steady-State Irradiation Behavior of Mixed Oxide Fuel Pins Irradiated in EBR-II, in: *Proceedings of the International Conference on Fast Breeder Reactor Fuel Performance*, Monterey, California, March 5–8, 1979, Rep. No. ISBN: 0-89448-105-3, pp. 2–15.
12a. L. A. Lawrence, J. W. Weber, and J. L. Devary, Fuel–Cladding Chemical Interaction of Mixed Oxide Fuels, in: *Proceedings of the International Conference on Fast Breeder Reactor Fuel Performance*, Monterey, California, March 5–8, 1979, Rep. No. ISBN: 0-89448-105-3, pp. 432–444.
12b. D. C. Fee and C. E. Johnson, Fuel–Cladding Chemical Interaction in Uranium–Plutonium Oxide Fast Reactor Fuel Pins, *J. Nucl. Mater.*, **96**(1981).

13. E. T. Weber, L. A. Lawrence, C. N. Wilson, and R. L. Gibby, In-Reactor Performance of Methods to Control Fuel–Cladding Chemical Interaction in: *Proceedings of the International Conference on Fast Breeder Reactor Fuel Performance*, Monterey, California, March 5–8, 1979, Rep. No. ISBN: 0-89448-105-3, pp. 445–459.
14. R. A. Karnesky, J. W. Jost, and I. Z. Stone, Cesium Migration in LMFBR Fuel Pins in: *Proceedings of the International Conference on Fast Breeder Reactor Fuel Performance*, Monterey, California, March 5–8, 1979, Rep. No. ISBN: 0-89448-105-3, pp. 343–352.
15. C. N. Wilson, R. L. Gibby, and E. T. Weber, Titanium Oxide Cesium Getters for Low O/M FTR Fuel Pins, paper presented at American Ceramic Society Basic Science and Nuclear Divisions Fall Meeting, October 14–17, 1979, New Orleans, Louisiana.
16. A. Boltax and A. Biancheria, Fuel–Cladding Mechanical Interaction Effects in Fast Reactor Mixed Oxide Fuel, paper presented at the IAEA International Working Group on Fast Reactors Technical Committee on Fuel and Cladding Interactions, Tokyo, Japan, February 21–25, 1977.
17. A. Biancheria, T. S. Roth, U. P. Nayak, and A. Boltax, Fuel–Cladding Mechanical Interaction in Fast Reactor Fuel Rods, in: *Proceedings of the International Conference on Fast Breeder Reactor Fuel Performance*, Monterey, California, March 5–8, 1979, Rep. No. ISBN: 0-89448-105-3, pp. 513–535.
18. J. J. Holmes and J. L. Straalsund, Effects of Fast Reactor Exposure on the Mechanical Properties of Stainless Steels, in: *Proceedings of the International Conference on Radiation Effects in Breeder Reactor Structural Materials*, Scottsdale, Arizona, June 19–23, 1977, M. L. Bleiburg and J. W. Bennett, eds., American Institute of Mining Metallurgaical, and Petroleum Engineers, Inc. (1977), pp. 53–64.
19. J. L. Straalsund, R. L. Fish, and G. D. Johnson, Correlation of Transient Test Data with Conventional Mechanical Properties Data, *Nucl. Technol.*, **25**, 531–540 (1975).
20. G. D. Johnson, J. L. Straalsund, and G. L. Wire, A New Approach to Stress-Rupture Data Correlation, *Mater. Sci. Eng.*, **28**, 69–75 (1977).
21. A. K. Miller, An Inelastic Constitutive Model for Monotonic, Cyclic, and Creep Deformation. Part I. Equations Development and Analytical Procedures; Part II. Application to Type 304 Stainless Steel, *J. Eng. Mater. Technol.*, **98**, 97–113 (1964).
22. D. R. Harries, Neutron-Irradiation-Induced Embrittlement in Type 316 and Other Austenitic Steels and Alloys, *J. Nucl. Mater.*, **82**, 2–21 (1979).
23. M. L. Grossbeck, J. O. Stiegler, and J. J. Holmes, Effects of Irradiation on the Fracture Behavior of Austenitic Stainless Steels, in: *Proceedings of the International Conference on Radiation Effects in Breeder Reactor Structural Materials*, June 19–23, 1977, M. L. Bleiburg and J. W. Bennett, eds., American Institute of Mining, Metallurgical, and Petroleum Engineers, Inc. (1977), pp. 53–64.
24. E. W. Hart and Che-Yu Li, Use of State Variables in the Description of Irradiation Creep and Deformation of Metals, in: *Proceedings of the International Conference on Zirconium in the Nuclear Industry*, A. Lowe and G. Parry, eds., Am. Soc. Test. Mater. Spec. Tech. Publ. 633 (1977), pp. 315–325.
25. G. Arnaud, A. Bernard, and P. Amman, Deformation of the Hexagonal Duct—Performance and Choice of Materials, Paper F14 in: Proceedings of the International Meeting on Irradiation Behavior of Metallic Materials for Fast Reactor Core Components, June 4–8, 1979, sponsored by CEA, Ajaccio, Corsica, France.
26. E. R. Gilbert, J. L. Straalsund, and G. L. Wire, Irradiation Creep Data in Support of LMFBR Core Design, *J. Nucl. Mater.*, **65**, 266–278 (1977).
27. *Proceedings of the International Conference on Radiation Effects in Breeder Reactor Structural Materials*, Scottsdale, Arizona, June 19–23, 1977, M. L. Bleiburg and J. W. Bennett, eds., American Institute of Mining, Metallurgical, and Petroleum Engineers, Inc. (1977).

28. *Proceedings of the International Meeting on Irradiation Behavior of Metallic Materials for Fast Reactor Core Components,* June 4–8, sponsored by CEA, Ajaccio, Corsica, France.
29. H. J. Bergmann, G. Knoblauch, D. Haas, and K. Herschbach, Examinations on Swelling and Irradiation Creep of the Austenitic Stainless Steel W-Nr. 1.4981CW, in: *Proceedings of the International Meeting on Irradiation Behavior of Metallic Materials for Fast Reactor Core Components,* June 4–8, 1979, sponsored by CEA, Ajaccio, Corsica, France.
30. M. M. Paxton, B. A. Chin, E. R. Gilbert, and R. E. Nygren, Comparison of the In-Reactor Creep of Selected Ferritic, Solid Solution Strengthened, and Precipitation Hardened Commercial Alloys, *J. Nucl. Mater.,* **80,** 144–151 (1979).
31. F. Rosa, P. S. Maiya, and R. W. Weeks, Effect of Grain Boundary Penetration of AISI Type 316 Stainless Steel by Cesium Oxides on Elevated Temperature Tensile Properties, in: *Proceedings of the Symposium on Elevated Temperature Properties of Austenitic Stainless Steels,* A. O. Schaefer, ed., Metals Properties Council Inc., ASME, New York, (1974), pp. 97–112.
32. R. F. Hilbert, E. A. Aitken, and P. R. Pluta, High Burnup Performance of LMFBR Fuel Rods—Recent G. E. Experience, in: Progress in Nuclear Energy. Proceedings of the European Nuclear Conference, Vol. 3 (April 1975), pp. 526–529.
33. D. Lee, R. A. Rand, and G. G. Trantina, Crack Nucleation in 316 Stainless Steel Fuel Cladding, workshop paper presented at the 4th International Conference on Fracture, Waterloo, Canada, June 19–24, 1977, pp. 723–739.
34. C. Bagnall and D. G. Jacobs, Relationships for Corrosion of Type 316 Stainless Steel in Liquid Sodium, WARD-NA-3045-23 (1975), Westinghouse Electric Corp.
35. O. K. Chopra, J. Y. N. Wang, and K. Natesan, Review of Sodium Effects on Candidate Materials for Central Receiver Solar-Thermal Power Systems, Argonne National Laboratory Report, ANL-79-36 (July 1979).
36. S. A. Shiels, A. R. Kerton, and R. P. Anatatmula, The In-Sodium Corrosion Behavior of Candidate Commercial Fuel Cladding and Duct Alloys, HEDL TME 77-71 (February 1978).
37. J. W. Weber, In-Reactor Corrosion Behavior of Stainless Steel Cladding in High-Temperature Sodium, presented at the International Conference on Liquid Metal Technology in Energy Production, Seven Springs, Pennsylvania, May 3–6, 1976.
38. J. W. Bennett, E. C. Norman, C. M. Cox, M. G. Adamson, A. Boltax, and J. H. Kittel, Advanced Alternate Breeder Fuels Testing in the U.S., in:*Proceedings of the International Conference on Fast Breeder Reactor Fuel Performance,* Monterey, California, March 5–8, 1979, Rep. No. ISBN: 0-89448-105-3, pp. 793–803.
39. E. Edmonds, W. Sloss, K. Q. Bagley, and W. Batey, Mixed Oxide Fuel Performance, in: *Proceedings of the International Conference on Fast Breeder Reactor Fuel Performance,* Monterey, California, March 5–8, 1979, Rep. No. ISBN: 0-89448-105-3, pp. 54–63.
40. A. L. Lotts, compiler, Fast Breeder Reactor Oxide Fuels Development—Final Report, *Oak Ridge National Laboratory Publication No. 4901* (November 1973).
41. L. A. Neimark, Argonne National Laboratory Progress Reports ANL-RDP35, December 1974, p. 5.12; *ANL-RDP38,* March 1975, p. 5.17; *ANL-RDP41,* June 1975, p. 5.6.
42. W. G. Johnston, J. H. Rosolowski, A. M. Turkalo, and T. Lauritzen, An Experimental Survey of Swelling in Commercial Fe–Cr–Ni Alloys Bombarded with 5 MeV Ni Ion, *J. Nucl. Mater.,* **54,** 24–40 (1974).
43. D. R. Harries, The UKAEA Fast Reactor Project Research and Development Program on Fuel Element Cladding and Sub-Assembly Wrapper Materials, in: *Proceedings of the International Conference on Radiation Effects in Breeder Reactor Structural Ma-*

terials, Scottsdale, Arizona, June 19–23, 1977, M. L. Bleiberg and J. W. Bennett, eds., Met. Soc. AIME (1977), pp. 27–40.
44. J. J. Laidler, J. J. Holmes, and J. W. Bennet, U.S. Programs on Reference and Advanced Cladding/Duct Materials, in: *Proceedings of the International Conference on Radiation Effects in Breeder Reactor Structural Materials,* June 19–23, 1977, Scottsdale, Arizona, M. L. Bleiburg and J. W. Bennett, eds., by Met. Soc. AIME (1977), pp. 41–52.
45. J. Lehman, J. M. Dupouy, R. Broudeur, J. L. Boutard, and A. Maillard, Irradiation Creep of 316 and 316 Ti Stainless Steels, in: *Proceedings of the International Meeting on Irradiation Behavior of Metallic Materials for Fast Reactor Core Components,* June 4–8, 1979, sponsored by CEA, Ajaccio, Corsica, France.
46. E. A. Little and D. A. Stow, Void Swelling in Irons and Ferritic Steels. II. An Experimental Survey of Materials Irradiated in a Fast Reactor, *J. Nucl. Mater.,* **87,** 25–39 (1979).
47. J. Erler, A. Maillard, G. Brun, J. Lehmann, and J. M. Dupouy, The Behavior of Ferritic Steels under Irradiation with Fast Neutrons, in: *Proceedings of the International Meeting on Irradiation Behavior of Metallic Materials for Fast Reactor Core Components,* June 4–8, 1979, sponsored by CEA, Ajaccio, Corsica, France.
48. J. J. Huet, A. Delbrassine, P. Van Asbroeck, and W. Vandermeulen, Radiation Effects in Ferritic Steels in: *Proceedings of the International Conference on Radiation Effects in Breeder Reactor Structural Materials,* June 19–23, 1977, Scottsdale, Arizona, M. L. Bleiburg and J. W. Bennett, eds., Met. Soc. AIME (1977), pp. 357–366.
49. E. A. Little, Void Swelling in Irons and Ferritic Steels. I. Mechanisms of Swelling Suppression, *J. Nucl. Mater,* **87,** 11–24 (1979).
50. G. W. Cunningham, Materials Development for Advanced Reactors, *Nucl. Technol.,* **28,** 301–304 (1976).
51. J. M. Dupouy, French Program on LMFBR Cladding Materials Development, in: *Proceedings of the International Conference on Radiation Effects in Breeder Reactor Structural Materials,* June 19–23, 1977, Scottsdale, Arizona, M. L. Bleiburg and J. W. Bennett, eds., Met. Soc. AIME (1977), pp. 1–12.
52. J. M. Kendall, LMFBR Steam Cycles—Is Efficiency the Ultimate Goal?, paper presented at 1979 Annual Meeting of ANS, Atlanta, Georgia, June 3–7, 1979; summary in: *Trans. Am. Nucl. Soc.,* **32,** 564–565 (1979).
53. J. O. Barner, T. W. Latimer, J. F. Kerrisk, R. L. Petty, and J. L. Green, Advanced Carbide Fuels—U.S. Experience, in: *Proceedings of the Topical Meeting on Advanced LMFBR Fuels,* Tucson, 1977, American Nuclear Society and Energy Research and Development Administration, pp. 268–298.
54. T. W. Latimer, R. L. Petty, J. F. Kerrisk, N. S. DeMuth, P. J. Levine, and A. Boltax, Irradiation Performance of Helium-Bonded Uranium–Plutonium Carbide Fuel Elements, in: *Proceedings of the International Conference on Fast Breeder Reactor Fuel Performance,* Monterey, California, March 5–8, 1979, Pub. No. ISBN: 0-89448-105-3, pp. 816–826.
55. J. F. Kerrisk, N. S. DeMuth, R. L. Petty, T. W. Latimer, J. A. Vitti, and L. J. Jones, Design and Performance of Sodium-Bonded Uranium–Plutonium Carbide Fuels, in: *Proceedings of the International Conference on Fast Breeder Reactor Fuel Performance,* Monterey, California, March 5–8, 1979, Rep. No. ISBN: 0-89448-105-3, pp. 804–815.
56. R. W. Stratton and L. Smith, The Irradiation Behavior of Spherepac Carbide Fuel, in: *Proceedings of the Topical Meeting on Advanced LMFBR Fuels,* Tucson, Arizona, 1977, American Nuclear Society and Energy Research and Development Administration, pp. 79–85.
57. K. Q. Bagley, W. Batey, R. Paris, W. M. Sloss, and G. P. Snape, U.K. Irradiation

Experience Relevant to Advanced Carbide Fuel Concepts for LMFBRs, in: *Proceedings of the Topical Meeting on Advanced LMFBR Fuels*, Tucson, Arizona, 1977, American Nuclear Society and Energy Research and Development Administration, pp. 313–325.
58. A. A. Bauer, P. Cybulskis, and J. L. Green, Mixed-Nitride Fuel Performance in EBR-II, in: *Proceedings of the Topical Meeting on Advanced LMFBR Fuels*, Tucson, Arizona, 1977, American Nuclear Society and Energy Research and Development Administration, pp. 299–312.
59. A. A. Bauer, P. Cybulskis, N. S. DeMuth, and R. L. Petty, He- and Na-Bonded Mixed-Nitrite Fuel Performance, in: *Proceedings of the International Conference on Fast Breeder Reactor Fuel Performance*, Monterey, California, American Nuclear Society, Rep. No. ISBN: 0-89448-105-3, pp. 827–841.
60. R. A. Murgatroyd and B. T. Kelly, Technology and Assessment of Neutron Absorbing Materials, *At. Energ. Rev.*, **15**, 3–74 (1977).
61. Compilation of B_4C Design Data for LMFBRs, HEDL-TME 75–19 (1975).
62. D. E. Mahagin and R. E. Dahl, Nuclear Applications of Boron and the Borides, in: *Boron and Refractory Borides*, V. I. Matkovich, ed., Springer Verlag, Heidelburg (1977), Chapter VII, pp. 613–632.
63. J. A. Basimajin, A. L. Pitner, D. E. Mahagin, H. C. F. Ripfel, and D. E. Baker, Irradiation Effects in Boron Carbide Pellets Irradiated in Fast Neutron Spectra, *Nucl. Technol.*, **16**, 238–248 (1972).
64. G. W. Hollenburg, J. L. Jackson, and J. A. Basmajian, In-Reactor Measurement of Neutron Absorber Performance, *Nucl. Technol.*, **49**, 92–101 (June 1980).
65. IAEA Specialist Meeting on Absorbing Materials and Control Rods for Fast Reactors, Dimitrovgrad, U.S.S.R., June 1973, Summary Report edited by R. E. Dahl and J. W. Bennett, HEDL TME 73–91 (1973).
66. G. W. Hollenburg, B. Mastel, and J. A. Basmajian, Helium Bubbles in Irradiated Boron Carbide, *J. Am. Ceram. Soc.*, **63** (7–8), 376–380 (1980).
67. C. E. Beyer and G. W. Hollenberg, Physically Based Model for Helium Release from Irradiated Boron Carbode, paper presented at American Ceramic Society Basic Science and Nuclear Divisions Fall Meeting, October 14–17, 1979, New Orleans, Louisiana.
68. G. W. Hollenburg and W. V. Cummings, Effect of Fast Neutron Irradiation on the Structure of Boron Carbide, *J. Am. Ceram. Soc.*, **60**, 520–525 (1977).
69. T. Inoue, T. Onchi, and H. Koyama, Irradiation Effects of Boron Carbide used as Control Rod Elements in Fast Breeder Reactors, *J. Nucl. Mater.*, **74**, 114–122 (1978).
70. A. E. Pasto and M. M. Martin, Eu_2O_3: Properties and Irradiation Behavior, ORNL 5291, Oak Ridge National Laboratory (1977).
71. A. E. Pasto and V. J. Tennery, Results of BICM-2 Irradiation Test of Eu_2O_3 in EBR-II, ORNL-5345, Oak Ridge National Laboratory (December 1977).
72. M. Levenson and E. L. Zebroski, A Fast Breeder System Concept: A Diversion-Resistant Fuel Cycle, *Nucl. Eng. Des.*, **51**, 119–132 (1979).
73. DOE Alternate Breeder Fuel Development Program, Reports of Task Groups (May 1977).
74. W. N. Bishop, Postirradiation Examination of Thoria–Uranium Fuel Rods, Babcock & Wilcox Publication No. 3809-7 (October 1969).
75. J. T. A. Roberts, Ceramic Utilization in the Nuclear Industry: Current Status and Future Trends, Part II, *Powder Metall. Int.*, **11**, 72–80 (1979).
76. J. H. Kittel, D. L. Johnson, W. N. Beck, and J. A. Horak, Metallic Fuel Systems for Alternative Breeder Fuel Cycles, in: *Proceedings of the International Conference on Fast Breeder Reactor Fuel Performance*, Monterey, California, 1979, Am. Nucl. Soc. Pub. ISBN: 0-89448-105-3, pp. 925–934.

77. J. H. Kittel and L. C. Walters, Development and Performance of Metal Fuel Elements for Fast Breeder Reactors, paper presented at the European Nuclear Conference, May 6–11, 1979, Hamburg, West Germany; abstract in *Trans. Am. Nucl. Soc.*, **31**, 177–178 (1979).
78. P. S. K. Lam, R. B. Turski, and W. P. Barthold, Performance of U–Pu–Zr Metal Fuel in 1000 MWe LMFBRs, in: *Proceedings of the International Conference on Fast Breeder Reactor Fuel Performance*, Monterey, California, 1979, Am. Nucl. Soc. Pub. No. ISBN: 0-89448-105-3, pp. 935–953.
79. C. M. Walter, G. H. Golden, and N. J. Olsen, U–Pu–Zr Metal Alloy: A Potential Fuel for LMFBRs, Argonne National Laboratory Report, ANL-76-28 (November 1975).
80. G. J. Fischer and R. J. Cerbone, The Fast Mixed Spectrum Reactor: Interim Report, Initial Feasibility Study, BNL-50976, Brookhaven National Laboratory (1979).

4

Fission Reactor Pressure Boundary Materials

The nuclear pressure boundary can be considered as encompassing those components of the primary and secondary systems of the power plant which contain pressurized coolant, including the reactor vessel, steam generators, steam and primary coolant piping, valves, pumps, nozzles, and assorted small components (refer to Figs. 1-1 to 1-3; Chapter 2). Steam generators are treated separately in Chapter 6.

The overriding requirement of the pressure boundary is maintenance of integrity over the design lifetime. Although there are secondary containment systems (outside the scope of this book) designed to accommodate a rupture, the primary system integrity is necessary both for proper functioning of the plant and provision of additional protection for the safety of the public. Accordingly, the component failure probabilities that are considered acceptable are very low. For instance, for the LWR, where the pressure vessel is considered to be the critical pressure boundary component, the U.S. Advisory Committee on Reactor Safeguards† has specified an acceptable failure rate of 10^{-6} failures per vessel year.[1] Similarly, for the LMFBR, where the primary piping rather than the vessel is the limiting component, an unreliability goal of 10^{-8} pipe ruptures per reactor year has been established.[2]

The emphasis in this chapter, then, will be on data and analyses developed with the primary objective of demonstrating that, under all anticipated and hypothetical (accident) service conditions, the probability of developing a through wall crack in a critical component is acceptably low; and, even if a through wall crack does form, a small leak results,

† The ACRS advises the Nuclear Regulatory Commission.

thus allowing sufficient time to take preventive action before a major failure occurs (i.e., the so-called "leak-before-break" concept). The components of interest, which will be discussed in the following sections, include the vessel and piping of LWRs and the piping in the LMFBR.

The combination of a very low failure probability and leak-before-break situation, when fully demonstrated, will not only be beneficial in terms of ensuring overall plant safety, but can lead to substantial increases in plant availability that would accrue from reductions in downtime for inspection and repair or replacement. The experience to date with LWRs illustrates such potential gains. Although vessel inspections have contributed little to total plant outage times, inspection and repairs of pipe and nozzle cracks in BWRs have had a significant impact that ranged from a high of 4.7% capacity loss in 1975 to 0.5% capacity loss in 1978.[3]

4.1. Design and Materials of Construction

The pressure–temperature–neutron flux conditions experienced by fission reactors is accommodated in a simple steel (Table 1-1) vessel geometry constructed from welded plate and forgings, which is expected to be nonreplaceable (i.e., a 30- to 40-year lifetime). The selection of piping and valves is equally straightforward. Either cast or wrought austenitic stainless steel or austenitic clad ferritic steel piping, 100 mm to 1100 mm in diameter, and cast or forged valve and pump bodies are considered standard. A nuclear plant involves miles of piping, constructed into complex configurations involving thousands of welds and numerous valves and pumps.

An interesting trend emerges from comparison of the basic design parameters of the reactor vessels. The LWR, for instance, has largely followed the chemical pressure vessel philosophy with thick-walled low-alloy steel pressure vessels, clad on the inside with austenitic stainless steel for corrosion resistance. The PWR operates at approximately twice the pressure of the BWR (15.2 versus 7.6 MPa), and the wall thickness is increased appropriately. On the other hand, the operating conditions of the LMFBR vessel are generally less severe (coolant pressures are a factor of 10 or more lower, neutron fluences are lower, while temperatures are over 200 K higher) and, although creep must be a consideration in design, a thinner-walled, austenitic stainless steel "tank" is considered adequate.

These differences in design philosophy should become evident in the discussions following.

4.1.1. LWR Vessel and Piping

Figure 4-1a shows a section through a typical BWR pressure vessel and the internal structure. A reactor vessel for a 1000-MWe system is about 6 m in diameter and 22 m high; the wall is approximately 165 mm thick. Figure 4-1b illustrates a typical PWR vessel and internals. A 1000-MWe class vessel is about 4.25 m in diameter and 12.2 m high, and the wall is about 230 mm thick.

The reactor vessel is the most critical component in the LWR pressure boundary from the viewpoint of safety. Vessel integrity is of particular concern in PWRs because of the higher coolant pressures and higher neutron doses. This concern with safety is exemplified by the attention devoted to design, materials selection, fabrication, inspection, and overall quality assurance in the ASME Code Section III and to continuing in-service inspection covered in ASME Section XI.

Typical reactor pressure vessel steels for both the BWR and the PWR are American Society for Testing and Materials (ASTM) A508 Class 2 forgings and ASTM A533 Grade B Class 1 plate. A508 Class 2 is a manganese–molybdenum steel modified with nickel and containing some chromium; A533-B Class 1 is the plate counterpart, similar in composition but with reduced chromium. Both steels are first austenitized at 1143 to 1173 K, followed by a quench and, finally, a temper treatment usually at 923 to 948 K for a minimum of 1 h per 25 mm of thickness.

The vessel is fabricated by welding. For corrosion resistance the vessel is clad (on the I.D.) with Type 308 stainless steel weld overlay, normally 5 mm thick. In the case of the BWR, only the portions exposed to liquid water are clad. After completion of these steps the entire vessel is given a final postweld heat treatment (PWHT) for 1 h per 25 mm of thickness. Usually the PWHT temperature is between 868 and 893 K. During fabrication, vessel components are also given short-time intermediate PWHT, and total PWHT time for some sections may be as much as 40 h.

Most of the primary coolant piping in recirculating systems of operating BWR plants is made of Type 304 stainless steel. Also present in smaller quantities are Types 316, 304L, 316L, and 347 stainless steels and Inconel 600 piping. More recently, the use of Type 316L stainless steel is becoming prevalent in small-diameter (100 to 152 mm) piping. Ferritic steels are commonly used for steam lines, feedwater lines, etc. Product forms include both seamless and welded pipe, and small quantities of cast pipe. The average carbon levels of Types 304 and 316 stainless steel used in piping are 0.06% and 0.05%, respectively (compared to 0.08% maximum allowable).

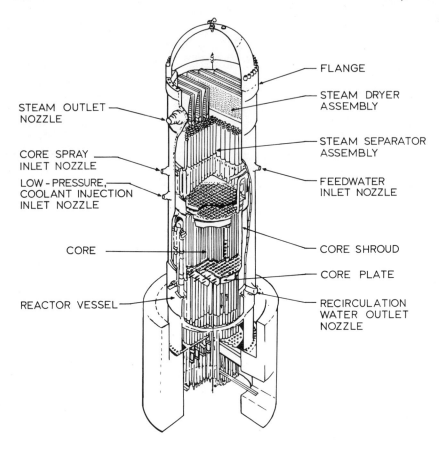

Fig. 4-1a. Drawing of a boiling water reactor illustrating details of the reactor pressure vessel and internals.

Due to problems with the austenitic steels (see Section 4.3.1.4) the use of carbon steel piping and components has increased significantly in the evolution of the BWR design. In the modern BWR-6, for instance, most of the piping systems, except for the main recirculation water lines and standby liquid control, are made of unclad SA 106 Grade B and SA 333 Grade 6 carbon steels.

The common PWR piping materials include Types 304 and 316 stainless steels. In particular, small-diameter pipe in the primary system and much of the pipe in other systems of PWRs are made of these alloys. Both the centrifugally cast form (CF8A, CF8M) and forged pipe of these alloys have been commonly used for large primary piping in Westinghouse

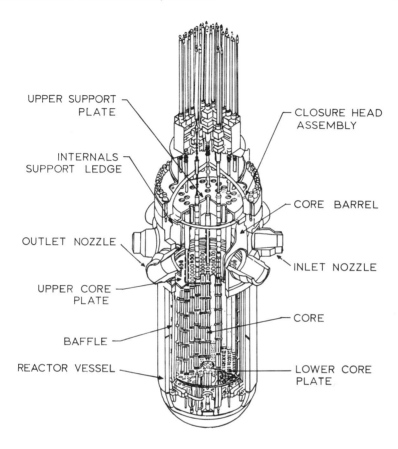

Fig. 4-1b. Drawing of a pressurized water reactor illustrating details of the reactor pressure vessel and internals.

plants. Other PWR suppliers use ferritic steel pipe (ASTM 516 Grade 70) clad with Type 308L austenitic stainless weld overlay.

The piping in a LWR is constructed into complex configurations, involving miles of pipe and thousands of welds. For example, there are approximately 6000 welds in stainless steel piping in a BWR; and the recirculation piping system alone contains 30 welds on 100-mm-diameter pipe, 60 welds on 246-mm-diameter pipe, 9 welds on 559-mm-diameter pipe, and 33 welds on 711-mm-diameter pipe, for a total of 132 welds. These welds (and similar welds in the PWR) are generally so-called "girth-butt" welds between similar-size pipes, or between pipes and valves or other fittings (Fig. 4-2).

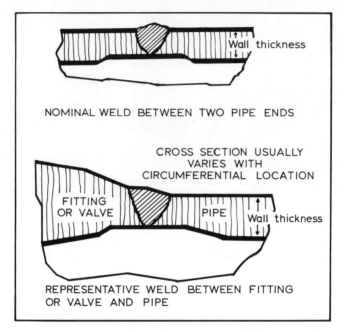

Fig. 4-2. Piping weld configuration (girth butt weld is the example).

4.1.2. LMFBR Vessel and Piping

The LMFBR vessel, shown in Fig. 4-3 (this example is the U.S. Clinch River Breeder Reactor, a 375-MWe prototype), is large in diameter (~6 m) and thin-walled (~75 mm thick) since internal pressures are very low, of the order of 1 MPa. It is constructed entirely of Types 304 or 316 stainless steel.

Type 304 stainless steel is selected for piping, but, in some designs, Type 316 stainless steel is used in the hot leg regions of the primary and secondary systems. A 0.06 to 0.10% carbon range is specified for long-term compatibility in sodium and adequate lifetime mechanical properties. Weld or filler metals include Type 308 (for welding Type 304) and Type 316 or 16–8–2 (i.e., 16 wt.% Cr–8 wt.% Ni–2 wt.% Mo) for Type 316 stainless steel.

Two pipe forms are in use: seamless pipe (ASME SA-376 requirements) that is solution treated following extrusion and cold reduction processes, and welded pipe either using gas tungsten arc (GTA) or submerged arc (SA) techniques. As for the LWR, a lengthy complex piping system is necessary, involving numerous pipes and fittings. Welds to fittings will likely be of the girth type made by GTA (Fig. 4-2). Shop welds

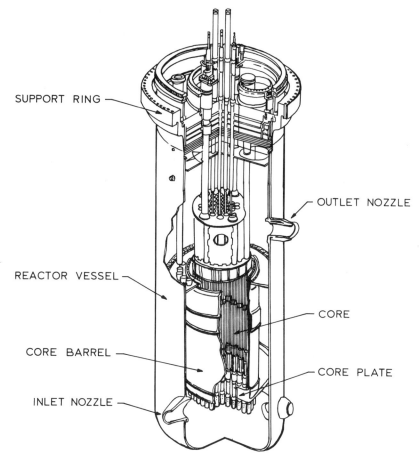

Fig. 4-3. Sectional view of pressure vessel internals and core of a prototype 375-MWe liquid metal fast breeder reactor.

will be annealed following fabrication, but field girth welds will probably not be.

4.2. Developments in Fracture Mechanics

While recognizing the importance of precluding extensive plastic deformation that might lead to distortion or buckling of components in the pressure boundary, particularly those thin-walled members operating at high temperatures (e.g., in LMFBR and CTR), it is the flaw–structure interaction process that is most likely to determine the integrity of both thin- and thick-walled components. Accordingly, flaw–structure interaction is the focus of this section.

Flaw–structure interaction analysis is a central feature of the Structural Integrity (SI) plan,[4] described in detail earlier in Chapter 1 (e.g., refer to Fig. 1-8). Assurance of pressure boundary integrity rests on the accuracy and reliability of the overall SI analysis. The reader will recall that the analysis must include all elements of a postulated failure, such as loadings (either real, caused by coolant pressure and thermal stresses, or specified hypothetical ones that might be experienced in accident transients), flaws present (real or specified), mechanical properties, and, of course, the flaw–structure interaction process.

Current approved methods, embodied in the ASME Section III and XI Codes, are based on linear elastic fracture mechanics (LEFM) and assume that the structure is elastic to failure, and, further, that failure occurs in a brittle and catastrophic manner. Although LEFM has proven very effective in describing subcritical flaw growth via fatigue and/or stress corrosion cracking, it can be highly conservative in its treatment of the growth of critical flaws for the steels of interest here. Steels used to fabricate nuclear pressure vessels and piping are ductile and will deform significantly before failure at operating temperatures, even in the presence of large flaws. Consequently, any flaws present in these materials may be blunted by plastic flow, and, if propagated, will do so in a ductile manner. Therefore, a significant underestimate of the failure conditions and the associated critical flaw size can result from the assumption of an elastic and brittle flaw–structure interaction. This is particularly the case in components operating at elevated temperatures which are covered by ASME Code Case 1592.

The following subsections will focus on recent developments to improve analytical procedures for the prediction of flaw growth in service, in both the elastic and elastic-plastic regimes. They will expand on the introductory description of structural design methodology contained in Section 1.3 of Chapter 1.

4.2.1. Elastic Fracture

Inasmuch as the only ASME Code-approved fracture mechanics analysis is LEFM, considerable effort is going into the refinement of this methodology. Recent developments in LEFM analysis of flaw–structure interaction can perhaps be classified into three broad groups of activities. The first is associated with further refinements in the methodology, particularly inexpensive two- and three-dimensional computer codes and the application of probabilities;[4, 5] the second, which will be discussed later in Section 4.3.1, is the development of statistically based K and da/dN vs. ΔK curves for use in the analysis;[4] and the third area is increased application of LEFM to real structural components.[4, 6–8]

Two recent examples of application of LEFM to real components, one LWR and one LMFBR, have been selected for more detailed presentation here to illustrate the types of problems being tackled and the different approaches being used.

BWR Nozzle Analysis. The feedwater nozzle in a BWR pressure vessel (refer to Fig. 4-1a) is a particularly important region from the standpoint of the in-service inspection requirements imposed by the ASME Section XI Code. The complex geometry makes in-service inspection difficult, and high stresses due to the stress concentration effect of the nozzle result in relatively small acceptable flaw sizes.

Several incidences of cracking in these nozzles have been reported[4, 5a] (see Section 4.3.1.3 for details). Currently, Section XI requires an evaluation (i.e., an SI analysis) of any flaw indication greater in depth than 2.5% of the wall thickness (i.e., ~4 mm) and stipulates that the predicted end-of-life flaw size shall be less than one-tenth of the critical flaw size for normal operation, otherwise flaw removal and repair procedures must be carried out.

In these particular instances, cracking is considered to arise from thermal fatigue. Two types of thermal loading are present: a high-frequency, very localized fatigue initiation phase, from coolant flow perturbations, and a low-cycle fatigue propagation phase, from normal startup and shutdown transient stresses. For the SI analysis end-of-life flaw sizes were computed on the basis of Eq. (1-8) and a life prediction analysis.

Integral to this type of analysis is the calculation of three-dimensional stress intensity factors. For a complex geometry such as a nozzle, which is subjected to steep stress and temperature gradients, this can prove to be a very expensive and time-consuming process using general-purpose finite element techniques, in some instances prohibitively so. However, recent development of boundary-integral-generated influence functions (BIGIF) for a series of geometries pertinent to nuclear systems facilitates the calculation of stress intensities in such components. The BIGIF method saves in computer costs and results in higher accuracy.[4, 7]

Besuner *et al.*[7] have reported on the use of BIGIF to compute stress intensities for various crack depths at various locations in the BWR nozzle, subjected to combined pressure and thermal loadings. The approach exploits the elastic superposition principle by breaking down a stress intensity solution for the arbitrary (and complex) crack situation into a solution of (a) the uncracked stress field and (b) a solution of the appropriate geometry with the crack, but with a much simplified loading that

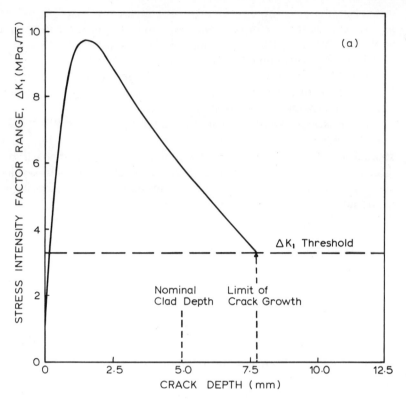

Fig. 4-4a. Computed stress intensity factor range vs. crack depth for short-range cyclic "initiation" stresses (Ref. 7).

is generally a unit pressure on the crack face. Thus, the complexity of the stress field calculation is separated from the complexity of the stress intensity calculation. In the approach of Besuner *et al.*, a general-purpose finite-element temperature and stress analysis method is used for the former calculation, while the new three-dimensional boundary-integral-generated influence function method is used for the latter calculation. The accuracy of this combination is limited essentially by the accuracy of the stress analysis.

Some results of the application of the influence function methodology to the nozzle problem are presented in Fig. 4-4. As noted earlier, the cracking is caused by a two-step mechanism. The first is attributable to rapidly cycling, high thermal stresses of very short range. The steep thermal gradients cause the stress intensity profile depicted in Fig. 4-4a. These stresses can initiate cracks, but, according to Fig. 4-4a, the extent of growth is limited to about 7.5 mm. However, normal startup (and shut-

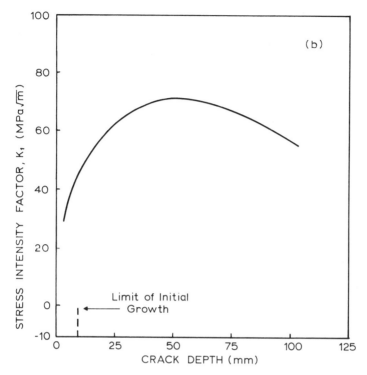

Fig. 4-4b. Computed stress intensity factor vs. crack depth for long-range thermal transient "growth" stresses (Ref. 7).

down) stresses can yield stress intensities (Fig. 4-4b) sufficiently high to propagate this small flaw through the nozzle wall by fatigue. Clearly, therefore, it is very important to have an accurate prediction of the stress intensity–crack depth relationship in order to determine the crack growth rates in these components and hence the time when repair procedures might be necessary.

LMFBR Pipe Analysis. Unstable fracture, either brittle or creep rupture, is highly unlikely under the expected operating conditions of the LMFBR vessel and piping. A more likely mode of failure is from the extension of subcritical flaws by fatigue (or creep-fatigue), and, accordingly, much effort has been expended to ensure that fracture mechanics techniques are available to treat elevated-temperature flaw growth behavior.

As noted in Chapter 1, K has also proven useful in characterizing fatigue crack growth at elevated temperatures presumably because the plastic zone size remains small under cyclic loading conditions. James[8] has applied this methodology to the LMFBR piping integrity question and has concluded that LEFM is adequately conservative for such situations. He assumed a hypothetical (worst case) situation that involved fatigue growth of preexisting large flaws (assumed to be the maximum that would remain undetected by NDE) in austenitic stainless steel elbows, which are the most highly stressed areas in the piping system. One example was a surface flaw on the inside surface of a cold-leg elbow operating at a temperature of about 700 K. This example was chosen to illustrate the rather inert behavior of the liquid-sodium environment surrounding such a flaw (see Section 4.3.2). The other example was a flaw located on the outside surface of a hot-leg elbow operating at a temperature of approximately 811 K. Since the environment surrounding this second flaw was impure nitrogen gas, this example illustrated the environmental and time-dependent considerations that must be made in such a case. All normal, upset, and emergency loadings over the reactor lifetime were accounted for in the analyses, and in the case of the cold-leg elbow, unlikely "abnormal" loadings were also considered.

In the cold-leg elbow, fatigue cycling was considered between two loading conditions, each involving thermal-expansion bending stresses, but, in addition, one having a severe thermal transient superimposed. Although large tensile stresses, approximately 207 MPa, were calculated for the thermal transient, and a service duty of 672 cycles was assumed, the estimated crack extension based on data in Figs. 4-27 to 4-30 (Section 4.3.2.2), is of the order of 0.2 μm, which is minimal. The abnormal loading increased crack growth to 0.25 mm, which is still not too significant. Crack growth in the hot-leg elbow, resulting from 27 different loading events, is approximately 1.72 mm, which is also considered acceptably low. In addition, sufficient membrane stress is present in both cases so that crack extension would proceed in the wall-thickness direction as well as along the surface. This would ensure that even if the magnitude of crack growth exceeded the predictions, wall penetration would occur (and hence produce leakage) long before critical crack lengths were approached. Therefore, as James[8] concluded, the admittedly conservative LEFM analysis serves the purpose of demonstrating, in this instance at least, the stability of large flaws in LMFBR pressure boundary components.

From the foregoing examples it is evident that LEFM can cope with small departures from linear elastic behavior in the immediate vicinity of the crack tip, such as in fatigue up to temperatures of ~0.5 T_m. However, as the degree of plasticity increases, the accuracy of K as a single-parameter description of the stress distribution ahead of the crack decreases. The presence of a plastic zone at the tip of a crack allows the two faces to move apart without any increase in crack length, and once these displacements become sufficient the elastic theory can no longer predict the local stresses and strains, and K_{Ic}, for example, can become very conservative as a failure criterion. The inherent toughness and ductility of the nuclear pressure boundary materials means that large strains must accumulate in the presence of a crack before fracture initiation occurs. Therefore, in order to predict realistically the conditions for failure in these components, a methodology that handles elastic-plastic and fully plastic fracture is necessary. This is of current importance in the LWR inasmuch as LEFM calculations of subcritical crack growth in pressure boundary components indicate attainment of critical crack sizes for unstable fast fracture (i.e., $K > K_{Ic}$). This is not as big an issue in the LMFBR because the admittedly conservative LEFM analyses show that flaws do not grow anywhere near to a critical depth under the expected duty cycle, and therefore at this point a more rigorous treatment is apparently not justified.

4.2.2. Elastic-Plastic and Fully Plastic Fracture

In the LWR industry, the justification for development of general yielding fracture mechanics (GYFM), which includes elastic-plastic and plastic fracture, was originally derived from concerns with pressure vessel integrity,[9, 11] but with the observations of cracks in both nozzles and piping (refer also to Section 4.3.1.4) momentum has now shifted in the direction of these components of the pressure boundary. In the LMFBR industry, interest in GYFM (and the related creep fracture mechanics) is growing, but reliance is still placed on LEFM analyses for design purposes. The basic principles involved in plastic fracture mechanics developed for the LWR are of general use to flaw–structure interaction analysis of ductile materials and should be equally applicable to the LMFBR steels.

The first experiments that clearly demonstrated the conservatism inherent in LEFM when applied to LWR vessels operating at or near actual service conditions were the large-scale pressure vessel tests performed in the heavy section steel technology (HSST) program.[10, 11] The results showed that, even with sharp flaws, 30% through the wall, vessels yielded significantly, and stable ductile crack growth preceded failure, at over three times the Code design pressure.[10] Table 4-1 summarizes two vessel tests, one with a flawed nozzle and the other with a flaw in the cylindrical section, along with the LEFM predictions for failure pressure and failure strain. As the table shows, LEFM is conservative in the prediction of failure pressure and greatly conservative in the prediction of failure strains.

As explained in Chapter 1, GYFM therefore has its origin in the need to account for large-scale yielding that allows significant amounts of stable crack growth to occur under rising load (or deflection) prior to general

Table 4-1. Upper-Shelf Intermediate Vessel Test Results (4)

Parameter	Vessel No. 5	Vessel No. 6
Flaw type	Nozzle	Cylinder
Crack depth (mm)	30.5	47.5
Failure pressure		
Actual (MPa)	183.4	220
LEFM prediction (MPa)	131.7	189.6
Error in prediction (%)	−17.70	−16.00
Failure strain		
Actual	0.25	2.00
LEFM prediction (%)	0.10	0.48
Error in prediction (%)	−150.00	−317.00

failure (refer to Figs. 1-9 and 1-10). Experimental studies show that the progress of ductile fracture from an existing sharp-tipped flaw may be separated into four regimes: the blunting of the initial sharp crack tip, crack growth initiation, stable crack growth, and unstable crack propagation.[12, 12a] For materials on the upper shelf (refer to Fig. 1-12) there is significant crack tip blunting and substantial stable crack growth. More importantly, the "resistance" of the material to further crack extension increases with increasing crack growth due to the work hardening and strain redistribution of the plastically deformed material in the vicinity of the crack tip. In this case, the assessment of the margin of safety of flawed structures based on the onset of crack extension is conservative, since the "toughness" of the material may increase significantly with relatively small crack extension.

It is important to note that stable crack growth can occur either under small-scale yielding conditions when the plastic zone size is small compared to crack length (Fig. 1-10b) or under large-scale yielding conditions when the plastic zone extends across the remaining ligament (Fig. 1-10c), but it is the latter condition that generally characterizes the behavior of pressure vessel steels on the upper shelf.

In contrast to LEFM, which requires only a criterion for crack initiation, GYFM requires criteria for both crack initiation and stable crack extension. It is also necessary to be able to predict the conditions for structural instability which is the limit of stable extension. The true margin of safety against rupture can only be defined when one can predict accurately the conditions for (1) stable crack extension throughout the wall of the components (leakage) or (2) structural instability after crack initiation (burst). As noted in Chapter 1, most effort on GYFM has gone into the selection of the global, geometry-independent parameters that will characterize crack initiation and extension. From the computation and measurement point of view, the viable candidates are the J integral and the crack opening displacement COD (δ) and angle COA (α).[4, 9, 12, 13]

Shih et al.,[12] on the basis of combined analytical and experimental investigations on A533B steels, support the characterization of the onset of flat fracture by either of the parameters J_{1c} or δ_{1c}, and stable flat crack growth by either the J-resistance or the COD-resistance curves under large-scale yielding (see Fig. 4-8, Section 4.3.1.1).

Paris and co-workers[13] had earlier proposed to characterize fracture toughness during stable growth by the dimensionless parameter T_j, the tearing modulus, where

$$T_j = \frac{dJ}{da} \cdot \frac{E}{\sigma_0^2} \qquad (4\text{-}1)$$

Shih et al.,[12] following on from the work of Paris[13] and Rice,[14] suggested COA as a potential stable-growth parameter, i.e., $\alpha = d\delta_t/da$, which, like J, can be restated in terms of a tearing modulus T_δ; thus

$$T_\delta = \frac{d\delta_t}{da} \cdot \frac{E}{\sigma_0^2} \qquad (4\text{-}2)$$

The details of the analytical developments can be found in Ref. 12 and in the papers by Hutchinson and Paris,[15] Paris et al.,[13] and Rice.[14] From this work, it appears that the material toughness associated with initiation can be characterized by the critical parameters J_{1c} or δ_{1c}, while the material toughness associated with crack growth and instability can be characterized by the dimensionless parameters T_j or T_δ. This two-parameter characterization of fracture toughness properties, namely, J_1 and T_j, or δ_1 and T_δ, is analogous to the characterization of deformation properties of a material by the yield stress and the strain-hardening exponent, and thus lends itself to the finite-element analytical approaches developed for elastic-plastic stress analysis. Current research is being directed at consolidating this position, in particular the range of validity of the COD-based tearing instability parameter, since the COD may characterize crack growth over a greater distance than the J integral which is limited to 6% of the remaining ligament.[12]

Presently, the calculations necessary to directly apply any of the ductile fracture parameters require large computer programs, skilled operators, and relatively large expenditures of time and money. These requirements would undoubtedly restrict the application of GYFM to a few engineering structures. To facilitate the application of GYFM, therefore, a ductile fracture handbook approach is being developed, which potentially reduces a plastic fracture analysis to simple graphical or semianalytic procedures yielding results with acceptable accuracy.[9] The basic tenet of this research, as reported by Marston,[9] is that any elastic-plastic fracture condition can be estimated [within engineering accuracy ($\pm 5\%$)] by the following scheme:

$$F^{eP} = \alpha_1 F^e + \alpha_2 F^P \qquad (4\text{-}3)$$

where F^e, F^P, and F^{eP} are fracture conditions under elastic, plastic, and elastic-plastic situations, respectively, and α_1 and α_2 are weighting parameters. This approach is sketched in Fig. 4-5.

To complete the structural integrity analysis, it is necessary to predict the end of stable crack growth, i.e., structural instability. This is particulary important to the establishment of an unambiguous "leak-before-break" criterion. The tearing modulus approach due to Paris[13] is one way; another is similar to the traditional LEFM R curve and is based on either the J-resistance or COD-resistance curves.[9] Figure 4-6 illustrates the expected usage of the latter approach. Experimental data in the form of a J-resistance curve are shown as a solid line in the plot of J versus crack length, with initial crack length a_0. For a given structure, the estimates of J^{eP} as a function of applied load (σ^∞/σ_0) and crack length a are plotted as dashed lines. The point A, at which the cracks meet tangentially, is a point of instability under load-controlled conditions and shows when the structure with the given load and geometry will exhibit unstable crack growth. A COD-resistance curve can also be used.

The ductile fracture analysis has been applied to the aforementioned BWR nozzle problem, with the intent of conservatively showing that the worst result of an undetected large nozzle flaw (even under accident situations) would be a stable leak-before-break condition.[16, 16a] Experiments and analysis showed that structural stability exists for even very large flaws (e.g., 60 to 70% of the way through the nozzle) and pressures well in excess of anticipated accident pressure (e.g., up to 150% of normal). The "safety issue" associated with nozzle cracks is, therefore, really a nonproblem. Another concern of the same type is to show the conditions for leak-before-break in flawed stainless steel piping (see Section 4.3.1.4a).

4.3. Material Characteristics

As illustrated in Fig. 1-8 in Chapter 1, SI analyses require knowledge of a number of materials properties including yield strength and plastic flow, fracture toughness, fatigue resistance, and stress corrosion cracking susceptibility. Two aspects of these properties have received most attention in recent years. The first is the treatment of the change in properties

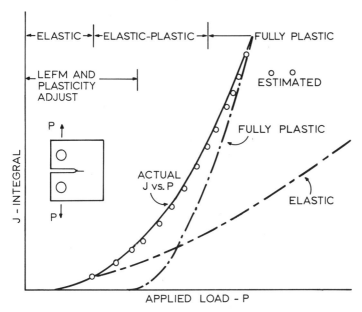

Fig. 4-5. Schematic of the elastic-plastic fracture estimation scneme (Ref. 9).

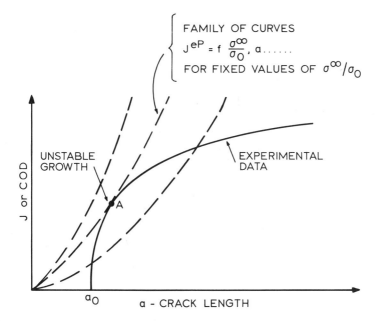

Fig. 4-6. The use of a *J*-resistance or COD resistance curve for predicting crack instability (Ref. 9).

during the life of the structure (due, for example, to the effects of irradiation or to microstructural changes brought about by long-term exposure at the service temperature) and how they are affected by changes in service conditions such as the temperature and chemistry of the service environment. The second is the treatment of variability in the properties for a given set of operating conditions due to heat-to-heat variations in material composition and processing history, coupled with data scatter associated with uncertainties in the test procedures used to measure the properties of interest.

The first problem has necessitated extensive testing as a function of all the variables considered likely to have an influence on the property of interest. Unfortunately, it is virtually impossible to study either the synergistic interactions between all variables or the complete range of all variables of interest. Shortcuts have been made, for example, either by developing empirical correlations for extrapolative purposes or by simulations of the environment.

The second problem, that of reducing the materials property distributions to a form suitable for SI analysis and structural design, has in the past been handled primarily through the use of lower-bound or upper-bound curves (e.g., Fig. 1-13). However, experience with the use of bounding curves has revealed cases of undue conservatism which have caused problems (particularly in LWRs where more experience exists). Therefore, there is a trend in the industry to move to the statistical treatment of a data set, in which the entire data set is described by a suitable distribution function and then statistical procedures are used to develop design curves that have an acceptably small probability of being transgressed.

The literature abounds with data on properties, which pursue one or both of these two directions. In the interest of conciseness, therefore, the following discussion will be limited to "properties of current interest," either because of an actual problem area (in LWRs) or a recognized potential problem area (in LMFBRs).

4.3.1. LWR Materials

Because of the ductile-brittle nature of ferritic vessel steels, researchers have been preoccupied with their fracture toughness, and, in particular, the potential for both a shift in ductile-to-brittle transition temperature and a reduction in upper-shelf fracture toughness resulting from irradiation (refer to Chapter 1, Sections 1.3.4.2 and 1.3.4.5). The situation is more serious for the PWR vessel since it operates under greater pressures, the section thicknesses are greater, and it is subjected to higher irradiation dose than the BWR vessel by a factor of 10 or so.

Another concern is the mechanism(s) by which subcritical flaws grow in service. As a rule, the Codes stipulate that the calculated end-of-life flaw size must not exceed some fraction (e.g., one-tenth) of the critical flaw size that is determined from the fracture toughness value at that time. This is one example of safety factors employed in design codes to account for various uncertainties in properties, analyses, etc. Corrosion fatigue appears to be the primary subcritical flaw growth mechanism in the vessel ferritic materials of PWRs and BWRs, while stress corrosion cracking is most prevalent in the austenitic parts of BWR vessel nozzles (e.g., safe ends) and in BWR austenitic steel piping.

The following subsections will review the state-of-the-art with respect to fracture toughness, fatigue crack growth, and stress corrosion cracking in these pressure boundary components.

4.3.1.1. Fracture Toughness

Fracture toughness, or more specifically the plain strain fracture toughness, K_{Ic}, is a material property whose magnitude is dependent on loading rate and temperature, and, most probably, on irradiation level. This latter effect, irradiation embrittlement, is a very important current issue of concern and will be treated separately in the following subsection.

In Chapter 1, Section 1.3.4.2, three fracture toughness parameters, K_{Ic}, K_{Id}, and K_{Ia}, were defined for different loading conditions and compared in Fig. 1-13. Also, Fig. 1-12 illustrated the ductile-to-brittle fracture transition that is characteristic of ferritic reactor vessel steels. The example shown there is a Charpy impact transition curve for a 150-mm-thick plate of A533B steel where the transition temperature is close to room temperature.[17] Vessels normally operate in the high toughness regime, well onto the upper shelf.

Accurate knowledge of the plain strain fracture toughness (K_{Ic}) is one of the key requirements in the ASME Section XI flaw evaluation analysis.[18] However, as was noted in Chapter 1, the direct measurement of K_{Ic} is not a viable method for developing the required toughness-temperature inputs to the LEFM analysis inasmuch as prohibitively large specimens are required to obtain valid measurements in the upper shelf, high toughness, regime.

There are methods available to conservatively estimate K_{Ic} from smaller fracture toughness specimens, some of which have been developed with the elastic-plastic fracture methodologies. For instance, measured values of the elastic-plastic initiation fracture toughness parameter J_{Ic} can be estimated from K_{Ic} through the relationship

$$\frac{EJ_{Ic}}{(1-\nu)^2} = K^2_{Ic} \qquad (4\text{-}4)$$

where E is Young's modulus, and v is Poisson's ratio. Direct experimental evaluation of the K_{Ic} vs. J_{Ic} relationship is in progress,[4] but it is not clear whether K_{Ic} can always be estimated from J_{Ic} measurements, especially in the lower transition region.

One method often employed to obtain conservative fracture toughness values is to measure K in dynamic tests, i.e., tests at high strain rate, to yield K_{Id}. Because of the increased dynamic yield stress, valid K_{Id} values up to 220 MPa$\sqrt{m}$ have been obtained on 203-mm-thick specimens, using loading rates nearly one million times higher than those employed to determine K_{Ic}.[11]

Another related toughness quantity is the value of the stress intensity at which a running crack no longer propagates, the so-called crack arrest toughness K_{Ia}. Arrest fracture toughness becomes important in postulated accident situations when crack initiation is predicted, and the resulting crack extension then determines the safety of the structure. Crack arrest is assumed to occur when the stress intensity factor at the propagating crack falls below the crack arrest fracture toughness. Extensive crack arrest testing has been undertaken recently, with the result that a reference crack arrest curve for actual vessel steels now exists, and the applicability of the material property, crack arrest toughness, to the analysis of propagating cracks in pressure vessel structures is justified.[4]

Overall it appears that the most practical method for obtaining K_I is to estimate it from Charpy impact (CVN) data since Charpy specimens are already in the reactor vessel irradiation surveillance programs to assess the radiation-induced shift in the ductile brittle transition temperature (refer to Fig. 1-21 earlier and Section 4.3.1.2 below).[19] There are a number of empirical Charpy-toughness correlations in the literature.[20] In addition, an analytically developed correlation has recently been published.[21] Some of the correlations are listed in Table 4-2, adapted from Stahlkopf and Marston.[20]

One of the obvious drawbacks of the straight CVN–K_I correlations is that the Charpy energy versus temperature relationship is assumed to be identical to the fracture toughness versus temperature relationship. Recently, Oldfield[22] has developed a new statistical approach to the treatment of fracture toughness data as a function of temperature. Rather than the standard single-parameter K_{IR} curve, drawn to the lower bound of K_{Ic}, K_{Id}, and K_{Ia} data, which was discussed in Chapter 1, they propose a more physically realistic four-parameter function (hypobolic tangent) to describe K_I, which can be used for either K_{Ic}, K_{Id}, or K_{Ia}, depending on the circumstances.

Thus

$$K_I = A' + B' \tanh \left(T - \frac{T_0'}{C'} \right) \qquad (4\text{-}5)$$

Table 4-2. Correlations between CV Energy and Fracture Toughness[a]

Correlations	Comments
Transition-region CVN correlations	
Barsom–Rolfe	$\sigma_y = 269\text{–}1696$ MPa
$K_{Ic}^2/E = 2(CVN)^{3/2}$	Static tests
$K_{Ic}^2/E = 5(PCVN)$	Precracked Charpy tests
Corten–Sailors	CVN = 7–70 J
$K_{Ic} = 15.5(CVN)^{1/2}$ or $K_{Ic}^2/E = 8(CVN)$	Static tests
$K_{Id} = 15.873(CVN)^{3/8}$	Dynamic (high strain rate) tests
Marandet–Sanz	Static tests
$K_{Ic} = 20(CVN)^{1/2}$	$T_{K_{Ic}}$ at $K_{Ic} = 100$ MPa$\sqrt{m}$
$T_{K_{Ic}} = 16.2 + 1.37 T_{28}$	T_{28} at CVN = 28 J
Upper-shelf CVN correlations	
Rolfe–Novak	$\sigma_y = 690\text{–}1696$ MPa
$[K_{Ic}/E]^2 = 5[CVN/\sigma_y]$	Static tests
Wullaert–Server	$\sigma_y = 345\text{–}483$ MPa
$K_{Jd} = 20(DVN)^{1/2}$	Dynamic J-integral initiation
$K_{Jc} = 2.1(\sigma_y\, CVN)^{1/2}$	All loading rates with appropriate σ_y
or	
$[K_{Jc}/\sigma_y]^2 = 4.41[CVN/\sigma_y]$	
Lawrence Livermore Laboratory	
$[K_{Jc}/E]^2 = CVN(9.66 + 0.04\sigma_y)$	$K_{Jc} = [EJ_{Ic}]^{1/2}$
	$K_{Jc} = [EJ_{Id}]^{1/2}$

[a] After Ref. 20.

where A', B', T_0', and C' are coefficients. To generate the coefficients, Charpy and fracture toughness data from the same heats are statistically analyzed. Conversion factors are generated within the analysis to transform the coefficients of the Charpy fit to corresponding coefficients for a fracture toughness curve estimate (i.e., Charpy curve-fit coefficients A, B, T_0, C, to fracture toughness coefficients A', B', T_0', C').

The fracture toughness–temperature response of the vessel material can be estimated statistically from the measured Charpy transition curve and the conversion factors developed. Conversion factors are available to develop average toughness estimates, as well as to develop statistical bounds on the estimated fracture toughness. This technique has been applied to data from 50 production heats of pressure vessel steel. Efforts are under way to produce a statistically based reference toughness curve as an alternative to the current ASME Code lower-bound curve. An early version is presented in Fig. 4-7.

Although current analyses are limited to the use of LEFM and K, as was discussed earlier, there are more appropriate ductile fracture toughness parameters emerging from the ductile fracture programs.[9, 12] To reiterate the important results from that work, it is felt that K_{Ic} should

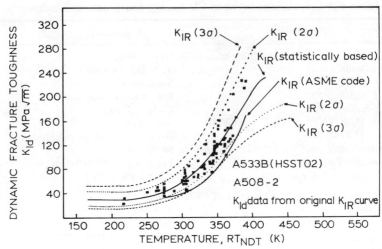

Fig. 4-7. Hyperbolic tangent model fit to dynamic fracture toughness data on A533B and A508-2 steels used in original ASME Section III Code K_{IR} curve (Ref. 4).

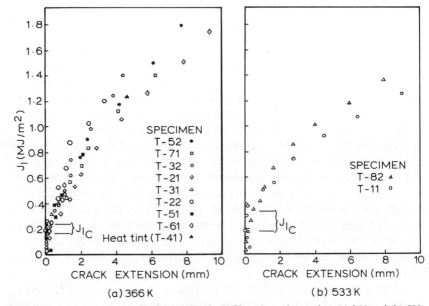

Fig. 4-8. J-resistance curves for SA533 Grade B Class 1 steel tested at (a) 366 and (b) 533 K (Ref. 12).

be replaced by the ductile crack initiation parameters, either the critical J integral (J_{Ic}) or crack opening displacement (δ_{Ic}). If one is interested in predicting structural instability on the upper shelf, J resistance (dJ/da) and crack tip opening angle ($d\delta_t/da$), which is a variant of COD, appear to be the most suitable parameters for characterizing the limits of stable crack extension. All these parameters are properties of the material which can be measured in experiments. Some typical results are shown in Fig. 4-8.[12] The tests are on 100-mm-thick compact tension specimens of A533 Grade B Class 1 nuclear vessel steel side-grooved to a depth of 25%. The plots are J-a, or J-resistance, curves for this material at 366 K (Fig. 3-8a) and 533 K (Fig. 3-8b). Both of these temperatures are on the Charpy "upper shelf," but the results indicate that the fracture resistance, dJ/da, still varies with increasing crack depth; these data are for crack extensions approaching 10 mm. A significant amount of ductile fracture testing has been completed but more will be required to support a Code revision.

4.3.1.2. Radiation Embrittlement

One of the principal concerns with LWR vessel steels has been the shift in ductile-to-brittle transition temperature as a function of fast neutron fluence in the high-flux beltline region. Figure 1-21 in Chapter 1 schematically indicates both a shift in transition temperature (so-called ΔTT or NDT shift) and a lowering of upper-shelf Charpy V-notch energy (so-called ΔUSE) at intermediate and high fluences. However, the actual situation is much more complicated. For example, there will be changes in neutron fluence and in damage as functions of vessel wall thickness, and the final level of damage will depend not only on fluence but on irradiation temperature, the amount and type of trace elements such as copper and phosphorus, and on the microstructure of the material.

An irradiation-induced transition temperature shift and decrease in upper-shelf toughness, whether manifested in CVN energy, K_{Ic}, or K_{Id}, is of significance to the predicted vessel failure probability as influenced by degradation of properties. The present upper-shelf Charpy impact limit for unencumbered LWR operation is 68 J (in Title 10, Code of Federal Regulations, Part 50, Appendices G and H). While the previous section showed that the currently used reactor vessel steels are quite tough and relatively insensitive to brittle failure in the unirradiated state, it is essential that there still exist a substantial safety margin in terms of toughness at the end of life. An example of a possible problem would be where relatively small flaws are undetected during initial inspection because they are below the detection thresholds of radiography or ultrasonics, or because they are oriented in a nonoptimum manner insofar as detection is concerned. These flaws would be far below critical size for the as-built vessel, even if the vessel were highly stressed near the ambient temperature. However, if neutron irradiation caused property degradation, exemplified by a pronounced shift in transition temperature together with a major loss in upper shelf energy, the initially subcritical flaw might become critical in size under some combination of loads. Surveillance programs, both those measuring neutron-induced changes in steels and those devoted to the periodic inspection of the pressure vessel, help to prevent the circumstances leading to such a postulated failure.[20, 23]

The radiation damage effect is more critical in LWRs than in LMFBRs and, in particular, in the PWR vessel, which is subjected to the highest wall fluence. Most early irradiation data were obtained in test reactors to achieve high fluences in relatively short time periods, permitting prediction of neutron damage in commercial reactors near end-of-life.[23] Typical end-of-life fluences for a 40-year operating period are 2 to 5×10^{23} n/m^2 ($E > 1$ MeV) for PWRs, and approximately 1×10^{22} n/m^2 ($E > 1$ MeV) for BWRs, which fluences could be achieved in test reactors in a few months at high fluxes. Thus, most criteria (and theories) for radiation damage effects were developed on the basis of such irradiations.

However, while such radiations supplied a great deal of valuable information, it appears that they were not truly representative of commercial reactors. Recent results from two of the older PWR surveillance programs[20] reveal a radiation embrittlement saturation effect. That is, normal manifestations of embrittlement (change in transition temperature, change in Charpy upper shelf energy, and change in flow properties) are affected only up to some level of exposure and any subsequent exposure does not affect these properties. This is in contrast to high-flux irradiations where the manifestations of embrittlement vary proportionately with fluence and no saturation is observed. The latter data are the sole basis for the current embrittlement trend curves, as contained in the NRC Regulatory Guide 1.99.1.

The low-flux, high-fluence surveillance results obtained to date are from the surveillance programs of the Point Beach Unit 1 and the Connecticut Yankee nuclear power plants, which have seen 7.5 and 10 effective full-power years of operation, respectively.[20] An example of low-flux irradiated materials is shown in Fig. 4-9; this is an SA302 Grade B plate with a copper content of 0.19 wt.% and a phosphorus content of 0.019 wt.%. The Charpy trend curves are shown in Fig. 4-9a and the NDT shift summary is shown in Fig. 4-9b. Also indicated in Fig. 4-9a is the scatter band for the unirradiated Charpy data. All of the irradiated data lie within a similar band as shown. It appears from these results that irradiation may not increase the scatter in the Charpy impact energy. The

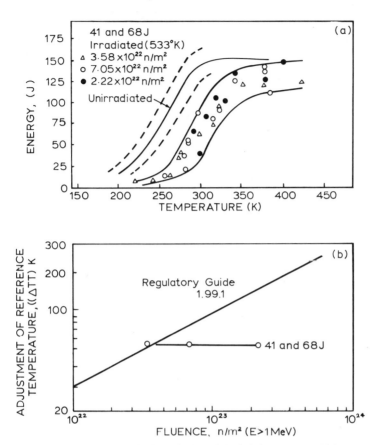

Fig. 4-9. Low-flux irradiation results for Point Beach Unit I plate (0.19 wt.% Cu, 0.019 wt.% P). (a) Charpy data (note unirradiated scatter band relative to irradiated data); (b) NDT shifts (ΔTT); actual shifts are compared with Regulatory Guide 1.99.1 predictions; note apparent saturation at 3×10^{22} nrt or below (Ref. 20).

saturation in NDT shift is clearly indicated in Fig. 4-9b at a fluence level of 3.6×10^{22} n/m² or less. A comparison of these results to those of the weld metal indicates that the level of saturation may be a function of not only chemistry but also microstructure.[20]

The significance of this saturation effect can be demonstrated by evaluating the current embrittlement issue in U.S. PWRs. There are presently several reactor vessels under scrutiny by the U.S. Nuclear Regulatory Commission because their predicted embrittlement levels will transgress the limit of 68 J in the future.[20] The same trend curves also indicate continual degradation of the properties; consequently in the future, several of these reactor vessels may require a thermal anneal treatment to restore properties. If this saturation effect can be demonstrated and proved, most, if not all, of the reactor vessels in question can be vindicated. The probability of ever thermal annealing a reactor vessel can be minimized. In addition to the transgression of the embrittlement limit imposed on Charpy upper-shelf energy, there is an operational problem associated with NDT shift since it determines the pressure–temperature limit curves for heat-up and cool-down of the plant. As Fig. 4-9b shows, the predicted (according to Regulatory Guide 1.99.1) shifts for PWR reactor vessels could exceed 450 K at end-of-life fluence conditions, which would make operation difficult if not impossible. However, the low-flux data indicate that the trend curves can overestimate the NDT shift by as much as 60 to 75% at end-of-life conditions. Consequently, the severe operational restrictions imposed by the large NDT shifts can be reduced if the saturation effect can be proven.

It is obvious from the data presented that there is a significant difference between embrittlement experienced in the low-flux irradiations typical of reactor vessel walls and high-flux irradiations in test reactors. One explanation for this difference is the occurrence of an *in situ* annealing at operating reactor temperatures, normally 560 K. This annealing is the result of defect diffusion and subsequent self-annihilation in the material. The defect diffusion rates in this material at 560 K are relatively low; consequently, the annealing process takes a significant amount of time. With the high-flux irradiations, the time required to reach the desired

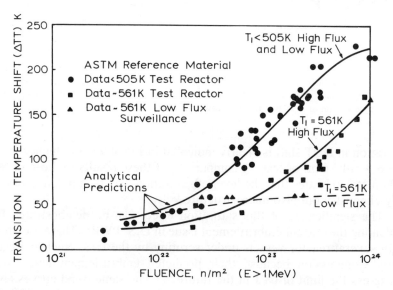

Fig. 4-10. Preliminary predictions of NDT shift for low-temperature, high- and low-flux irradiations (Ref. 24).

fluence levels is too short for this annealing to occur. In other words, the phenomenon of irradiation saturation is a "time at temperature" effect.

A model of the NDT shift, incorporating the self-annealing phenomenon, has been developed by Odette.[24] In Fig. 4-10 analytical predictions for the NDT shift as a function of fluence are shown for four flux/temperature combinations. At irradiation temperatures below about 505 K, the diffusion rates are so low that little or no thermal annealing occurs and, as a consequence, the transition temperature shift is predicted to be independent of flux. In contrast, the rate of annealing at 561 K is such as to lead to a very substantial difference in the behavior predicted for high-flux, test reactor irradiation as opposed to low-flux, power reactor irradiation. As seen in Fig. 4-10, the agreement between the analytical predictions and actual experimental data on the change of transition temperature as a function of fluence is surprisingly good. Clearly more data are necessary to justify alteration of the current trend curves contained in U.S. NRC Regulatory Guide 1.99.1 to account for this self-annealing effect. An accelerated program of measurements on specimens taken from surveillance capsules in several reactors (principally, PWRs because of the higher fluences) has therefore been implemented by the industry.[4, 20]

One of the reasons why a large data base must be assembled on the effects of irradiation on pressure vessel steels is the variability in properties that is apparently produced by residual elements. Figure 4-11 is a typical scatter band covering a spectrum of residual element concentrations as they affect the NDT shift. Evidence suggests a synergism between the various trace elements and neutron flux.[23]

The fundamental damage mechanisms responsible for the effects that small quantities of residual elements have on the degree of radiation damage are still undetermined. Elements of concern include phosphorus, sulfur, copper, vanadium, and nickel in the case of weldments. Prior fabrication history is also significant since forgings, plate, and weldments respond differently to residual elements and neutron fluence.[23]

Bush,[23] in reviewing the experimental observations, concludes that: (a) at low temperatures (506 K), neutron irradiation will cause damage ($\Delta\sigma_y$, NDT shift) independent of the level of residuals in the steel; (b) irradiations at 560 K cause little or no damage in low-residual steels while there is substantial damage in high-residual steels; and (c) the difference in damage between low- and high-residual steels at 560 K indicates that a dynamic recovery process, most probably controlled by vacancy diffusion, must be responsible.

The effects on radiation embrittlement sensitivities of additions of copper, copper plus vanadium, sulfur, and phosphorus plus sulfur have been studied, and the most significant change occurs with the addition of copper.[25-28] The proposed mechanism to explain greater radiation

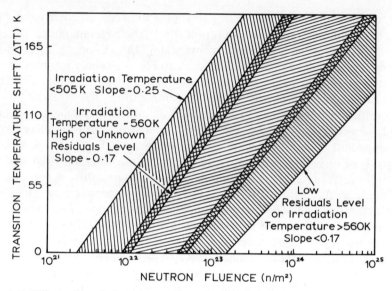

Fig. 4-11. Effects of irradiation temperature and residual element concentration on irradiation-induced 41-J Charpy-V transition temperature shifts in A302-B and A533-B steels (Ref. 23).

damage in steels with a high level of copper compared to steels with a low level of copper is that a stable copper-vacancy defect forms, which modifies the nucleation of defect aggregates so as to increase the yield strength and, in turn, increase the transition temperature. These aggregates are barriers to dislocation motion similar to a precipitation-hardened alloy.

A combination of experimental, analytical, and theoretical studies[23] has partially validated the copper model, but more experiments are required, in particular to understand the difference in response of forgings, weldments, and plates. Nevertheless, the need to minimize residual elements in vessel steels and weldments already is obvious and steps are being made to achieve cleaner steels (see Section 4-4).

4.3.1.3. Fatigue Crack Growth

Experience with nonnuclear structures indicates that fatigue and corrosion fatigue are the most likely mechanisms by which small flaws or cracks might grow to a critical size if they remained undetected by periodic in-service inspection. More emphasis has recently been attached to this property in nuclear systems following the release of the Marshall report[29] on PWR pressure vessel safety, which, in its recommendations, identified the need for further corrosion fatigue work.

Fatigue (subcritical) crack propagation is a localized phenomenon dependent upon the temperature, rate of load application, environment, material properties, and stress corrosion conditions. As noted in Section 1.3.3.2 earlier, the LEFM application to fatigue crack growth is well established [Eq. (1-8)]. Appendix A to ASME Code Section XI sets a procedure for evaluating the growth of a crack found during in-service inspection for the remainder of life of the vessel.[18] Reference fatigue crack growth curves, plotted according to Eq. (1-8), were shown in Fig. 1-16. These two curves are meant to be applied to ferritic pressure vessel steels with minimum yield strengths of 345 MPa and less. One curve is used for determining growth of flaws in an air environment,[30-32] and the second curve, which is considerably more severe, is for crack growth calculations for flaws exposed to the LWR water environment.[33,34]

Since the original formulation of the reference fatigue crack growth curves, a great deal of data has been obtained by a number of investigators. New data[35] have confirmed the behavior specified in the reference curves for air environment, but the reference curve for water environment was based on very little data (see Fig. 1-16), and later results[36-38] have shown that the actual behavior is considerably different. Recently, Bamford[39] and Scott[40] independently reviewed the data set now available, and the former worker proposed an alternative reference curve for fatigue crack growth in water environments. The major conclusions from these reviews are summarized below:

1. Medium-strength carbon and low-alloy steels (i.e., minimum yield strength < 450 MPa) have very similar behavior in a water environment.[39]
2. Crack growth will be somewhat slower in weldments and so the base metal reference curve is adequately conservative for a welded component.[39]
3. There is no difference in behavior between PWR and BWR environments; thus the water chemistry is not a dominant variable (at least in the results to date).[39]
4. Crack growth rates depend on loading frequency; growth rate increases with decreasing cyclic frequency, reaching a maximum acceleration at a frequency between 0.0017 and 0.017 cps. Results thus far indicate that frequencies lower than 0.0083 cps produce slower growth rates.[39] Mechanistically this observation is explained in terms of crack tip blunting by corrosion,[40] which appears to be more efficient in the oxygenated BWR water environment than in the very-low-oxygen PWR environment.
5. The effect of "hold time" during fatigue cycling does not appear to be significant.[39,40] In fact there are indications that the ramp and hold-time data may be less accelerated than sinusoidal or

positive sawtooth data obtained under the same load range and environment conditions.

6. The ratio of minimum load to maximum load, the so-called R ratio, is the most significant factor affecting crack growth in water environments.[39,40] Higher mean stress levels (i.e., higher R ratios) enhance crack growth at a given level of ΔK, at least over the range $0 \leq R \leq 0.75$. Scott[40] has proposed a hydrogen embrittlement corrosion fatigue mechanism to specifically account for the adverse effect of increasing R ratio.

7. No information is yet available on fatigue crack growth rates of irradiated carbon steels in water, but irradiation has little or no effect on crack growth in air.[41]

Figure 4-12, taken from Bamford's paper, illustrates the general trends in fatigue crack growth in LWR environments, which result from the foregoing conclusions. Bamford's proposed improved reference equa-

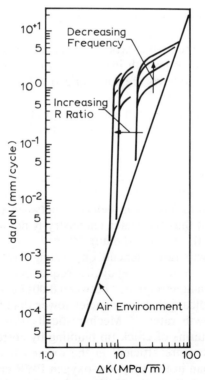

Fig. 4-12. Schematic of frequency and R ratio effects on fatigue crack growth in pressure vessel steels in LWR water environment (Ref. 39).

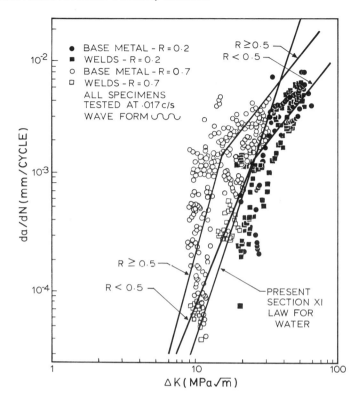

Fig. 4-13. Alternative reference law for fatigue crack growth; 95% global confidence limit on the mean data (Ref. 39).

tion[39] relates the three most important variables, viz crack growth rate da/dN, range of applied stress intensity factor ΔK, and R ratio, utilizing 600 to 700 data sets. Both a statistical (mean) treatment and a graphical (bounding) treatment of the data were reported. The former is shown in comparison to the current Section XI curves in Fig. 4-13.

Application of the two alternative relationships and the Section XI law to an example problem showed that the predicted crack growth was similar to that predicted by the Section XI law, but slightly lower for both cases.[39] As would be expected, the mean curve was less conservative than the bounding curve. Because it does not account for R ratio effects, the present Section XI law becomes less conservative as the proportion of high R ratio transients is increased. Both the alternative relationships retain their conservatism under those circumstances, and as such both are considered preferable to the present law. However, the decision as to which alternative law is preferable depends on the degree of conservatism desired.

To consolidate the position on corrosion fatigue behavior in LWR pressure boundary materials, research in this area is currently concentrated on (1) obtaining more data in water at high R values (0.75 to 0.95) and in the low ΔK regime; (2) extending the range of applicability to higher and lower strength steels (e.g., piping materials); (3) the effects of irradiation in water environments; and (4) the synergistic interactions which could affect corrosion fatigue crack growth rates in actual reactor operation.[4]

With respect to the latter aspect of fatigue behavior, however, it is conceivable that real crack growth in an operating plant structure would be much lower than predicted in most cases. The review by Bamford[39] indicated a number of factors which occur in operating reactor systems which would tend to slow down the growth rate of a crack. For instance, it has been experimentally observed[42] that crack growth in this environment can be slowed down and even arrested by load changes, particularly load increases, which are to be expected in operating systems. Crack growth in welds is also known to exhibit small arrests as the crack grows along the weld. Also, there is some indication that hold times, which are typical of reactor operation, tend to result in slower growth rates in this temperature regime.

Despite the concern over fatigue cracking in LWR vessels, to the author's knowledge, the only incidences attributed to fatigue have been on the inside surfaces of some BWR feedwater nozzles (refer to Fig. 4-14 for location), primarily on the nozzle corner, but, in some instances, the crack was sufficiently deep to penetrate the austenitic stainless steel cladding and extend into the base material of the nozzle. To date, the maximum depth of cracking, including the cladding, has been approximately 19 mm. As noted earlier in Section 4.2.1, thermal (corrosion) fatigue is considered to be the principal contributing factor,[43] based on the extensive nature of the cracking, the orientation of the cracks, and the duty cycle analysis.

The present method of dealing with nozzle and bore cracks, if they occur, is to remove them by grinding. However, if the crack is very deep or if a crack is reinitiated on the site of a previously ground area, there is a possibility of removing enough wall thickness to violate the requirements of the ASME Codes. In such a situation the only solution is to build up the thinned area with a repair weld. Although repair welding procedures with no post-weld heat treatment for stress relief are approved by the ASME Code, there is very little experience with the fracture, fatigue, and corrosion properties of this type of weld. Research is under way to determine these properties for the code-accepted nonstress relieved weld and also to explore alternate weld repair techniques.[4, 43a]

Two actions are being taken to eliminate the thermal fatigue cracks: first is the removal of all stainless steel cladding from the nozzle corners and bore, and second, a redesign of the nozzle configuration to minimize the fatigue stresses.[43]

During 1979, an unusual quantity of cracks or crack "indications" have also been reported in the carbon steel (A106B) feedwater pipes and nozzles of the secondary side of steam generators in some Westinghouse PWRs. These "1979" PWR cracks are in areas that may be considered roughly equivalent to those that are subject to cracking in BWRs, i.e., since a BWR acts as its own steam generator, the feedwater nozzle area of a PWR steam generator is comparable to the nozzle of a BWR vessel. Early indications are that the PWR cracking could also be thermal-fatigue initiated due to temperature transients, but, at this writing, no details are available. This is, however, a less serious problem than the BWR situation because smaller sections are involved and, being on the secondary side, they are more readily accessible for repair or replacement. Also, as for the BWR nozzle, a design solution to minimize the temperature/stress cycles appears more appropriate than a material change.

4.3.1.4 Stress-Corrosion Cracking

The susceptibility of the Fe–Ni–Cr steels to stress-corrosion cracking in aqueous environments has been observed in both the vessel nozzle *safe-ends* and piping of BWRs (Fig. 4-14). Together these incidents have provided the single largest source of capacity loss in the LWR pressure boundary,[3] and, accordingly, a significant research effort has been mounted to understand and remedy the problems.

4.3.1.4a. Stress-Corrosion Cracking of Piping. Leaks and cracks in the heat-affected zones (HAZs) of welds that join austenitic stainless steel piping and associated components in BWRs were first observed over 10 years ago. A few instances have since been reported each year. In the U.S. most of the affected piping is Type 304 stainless steel with diameters of 203 mm or less, in recirculation bypass, core spray lines, reactor-water-cleanup lines, and control-rod drive return lines. Cracking locations are shown in Fig. 4-14.

Since plant safety was not jeopardized, the remedial action in most instances has been to replace the affected lines using materials less susceptible to cracking. The motivation to understand and fully resolve this problem has increased, however, with the observation of throughwall

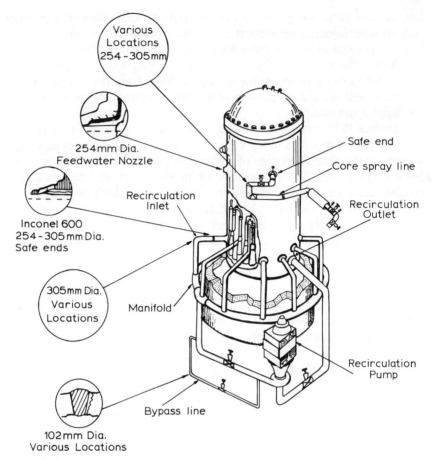

Fig. 4-14. Incidences and locations of stress corrosion cracking in the BWR pressure boundary. Total 191 (courtesy R. Smith, EPRI).

cracks in 305-mm-diameter pipes in three Japanese BWRs and in a 660-mm-diameter pipe in the recirculation loop in a single system at one operating plant in Germany.

A detailed review of the situation by a USNRC study group[44] concluded that, although cracking in large-diameter pipes in BWRs could not be ruled out, the mode of crack propagation would not be unstable and lead to excessive loss of coolant. That is, a leak-before-break condition would always prevail, allowing time for detection and repair. Nevertheless, from a utility's viewpoint, the downtime associated with increased surveillance and repair and replacement of piping systems represents an unacceptable loss in plant availability and, consequently, provided ample

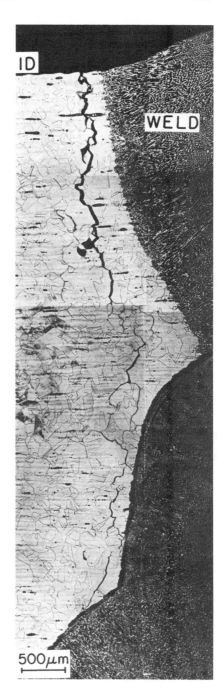

Fig. 4-15. Cross section of U-shaped crack in loop A weld 10K14 (J. Y. Park, S. Danyluk, R. B. Poeppel, and C. F. Chang, Metallurgical Examination of Cracks in the Dresden-2 BWR Emergency Core-Spray System 10-inch-Diameter Piping, Argonne National Laboratory report prepared for Commonwealth Edison Co., April 1976).

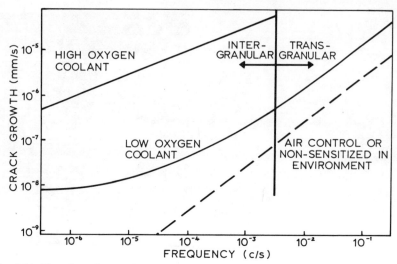

Fig. 4-16. Plot of crack growth rate versus cyclic frequency for Type 304 stainless steel piping (Ref. 45).

justification for an intensive research campaign to remedy the problem, which started four years ago.

It is well known that the austenitic stainless steels are susceptible to stress corrosion cracking if the proper metallurgical, stress, and environmental conditions are present. Oxygen is the most important environmental parameter, and this explains in large part why cracking is not seen in either PWR piping (the use of hydrazine additives and a hydrogen overpressure limit the oxygen in the coolant to very low levels) or LMFBR piping (sodium coolant has typically very low oxygen contamination).

Unlike the situation encountered in stainless steel clad LWR fuel rods (Chapter 2, Section 2.4), all of the stress corrosion cracking (except that resulting earlier from chloride contamination) observed in the stainless steel BWR piping has been identified as intergranular-stress-corrosion cracking (IGSCC) (Fig. 4-15) and is related to some degree of sensitization of the material.[45] This general phenomenon also influences the integrity of nickel-base alloy tubes in PWR steam generators and will be discussed in some detail in Chapter 6.

The trends in crack propagation data for sensitized stainless steel are summarized in Fig. 4-16,[46] plotted as da/dt versus cyclic frequency. The important conclusions that can be drawn from this figure are: (1) crack growth rate increases as cyclic frequency increases; (2) above 0.003 cps transgranular cracking is obtained, and below this value intergranular cracking is obtained; (3) slower growth rates occur in low oxygen coolant;

and (4) nonsensitized material always results in transgranular behavior. It should be noted here that transgranular SCC in Type 304 stainless steel is not considered to be a BWR concern.

The data pattern is in agreement with available stress corrosion models, in that three stages of growth are suggested by the behavior (refer to Fig. 1-14 earlier). Also the indications are that Stage 1 behavior is prolonged with the application of cyclic loads and that most of the crack propagation will occur when cyclic loads with wave shapes that promote stress corrosion (trapezoidal) are applied.[46]

Recent research into IGSCC has therefore concentrated on quantifying the characteristics of the three main conditions, namely susceptible material, stress, and corrodant, with respect to crack initiation and growth behavior.[45] The interactions between these three parameters necessary to produce cracking are subtle, as is demonstrated by the very low number of IGSCC incidents (191)† occurring in 17,000 welds in the 30 operating BWRs.[47] However, significant progress has been made in the understanding of this phenomenon,[44,45,47] and some key recent results are described below. Potential material improvements derived from this research are discussed in Section 4.4.2.

Sensitization. Austenitic stainless steels, such as Types 304 and 316, can become susceptible to intergranular attack if exposed to heat treatments that precipitate chromium-rich carbides along the grain boundaries.[48] The formation of such carbides may deplete the chromium adjacent to the grain boundaries to levels below those needed for corrosion protection by passivation (approximately 12% chromium), and these depleted zones then become susceptible to preferential attack by a corrosive environment. Measurements have been made of chromium depletion across austenitic grain boundaries adjacent to carbides.[49] Results showed that the grain-boundary chromium concentration dropped from 18.6% in the matrix to 10.8%, confirming Cr depletion of the grain boundary to levels below those required for passivation. Other work using Auger spectroscopy has shown high accumulations of sulfur and phosphorus at the grain boundaries,[50] which elements can affect sensitization and electrochemical characteristics at the grain boundary.

This "sensitization" (of grain boundaries) occurs when the steel is cooled slowly through the temperature range of 1143 to 679 K. The degree of sensitization depends greatly on the alloy composition and thermal-mechanical history.[48] The most significant factor is the carbon content; the higher the carbon content, the more susceptible the alloy is to sensitization. Low-carbon-grade stainless steels (0.03% maximum) have significantly lower susceptibility than do regular-grade stainless steels

† As of July 1979.

(0.08% maximum carbon) and, for practical purposes, are considered immune to IGSCC in a BWR environment when properly welded. Alloying elements have a less, but still noticeable, effect; molybdenum and chromium decrease susceptibility to sensitization somewhat, whereas nickel tends to increase susceptibility. Stabilization of austenitic stainless steel by the addition of carbide-forming elements (such as boron, titanium, and niobium) is an effective means of preventing sensitization. In BWR experience, no cracks have been reported in Type 304L or in Type 316 (not furnace-sensitized), nor in stabilized austenitic stainless steel used in some German reactors.

The "mill-annealed" pipe has a "nonsensitized" microstructure; therefore, the welding process is the primary cause of sensitization.[44,45] The degree of sensitization will vary depending on the heat input, interpass temperature, size of component, and other variables. Relatively small piping is more prone to weld sensitization because a smaller heat sink is available, and successive weld passes are made more frequently, keeping the interpass temperature higher. In a weldment, sensitization occurs in the base material within approximately 6 mm on either side of the heat-affected zone (HAZ) of the weld. Carbides are precipitated during the thermal and strain transients during multiple-weld passes. An important recent finding has been the synergistic effect of the superimposed temperature and strain cycles on the inner surface of multipass welded pipe, thus increasing the kinetics of sensitization.[51] The strain cycles apparently produced nucleation sites for carbide precipitation, which resulted in a faster sensitization rate than would have been expected based on total time and temperature alone.

Recent results also indicate that the degree of sensitization in the HAZ of a weld might increase significantly at BWR operating temperatures after a 10-year exposure.[52] Since this type of sensitization occurs below the normal sensitization temperatures of solution-treated and isothermally exposed material, it is called low-temperature sensitization (LTS). Prior treatment capable of nucleating carbides is required to obtain LTS. Research has shown that the thermal/strain cycle from welding can nucleate carbides that subsequently grow during LTS to increase the degree of sensitization and susceptibility to IGSCC. There is a corresponding depletion of the Cr at the grain boundaries, in one study down to 8.2% from 10.8%, which is well below "passivation" levels.[53]

Although the basic cause of sensitization is understood, the observed heat-to-heat variability of several heats of 100-, 254-, and 660-mm piping, both in the as-received and in the as-welded condition, is a result of factors that are not well understood, for example, thermomechanical history during fabrication or subtle compositional variations. Therefore, it is important to be able to detect a sensitized material both before and during service. The electrochemical potentiokinetic reactivation (EPR) technique has recently been developed for this purpose.[54] The technique relies on the fact that the electrochemical behavior of sensitized stainless steel differs from that of nonsensitized stainless steel. Sensitized steels are easily activated and they show relatively high current flow during the potential sweep, as compared to the flow shown by nonsensitized steels. EPR measurements have been correlated to different degrees of sensitization, either in welds or in as-received material.

Stress. Only tensile-type stresses will cause stress corrosion cracking; such stress may include residual stresses in the metal from fabrication and welding as well as tensile stresses imposed by operating load conditions. The variability of these stresses at and in the vicinity of welds is a major reason why only about 1% of the welds have cracked in service.

Weld configurations are of particular significance to stresses in IGSCC. Pipe, fittings, and valves are not precision-made products; the relatively large tolerances on diameters and wall thickness cause problems in achieving an adequate fit between two pipes (or pipe-to-fitting or pipe-to-valve) for butt welding (refer to Fig. 4-2). Complex stress patterns are therefore imposed by the various joining processes required to build the pipe configurations.

Data are available which show that residual stresses in laboratory test welds on 100-mm, 254-mm, and 660-mm Schedule 80 Type 304 stainless steel pipe-to-pipe welds can be significant.[44,45,55] Shack et al.[55] were also able to measure residual stresses on welds in a 254-mm pipe that was removed from BWR service. From the standpoint of IGSCC, the most significant residual stress is probably the tensile axial-direction stress on the inside surface within an axial distance of about 6 mm from the weld, and therefore detailed measurements were made of it. The data of Shack et al.[55] show the dependence of internal stress on pipe size; the maximum tensile, axial-direction stresses were:

   100-mm-diameter pipe welds: 262, 317, and 352 MPa for three tests
   254-mm-diameter pipe welds: 414 MPa
   660-mm-diameter pipe welds: 193 MPa

The lower residual stresses in the largest diameter pipe may well account for there being so very few instances of IGSCC in large pipes.

Additional surface residual tensile stresses on the order of 690 MPa can be produced by postweld grinding.[44] Grinding also produces a thin surface layer of cold-worked material.[45] This material often consists of deformation-induced martensite, which is believed to contribute to the initiation of IGSCC.[51] A highly cold-worked surface layer is not capable of much plastic deformation and will exhibit brittle fracture on the surface at relatively low levels of strain.

The magnitude of operation stresses is limited by the ASME Section III Code under which the piping systems are designed. Interestingly, the Code, although using a rather complex set of stress limits, does not yet include allowances for metal deterioration such as IGSCC, nor fabrication (residual) stresses. For the design code, the basic stress limit in straight pipe is the hoop membrane stress due to internal pressure, and is limited to 0.6 of the material yield strength. Bending stresses (calculated on an elastic basis) are limited to twice the material yield strength. Highly local stresses are limited only by fatigue considerations. It is important to note that in many piping systems the operation stresses do not exceed the yield strength. Further, even in a specific piping system in which some weld has operating stresses above the yield strength, it is likely that many of the welds in that piping system will have operating stresses below the yield strength.

General Electric has established a criterion that identifies those stresses significant to IGSCC and establishes appropriate procedures for calculating them.[44,45] Table 4-3 describes the design stress rule.

A stress rule value greater than 1.0 indicates a stress condition sufficient to facilitate IGSCC. However, a stress rule value greater than 1.0 does not imply that IGSCC will definitely occur within a given time frame. Figure 4-17 presents the calculated stress rule value for 93 incidents of IGSCC in BWR piping. For each incident, the stress rule value is greater than 1.0. To date there are no known cracking incidents for which the stress rule was less than 1.0.

Table 4-3. General Electric Design Stress Rule[a]

Premise .... Stress corrosion can be avoided if stresses are maintained below 0.2% offset yield stress.

Rule .... $\dfrac{P_M + P_B}{S_y} + \dfrac{Q + F + \text{Residual}}{S_y + 0.002E} < 1$

Definitions .... $P_M + P_B$ = primary membrane and bending stresses
S_y = ASME code 0.2% yield stress at applicable temperature
Q = secondary stress (includes thermal)
F = peak stress
E = ASME code elastic modulus at applicable temperature
Residual = sum of all sources of residual stress (including weld residual stress and stress resulting from compressive transients)

[a] After Refs. 44, 47.

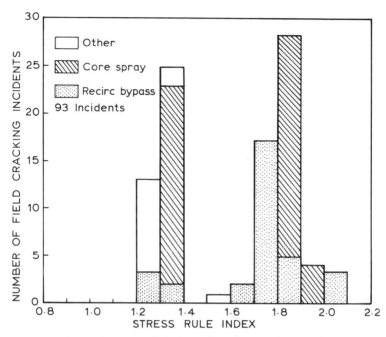

Fig. 4-17. General Electric stress rule index evaluated for 93 pipe cracking incidents (Ref. 46).

While recognizing the empirical nature of the GE stress rule, it is nonetheless a potentially useful tool for evaluating BWR piping systems and for identifying target pipe welds in existing BWRs that may be vulnerable to IGSCC. Improvements to the design rule are therefore continuing, as are efforts to characterize the stress system to which piping is subjected in service, and development of a more quantitative failure criterion. As discussed earlier, proving the leak-before-break concept is basic to the safety concerns over piping integrity.

Corrodent. Stress-corrosion cracking will proceed only if the environment is such that an electrochemical reaction can occur. Clarke and Gordon[56] substantiated the importance of oxygen (0.2 to 100 ppm) in the coolant to IGSCC through a correlation between oxygen and time-to-failure, and, in more recent work, Burley[57] has shown that this trend continues to even lower oxygen levels, in the ppb range.

The residual oxygen levels in a BWR coolant during operation result from radiolysis, which produces both oxygen and hydrogen peroxide in the reactor coolant. At temperatures above 448°K, hydrogen peroxide decomposes rapidly to hydrogen and oxygen.[58] During shutdown, when the reactor coolant is exposed to the air, oxygen levels build up rapidly to approximately the room temperature saturation level (8 ppm). On startup, the boiling process reduces oxygen levels, but the levels remain above steady state (0.2 to 0.4 ppm) until the reactor coolant temperature is well above the normal boiling point of water.[58] Levels of 0.5 to 1.0 ppm have been measured to temperatures up to 448 K.

Ford and Povich[59] utilized the constant extension rate test (CERT) to determine the temperature/oxygen combinations that will facilitate IGSCC and those that will not. Their data are shown in Fig. 4-18 with the BWR water chemistry data[58] superimposed. The practical significance of this plot is that IGSCC can occur (under dynamic straining conditions) under temperature/oxygen combinations likely to occur during normal BWR startup and that time spent in this susceptible region could be

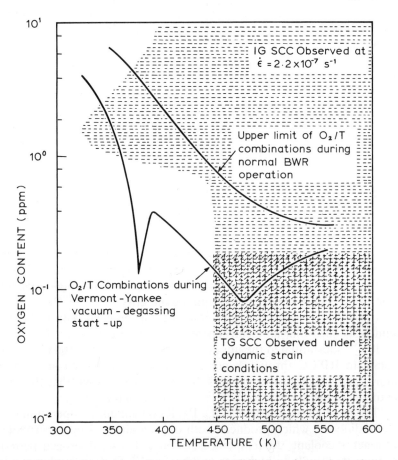

Fig. 4-18. Regions of stress corrosion cracking susceptibility superimposed over the temperature/oxygen combinations likely to occur during BWR start-ups (Ref. 58).

minimized by oxygen control procedures. Oxygen control, particularly during shutdown and startup, has therefore been recommended to the utility industry by General Electric.[44,45] However, since intergranular failure of sensitized Type 304 stainless steel can occur at operating temperatures under both high oxygen and extremely low oxygen levels, some extreme measures such as nitrogen blanketing of the reactor coolant or hydrogen injection (to continuously recombine with the oxygen) might be necessary. The Swedish reactors have adopted these approaches and no cracks have developed to date in their piping.[44,45]

In an attempt to integrate all the results that are emerging from the BWR pipe crack research, Hanneman *et al.*[45,60] have developed a working mechanistic model of IGSCC in the HAZ of BWR piping based on the three-stage growth process (summarized in Fig. 1-14 earlier). The model treats the crucial initiation and propagation stages separately. Three processes, acting separately or in combination, are considered for the former: (1) a relatively rapid process such as mechanical fracture; (2) a more moderate rate process such as fatigue; and/or (3) a normally very slow process such as localized dissolution along sensitized grain boundaries. "Initiation" or Stage I proceeds until the crack reaches a critical depth, a_{scc}^*, for "stable" IGSCC propagation, i.e., Stage 2 (i.e., $K > K_{Iscc}$). In the case of mixed-mode initiation of IGSCC, mechanical fracture could occur to only a fraction of the depth of a_{scc}^* (as when only a thin surface damaged layer is present) and then the process could switch to one or more slower modes of crack tip advance, such as fatigue or localized chemical dissolution along the grain boundary, until the stable propagation depth a_{scc}^* is reached. The value of a_{scc}^* will be dependent upon the alloy composition, its metallurgical condition, degree of sensitization, temperature, environment, and local total stress and strain rate.

The propagation of initiated IGSCC cracks will depend on the three prime variables discussed above, namely sensitization, stress (that is, local stress intensity factor K, and the loading cycle and resultant strain–time behavior), and environment. For a given initial flaw size the crack length a will advance with time on a curve that is dependent on both the microstructural and environmental states. Figures 4-19a and 4-19b show the schematic form of advance of an already initiated crack versus time for various levels of sensitization and also versus dissolved oxygen level, for a given total applied stress. Both static and cyclic (fatigue) crack growth are considered.

Although still being developed, this model is proving useful in interpreting data and field observations, and in guiding remedy research (see Section 4.4). For instance, in addition to the larger a_{scc}^* for a 600-mm versus a 100-mm pipe, the model[60] predicts that the Stage I crack growth rate of the larger pipe would be reduced at a given depth because of its

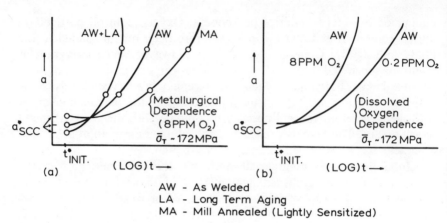

Fig. 4-19. Schematic of crack advancement versus time in Type 304 stainless steel for various metallurgical and environmental conditions (Ref. 60).

lower tensile subsurface residual stresses. Therefore, statistically, large pipes should crack much less frequently and take a longer time to show any detectable size of intergranular stress corrosion cracks. This is in agreement with field observations to date and provides encouragement for the future. Furthermore, subsurface residual stress gradients in the 600 mm normally welded case indicate the possibility of eventual crack arrest in the weld metal. By contrast, in the smaller-diameter pipes the value of a_{scc}^* will be small enough and the indicated Stage I da/dt value large enough to account for the much more frequent instances of detectable cracks in BWR service.

An LEFM analysis of stress-corrosion crack growth in sensitized pipe weldments, reported by Harris,[61] also indicated that the incidence of pipe cracking in the smaller lines should be several orders of magnitude more frequent than in larger lines. However, although Harris' analysis did show that failure of the large-diameter pipe is unlikely, it predicted a considerable probability of encountering long, shallow part-through cracks in them. The largest single factor influencing the calculated failure probabilities was the presence of residual stresses; parameters that describe the crack growth characteristics (e.g., Fig. 4-19) were found by Harris to have only a secondary influence. The reason for this major conclusion is interesting: owing to the axisymmetric nature of residual stresses in the large pipe, compressive residual stresses exist, and these are very effective in retarding (or arresting) crack growth. On the other hand, the residual stresses in the smaller line are not axisymmetric, and therefore can be (and are) completely tensile through the pipe wall at certain locations, thereby accelerating crack growth.

4.3.1.4b. Stress-Corrosion Cracking in Nozzle Safe-Ends. As shown in Figs. 4-1a and 4-14, the BWR vessel is fitted with nozzles for attachment of the piping. Inconel 600 or Type 316 stainless steel "safe-ends" are attached to the reactor vessel to serve as transition pieces to facilitate welding of the stainless steel piping to carbon-steel pressure vessel nozzles. Cracks have been found in these safe-ends.[43,44]

Cracking in the safe-ends usually resulted from material sensitization that occurs when the carbon steel components to which they are attached are given a postweld treatment. Inconel 600 is sensitized by grain-boundary precipitation of chromium carbides, in very much the same way as is stainless steel. Heat treatments within a band of temperature in the vicinity of 923 K reduce the resistance to IGSCC in oxidizing media; this is because the depleted regions overlap and form a continuous network. Experimental evidence indicates that such cracking only occurs within creviced areas. The role of a crevice is undoubtedly to promote localized corrosion. Solution heat treatment to redissolve carbides, although improving SCC resistance in aerated solutions, will not necessarily make the material immune in creviced areas (refer to Chapter 6 for further details).

Recently, cracking has also been observed in the recirculation inlet-nozzle safe-ends (refer to Fig. 4-14) at one U.S. BWR, which resulted from a unique crevice geometry associated with the thermal sleeve attachment.[44] This is the first incident of this type and crack initiation in this safe-end is believed to have resulted from a combination of high residual and operating stresses from the thermal-sleeve attachment weld, the oxygen in the coolant, and a chemical environment resulting from a crevice formed by the safe-end/thermal-sleeve-attachment configuration. There is insufficient evidence to indicate that sensitization of the Inconel 600 alone is a factor that contributed significantly to crack initiation or propagation.

In an attempt to mitigate this problem, in-service inspections are undertaken to examine the welds and surrounding areas of the nozzle safe-end regions to ensure that cracking does not go undetected during service, even in plants where the weld attachment configurations do not form crevices; and thermal-sleeve attachment configurations that form crevices will be avoided in new plants.

4.3.1.5. Closing Remarks

It should be evident from the data and analyses presented above that the single biggest concern in the LWR pressure boundary is demonstrating the inherent toughness of the vessel and piping steels. The state-of-the-art has advanced rapidly towards confirmation of the leak-before-break criterion; now the appropriate data and data treatments must be included in the codes that govern pressure boundary design and operation. The revised fracture toughness and fatigue correlations warrant early acceptance; the radiation embrittlement saturation effect looks very promising but requires consolidation.

Whereas for the PWR vessel the problem is one of eliminating hypothetical concerns through improved SI analysis, for the BWR piping the fact is that through-wall cracks do occur. Because of this "proven failure," the pipe cracking problem is perhaps the most crucial "unresolved item" in the LWR pressure boundary. Although the mechanistic understanding is advanced and the remedies (to be described later in Section 4.4) have a high probability of success, it will be several years before it is known whether the IGSCC phenomenon has been successfully eliminated from the BWR.

4.3.2. LMFBR Materials

Inasmuch as the austenitic stainless steels are the prime candidate materials for the LMFBR vessel and piping, the property data generally do not discriminate between components. The extensive data base that has been obtained on Types 304 and 316 stainless steel is documented in the *Nuclear Systems Materials Handbook* (NSMH)[62] referred to earlier in Chapter 3.

Needless to say, until more experience is gained with operation of LMFBRs, the true performance-limiting material characteristics must be assumed. However, a consensus does exist in the industry. With respect to the reactor vessel and piping, properties of interest for Type 304 stainless steel are essentially in the "subcreep" regime, i.e., low-cycle fatigue and crack propagation, because vessel and cold leg piping (below the reactor core) temperatures are relatively low (e.g., 643 to 753 K). In the upper, hot leg piping regions, where the temperatures can be substantially higher (e.g., to 923 K) time-dependent creep and corrosion effects must also be considered to interact with fatigue; here Type 316 stainless steel is favored over Type 304 due to its higher strength. Figure 1-11 earlier indicates these two operating regimes.

In the ASME Code Section III, time-independent (or subcreep) fatigue damage is represented as the number of cycles to failure over a cyclic strain range that is specified for the expected service life of the component. The relatively mature state of design for Types 304 and 316 stainless steel may be observed in Fig. 4-20. The new fatigue curve proposed by the ASME Subgroup on Fatigue Strength differs by only a few percent from the existing Code Case 1592-7 curve.

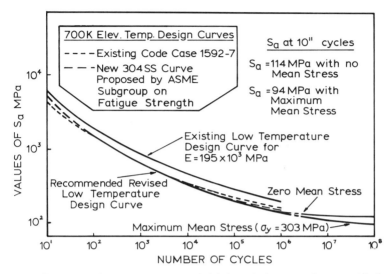

Fig. 4-20. ASME Code Case 1592 recommended fatigue design curve for austenitic Steels.

Unfortunately, the situation is not as clear at elevated temperatures, where, as noted in Chapter 1, Section 1.3.4.4, creep and/or environmental effects interact with fatigue. The latter are not explicitly accounted for in the current Code Case 1592,[63] and creep-fatigue damage assessments are presently performed on the basis of cycle and time-fraction summations [refer to Eq. (1-11)].

The sum of creep and fatigue damage fractions is restricted for Types 304 and 316 stainless steel to a limiting value of D. In the so-called "linear-damage rule" this limiting value is taken to be $D = 1$, whereas in the "bilinear ASME approach" more conservative values are used for the failure criterion D. As shown in Fig. 4-21, however, limited data infer that even the "more conservative" criterion may not be conservative for austenitic stainless steels.[64] Such problems have provided the impetus for the alternative criteria, based on frequency separation (FS) or strain-rate partitioning (SRP),[65] which were introduced in Chapter 1 and which will be discussed further in this section in connection with the growing experimental data base on this phenomenon.

The fatigue curves used in either temperature regime are based on test data of smooth, hourglass shaped uniaxially load specimens with constant conditions (temperature, wave shape, frequency, etc.) throughout each test. Appropriate safety factors are applied to a best fit of the data to obtain the design curve. The codes contain engineering treatments (largely empirical) to apply the laboratory data to the multiaxial fatigue situation expected in-service.

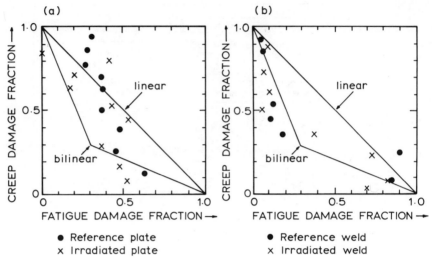

Fig. 4-21. Creep-Fatigue damage functions, comparing data on unirradiated and irradiated austenitic stainless steel plate (a) and welds (b) to ASME Code Case 1592 criteria (adapted from Ref. 64).

Coffin[65] and others[66] have shown that failure of the uniaxial test specimens used to generate the code fatigue data can be considered "crack initiation data" when applied to a prototype structure. This is due to the fact that for smooth specimens, the number of cycles required to initiate a crack ($\sim$ 3 mm long) is the major part of the total cycles to failure. As noted earlier for the LWR situation, this is usually not the case for vessel or piping components, which contain discontinuities and stress concentrations, since with localized strain concentrating effects there can be a significant period of stable crack growth even after crack initiation has occurred.[66] Thus, for a real structure, exceeding the number of allowable design cycles does not imply imminent failure but rather is an indication that fatigue damage is progressing with increasing chance of subsequent failure. Recognition of this potentially large design conservatism has provided impetus for the use of crack growth rate data as an indication of the rate of fatigue damage. A significant fatigue crack growth data base has therefore been developed for Types 304 and 316 stainless steel as part of the process to incorporate this feature into the high-temperature design codes, and a discussion of this work will form the second part of this section.

4.3.2.1. Elevated-Temperature Low-Cycle Fatigue

Of principal concern to the LMFBR pressure boundary are the long-term effects on the fatigue life of loading history, sodium exposure, temperature (i.e., thermal aging), and, to a lesser extent, irradiation [the presence of blanket and reflector regions in the core (Fig. 3-5) and a nonload-bearing guard vessel around the core (Fig. 4-3) limit the fast neutron fluence at the vessel wall].

Loading History. The effect of loading wave shape on the elevated-temperature, low-cycle fatigue behavior of Type 304 and 316 stainless steels is the subject of extensive study because the essential first step for a time-dependent fatigue damage criterion is to explain this behavior. Ever since fatigue experiments in vacuum demonstrated the existence of a pure creep-fatigue interaction,[65,67] studies have attempted to quantify the effect.[68–70] Extensive testing has shown that, in Type 304 stainless steel, hold times in tension (either at constant total strain or stress) reduce the life of the test sample severely, whereas hold times in compression or cycles involving equal hold times in tension and compression (symmetrical hold) are not nearly as damaging.[67,69,70] This difference in lifetime is also reflected in the modes of fatigue fracture of the specimens. Thus, the fracture modes of the specimens tested under compressive and symmetric hold times are transgranular with striations occurring on the fracture surface, whereas, in general, the specimens tested under tensile hold fail in an intergranular manner.[69]

The reader will note from Chapter 1, Section 1.3.4.4, that this wave-shape effect could not be explained by the original Coffin frequency-modified fatigue model and it prompted the consideration of frequency separation (FS).[65] The Manson strain range partitioning (SRP) approach, on the other hand, can in principle explain this behavior because it separates the multiaxial strains into elastic, plastic, and creep components[65] [Eq. (1-12)]. However, Majumdar and Maiya's damage-rate model[69] has been specifically applied to this phenomenon, and therefore, promises a broader application.

They assume that creep-fatigue damage progresses by a combination of crack growth ("fatigue crack damage") and grain-boundary cavity growth ("creep damage"). The growth rates are considered to follow the following form, where $|\epsilon_p|$ and $|\dot{\epsilon}_p|$ represent the current absolute value of plastic strain and strain rate, respectively:

For fatigue crack growth,

$$\frac{1}{a}\frac{da}{dt} = \begin{cases} T|\epsilon_p|^m|\dot{\epsilon}_p|^k & \text{for tension} \\ C|\epsilon_p|^m|\dot{\epsilon}_p|^k & \text{for compression} \end{cases} \quad (4\text{-}6)$$

and for creep cavity growth,

$$\frac{1}{c}\frac{dc}{dt} = \begin{cases} G|\epsilon_p|^m|\dot{\epsilon}_p|^{k_d} & \text{for tension} \\ -G|\epsilon_p|^m|\dot{\epsilon}_p|^{k_c} & \text{for compression} \end{cases} \quad (4\text{-}7)$$

where T, C, G, m, k, k_c, and k_d are material parameters that are functions of temperature, microstructure, and environment.

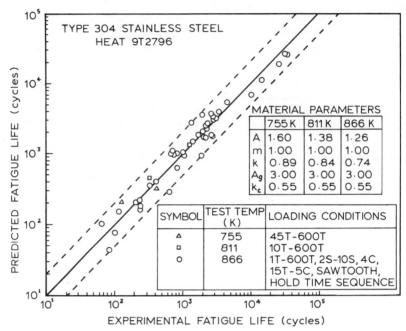

Fig. 4-22. Comparison of experimentally observed and predicted (by damage rate model) fatigue lives for Type 304 stainless steel under loading of various waveforms (Ref. 69). T is tensile hold, S is symmetrical hold, and C is compressive hold.

Fatigue damage and creep damage are assumed to accumulate independently, and the relationship

$$\frac{\ln a/a_0}{\ln a_f/a_0} + \frac{\ln c/c_0}{\ln c_f/c_0} = 1 \qquad (4\text{-}8)$$

governs failure of the low-cycle fatigue specimen. One can easily see that this equation is a form of summation rule similar to Eq. (1-11) in Code Case 1592, but is derived from actual phenomenological observations.

Majumdar and Maiya's procedure allows one to take into account the effect of various wave shapes on the low-cycle fatigue life. Figure 4-22, taken from their paper,[69] compares the predicted versus experimentally observed hold-time fatigue life in the temperature range 755 to 807 K. In most cases, the predicted life lies within a factor of 2 (dashed line) of the experimentally observed life, giving some credence to the various assumptions in their model. In fairness, however, it should be noted that both the linear damage rule and the strain range partitioning method would also predict this particular data set reasonably well. The linear damage rule yields nonconservative results for loading cases that involve a sawtooth waveform, but otherwise, at this juncture, there is little difference between the various treatments of creep-fatigue upon which to base a judgment. Analyses of the environmental interactions of sodium, aging, and irradiation, discussed below, will likely reveal the most appropriate method (e.g., Fig. 4-21), but relatively little work has been done on this to date.

Sodium. Almost all the tests on the effects of sodium have been continuous cycling in nature and at high temperatures (≥ 870 K).[71] Fatigue endurance of the austenitic steels is generally higher in sodium than in air, but the "enhancement" is a function of both oxygen and carbon potentials in the sodium. In one series of tests in high-carbon sodium, for example, the fatigue lives were comparable to air data.[71]

The potential for reduced fatigue endurance due to carburization recently led Zeman and Smith[72] to conduct a series of tests on Types 304 and 316 stainless steel at 823 K, in sodium containing 1 ppm oxygen and 0.3 ppm carbon, which conditions are slightly carburizing. They found that the fatigue lifetime of annealed Type 316 stainless steel is a factor of 3 to 4 greater when tested in sodium than when tested in air, whereas the fatigue lifetime of annealed Type 304 stainless steel does not differ substantially in the two environments. These results are in agreement with the earlier data at higher temperatures, up to ~970 K. In Fig. 4-23, $n = 0.5$ in the Coffin–Manson equation [Chapter 1, Eq. (1-10)] adequately represents the data in sodium. The fatigue life of Type 304 in sodium was also determined after thermally aging the material in 823 K argon for 18 Ms (5000 h). The fatigue lifetime of the thermally aged material tested in

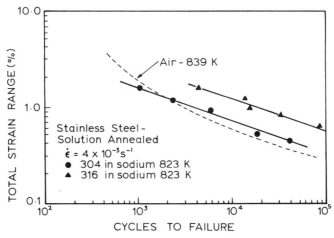

Fig. 4-23. Total strain range versus cycles to failure for solution-annealed Types 304 and 316 stainless steel in sodium at 823 K (Ref. 71).

sodium is slightly longer than that of the annealed material tested in sodium. An 18-Ms (5000-h) preexposure to the slightly carburizing sodium at 823 K, on the other hand, affected the fatigue lifetime of Type 304 more than that of Type 316 stainless steel. The sodium-exposed Type 304 had a shorter fatigue life than the annealed material at strain ranges above $\Delta\epsilon_t = 0.8\%$ and a longer fatigue life at strain ranges below this value. The data for Type 316 indicate little effect of pre-exposure in sodium under these conditions. This difference may be due in part to the substantially greater carbon penetration depth observed for Type 304 stainless steel.

From these data (and other data on crack growth in sodium to be described later) one would conclude, as Coffin[65] did, that there is an environmental component to elevated-temperature fatigue. Under continuous cycling fatigue, except when carburization occurs, sodium is a less aggressive environment than air. However, preliminary data from fatigue tests where hold times are included indicate that behavior in sodium is identical to that in air.[71] This result is not understood.

Temperature (and Thermal Aging). The importance of both creep and environmental effects to fatigue life as temperature is increased was mentioned earlier. Fatigue life is sharply reduced at temperatures above $0.5T_m$ and intergranular fracture predominates. However, for pressure boundary components that operate for the most part below $0.5T_m$, it is the indirect, long-term effects of temperature that are of more importance. The "thermal aging" process is manifested in a variety of ways—oxidation of inclusions or precipitates (e.g., carbides) to form voids, grain growth, coalescence, and growth of precipitates, etc.—that can influence the initiation and subsequent growth of fatigue cracks.

Although there is a lack of data on fatigue life of the austenitic steels as a function of long-term aging, a number of studies indicate potentially important effects. One such effect, that of carburization, was mentioned above. A second effect, that of grain size and precipitate distribution interaction, was seen by Maiya and Majumdar[73] in elevated-temperature, low-cycle fatigue tests on three different heats of Type 304 stainless steel. The three heats of steel showed approximately the same continuous-cycling low-cycle fatigue behavior as that of the reference heat (used for all U.S. LMFBR test programs) despite having different microstructural features. However, the three materials showed improved fatigue strength during tensile hold-time conditions where significant creep occurs (Fig. 4-24). On the basis of microstructural observations, the improved creep-fatigue resistance of the three heats can be attributed to a smaller grain size and a more closely spaced $(CrFe)_{23}C_6$ carbide precipitate distribution at the grain boundaries, both of which result in increased resistance to

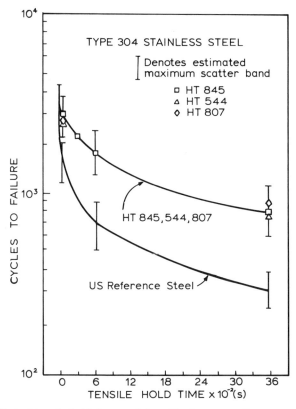

Fig. 4-24. Effect of tensile hold time on fatigue life for three different heats of Type 304 stainless steel at 866 K. $\Delta\epsilon_E = 1\%$ (Ref. 73).

grain boundary sliding. The more extensive data base on the effect of aging on fatigue crack growth will be discussed later.

Irradiation. The effects of fast neutron irradiation on the low-cycle fatigue performance of austenitic stainless steels have been the subject of considerable interest from the earliest days of the LMFBR program. However, interestingly, none of the three fatigue models described earlier treat the effects of radiation damage. The reason given is that the effects will be negligible for the low fluences likely to be experienced over the service life. While this is certainly true for the piping, the effect of irradiation on vessel fatigue life (particularly the welds) should be considered, inasmuch as the recent data of van der Schaaf et al,[64] shown earlier in Fig. 4-21, indicate a marked reduction in average total damage factor D from 0.86 to 1.0 in the reference condition to 0.42 to 0.47 in the irradiated condition. Further, none of the conditions satisfied the linear damage rule, and the bilinear rule could be considered a lower bound only for the plate material.

Michel and Korth[74] recently reviewed the status of low-cycle fatigue data obtained from post-irradiation tests on materials irradiated in EBR-II. Type 304 and 316 stainless steels and Type 308 weld metal specimens were tested after fast neutron fluences in excess of 10^{26} n/m² ($E > 0.1$ MeV) in the temperature range 673 to 973 K (i.e., covering both fatigue regimes). These data should be treated with caution inasmuch as properties measured in post-irradiation tests will not correctly reflect the effects of *in-situ* irradiation. Post-irradiation data should in fact be a conservative measure of property degradation because the enhanced in-reactor stress relaxation effects are absent.

Figure 4-25 shows the trends of fatigue life reductions for cycle lives up to 5000 cycles-to-fail as a function of neutron fluence. A more detailed analysis of the data used in Fig. 4-25 shows that when the test temperature and irradiation temperatures were identical there was very little effect of irradiation on the fatigue behavior of either base metal or weld metal. However, when the test temperatures (T_t) and irradiation temperatures (T_i) differed, in either direction, fatigue life was reduced; reduction factors ranged from 1.4 to 1.8 when $T_i > T_t$ and from 1.5 to 2.5 when $T_t > T_i$.[74]

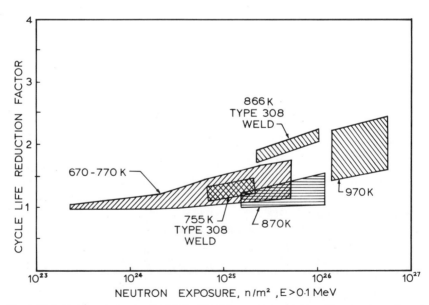

Fig. 4-25. Reductions in low-cycle fatigue life of Types 304 and 316 stainless steel as affected by neutron exposure and determined in post-irradiation tests (Ref. 74).

The addition of tensile hold times to investigate creep-fatigue interaction produced marked effects on fatigue life, as expected. A 866 K test on Type 316 stainless steel, for example, produced a reduction factor of 37 when a 0.5-h tensile hold period was added to each cycle, compared to only 1.5 for the cyclic fatigue case. This observation is similar to the results shown in Fig. 4-21, which were obtained at 823 K after a 7×10^{22} n/m² exposure.[64] The reduction in D was attributed to the early formation of intergranular cracks at low creep damage fractions and accelerated growth of these cracks enhanced by helium formation at the grain boundary. In similar tests for Type 316 at 973 K, however, the reduction-factor difference was only 2.7 versus 1.8 when a 0.1-h tensile hold time was employed.[74] Either damage must be annealing out in the higher temperature tests, or grain boundary sliding must be facilitating stress relaxation during the hold periods. The analogy between temperature-enhanced diffusion-controlled processes and *in-situ* irradiation-enhanced processes has been widely used. Therefore, one could well argue here that in-reactor creep-fatigue behavior might be similar to these higher-temperature observations.

The effect of prior cold work becomes pronounced in post-irradiation creep-fatigue tests.[74] At 866 K, 11% cold-worked Type 316 stainless steel exhibited a significantly reduced fatigue life particularly at the higher end of the neutron fluence range. However, Type 304 steel was not significantly affected by cold work. At the higher test and irradiation temperatures, both cold-worked Types 304 and 316 stainless steels exhibited significant reductions in cycle life over the annealed materials. The reason cold work increases the irradiation damage effect when testing in creep-fatigue and not necessarily so in fatigue testing without hold times is not adequately understood.[74]

It must be noted that the reductions in fatigue life due to irradiation have only been confirmed in the low-cycle regime (up to ~5000 to 10,000 cycles to fail). Above this regime, limited tests have indicated a reversal in the situation, with high-cycle properties being enhanced by the neutron exposure. A possible explanation for this effect has been advanced by Korth and Harper,[75] utilizing the strain partitioning equation, Eq. (1-13) (and Fig. 1-15) earlier. From this equation, it can be seen that when the plastic component is dominant (low-cycle regime), ductility will be the overriding mechanical property that will influence fatigue behavior, and reductions in ductility due to irradiation will therefore reduce the fatigue life. However, if the total strain range decreases to the point that elastic strains are dominant (high-cycle regime), then ultimate strength will be the overriding factor, and irradiated material, which has a higher ultimate strength, will show an increased life when compared to unirradiated material. Using the universal slopes equation, which is a form of Eq. (1-13),

Brinkman et al.[76] has demonstrated the crossover of irradiated fatigue data, as shown in Fig. 4-26.

Thus, the effects of irradiation on fatigue behavior should be predictable from knowledge of the changes in ultimate strength and ductility, at least for the post-irradiation condition. The fraction modification method was developed to provide such a prediction;[74] Eq. (1-12) becomes

$$(\Delta\epsilon_t)_{irrad} \simeq \phi_u A N_f^a + \phi_t B N_f^b \qquad (4\text{-}9)$$

where

$$\phi_u = \frac{(\phi_{UTS})_{irrad}}{(\phi_{UTS})_{unirrad}}$$

and

$$\phi_t = \left[\frac{D_{irrad}}{D_{unirrad}}\right]^{0.6}$$

and where σ_{UTS} and D are defined as ultimate strength and ductility, respectively, and the exponent 0.6 comes from the universal slopes equation.

Using this fraction modification method, very good correlation has been observed between predicted and actual irradiated fatigue data, with the predictions being slightly conservative.[77,78] Whether one can extend this approach to the *in-situ* irradiation condition is questionable, however.

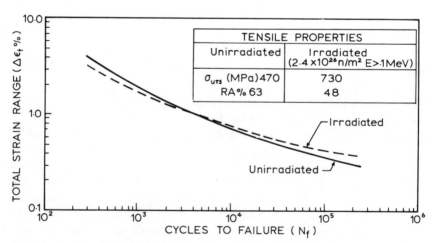

Fig. 4-26. Estimated irradiated fatigue properties of Type 316 stainless steel based on tensile properties at 703 K (Ref. 74).

In the case of creep-fatigue interaction, correlations and predictions have not developed to the level as have the methods for straight fatigue. The strain range partitioning (SRP) approach has been applied to irradiated stainless steels,[79] but insufficient data are available to determine whether correlations developed from short-term data accurately predict long-term behavior. Nevertheless, Michel and Korth[74] preferred the SRP approach to either the extrapolation or summation approaches. Extrapolation, in particular, is considered much too conservative due to a tendency toward saturation in fatigue damage at long times.

4.3.2.2. Fatigue Crack Growth

As was the case for ferritic steels, fatigue crack growth in austenitic steels in the subcreep regime has been very satisfactorily treated by using the tools of LEFM. The modeling of the process is well understood and very widely used (as noted from the LMFBR example in Section 4.2.1.). Even in the presence of such factors as corrosive environmental effects and temporary healing effect due to overload cycles, fracture mechanics seems to provide adequate modeling.[80]

The effect of the cyclic load profile and multiaxial stresses on the fatigue crack propagation has not been adequately studied, however. Regarding the load profile, in practice the tendency has been to use a linear cumulative damage rule. As for the effect of combined stress state, the very limited amount of results obtained thus far indicates that it is of secondary importance. Hence, in practice, the effect of load biaxiality is ignored and the maximum principal stress (or the stress intensity factor based on the maximum stress) is assumed to be the controlling load factor.[81]

The LEFM treatment of fatigue crack growth has, to date, proven adequate for most of the LMFBR pressure boundary, since creep effects are generally minimized through design.[81] Crack growth-stress intensity factor K correlations, appropriately parametrized for the effects of temperature, stress ratio, frequency, etc., have found success in predicting crack growth well into the creep regime. As will be discussed in more detail below, the Carden parameter can correlate Type 304 fatigue crack growth data in air to ~ 923 K.[82] Wave shape was not a factor in this algorithm, however, and it is not clear just how valid LEFM will be for situations involving significant hold times, which would allow the gross region in the vicinity of the crack to become fully plastic. The limitations of fracture mechanics analyses to the creep crack growth situation were discussed in Chapter 1. The developing fields of GYFM and creep fracture mechanics might therefore see application in SI analyses in this regime.

Notwithstanding these latter comments, however, the data that will be described below support the use of the single parameter, K description of fatigue crack growth, and the resulting linear plots of log da/dN versus ΔK over a wide range of conditions. Following the lines of the low-cycle fatigue discussion, fatigue crack growth data will be presented for Types 304 and 316 steels as a function of loading history, sodium environment, temperature (including aging), and irradiation. Extensive use will be made of the excellent reviews published by James over the past four years.[80,83,84] In reviewing these data, in comparison to the low-cycle fatigue data, some interesting points emerge: (1) the trends of fatigue life and crack growth rate with several of the aforementioned parameters are similar; that is, as fatigue life increases, crack growth rate decreases, and vice versa; (2) notwithstanding point (1) there are some differences evident in the two data bases, which must be due to the complication of considering three stages incorporated in the fatigue life data versus only one stage in the crack growth data; (3) perhaps because of point (2) the crack growth data from several sources appear to be in less conflict and more amenable to interpretation; and (4) the effects of corrosion appear to dominate over creep in elevated-temperature crack growth data, presumably due to the strong effect of environment on conditions at the crack tip.

Loading History. The effects of cyclic frequency, waveform (hold time), and stress ratio R on fatigue crack growth in austenitic stainless steels have been studied primarily in an air environment.[80]

The first two variables are interrelated in that decreased frequency is generally associated with increased hold time. This fact may account for the reported differences in behavior between (a) different fatigue crack growth studies and (b) fatigue crack growth and fatigue life studies. Nevertheless, the data do allow two broad generalizations that are similar to those noted by Bamford[39] and Scott[40] for the ferritic steels. These are: (a) fatigue crack growth rate increases as cyclic frequency is decreased over a broad temperature range (in one study[85] by a factor ~8 when cyclic frequency is decreased by a factor of ~50,000); and (b) the imposition of hold time upon the fatigue cycle does not appear to be detrimental beyond that effect associated with the decrease in cyclic frequency; in fact, hold time may be less damaging, at least at 811 K.[86]

The LMFBR *Nuclear Systems Materials Handbook*[62] employs a frequency/temperature correction factor to obtain the desired crack growth relationship for that frequency/temperature combination. The crack growth data are normalized to data at a standard frequency of 0.67 cps. This approach has been applied to the austenitic steels by James.[87] A representative plot of relative crack growth behavior versus cyclic frequency for Types 304 and 316 stainless steels is shown in Fig. 4-27, along with the resulting correction equation established for Type 304:

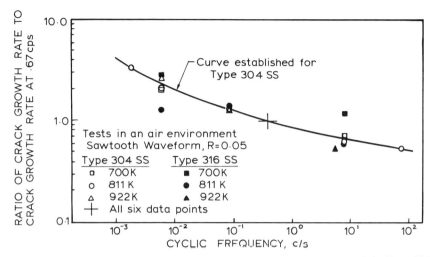

Fig. 4-27. Comparison of relative frequency effects on fatigue crack growth in Type 316 stainless steel with behavior of Type 304 stainless steel over the temperature range 700–922 K (adapted from Ref. 86).

$$Y = 0.295075 - 0.226644(X) + 0.0321061(X)^2 - 0.00338085(X)^3$$

(4-10)

where X is log (frequency in cpm) and Y is log (crack growth ratio). It is seen that the correction factor for Type 304 can also account for the trend in behavior of Type 316, although there is considerable scatter. James[87] cautions that this relation should not be extrapolated beyond the frequency range for which data were obtained. Although promising, this approach clearly requires more data to provide statistical significance to the correlation.

The data discussed above were obtained at constant R ratio. Data for annealed Type 304 stainless steel tested in air at 811 K show that the crack growth rate increases with increasing values of R over the range $-0.15 < R < 0.75$, which is similar to the behavior for ferritic steels, summarized in Fig. 4-12 earlier.

The use of the effective stress intensity factor, K_{eff} in place of ΔK has been proposed to account for the R effect; thus the crack growth law of Eq. (1-8) becomes

$$\frac{da}{dN} = C(K_{\max}[1 - R]^m)^n$$

(4-11)

where $K_{\text{eff}} = K\max[1 - R]^m$, and the constant m is determined empirically and is dependent upon the material and temperature. James[88] has demonstrated the applicability of this correlation over the range $0.063 < R < 0.807$. However, this so-called Walker parameter was recently rejected by Bamford[39] in the LWR vessel study reported earlier as being much too conservative, since it causes crack growth to increase exponentially as the R ratio approaches unity and therefore does not recognize the observed saturation effect at $R \approx 0.85$ in ferritic steels. Most probably the LMFBR components will not experience the R ratios that the corresponding LWR components are designed for, but this particular point should be considered for the austenitic steels in order to avoid unnecessary conservatism in SI analyses.

The Carden parameter[65,80,82] P attempts to combine the frequency (ν)/temperature (T) effect with the stress ratio (R) effect, and also accounts for the presence of a crack growth threshold. The equation is a very general one:

$$da/dN = MP\alpha \qquad (4\text{-}12)$$

where

$$P = \{[A_{1101} \exp(-\Delta H_1/RT) + A_{1102} \exp(-\Delta H_2/RT)](1/\nu)^k$$

$$+ C_{1101} \exp(\Delta H_3/RT)(1/\nu)\}[(K_{\max}^2 - K_{\text{th}}^2)(1 - R)^{2n}]^m \qquad (4\text{-}13)$$

and where M, α, A_{1101}, A_{1102}, C_{1101}, K, n, m, ΔH_1, ΔH_2, and ΔH_3, are constants that are independent of temperature; $K_{\max}$ is maximum stress intensity; and K_{th} is threshold stress intensity below which crack growth does not occur. Although this equation contains many constants, they do not change with temperature or frequency; once determined, they are supposedly valid for all conditions. Figure 4-28 shows the results obtained by Carden[82] in correlating data on Type 304 stainless steel from several sources. The slope of crack growth rate versus P is 45, so that $\alpha = 1$, and M can be determined from any single point on the curve. The agreement here is very good over a wide range of temperature and frequency.

Overall, it appears that the understanding of loading history effects on fatigue crack growth in Types 304 and 316 stainless steel would benefit from more data at lower frequencies and higher R ratios, with emphasis on sodium and irradiation environments.

Sodium. The fatigue-crack growth rate for Types 304 and 316 stainless steel in a sodium environment has generally been considered to be lower than in air environment at the same temperature.[80] In fact, it will be noted in Fig. 4-29 that the crack growth rates for Type 304 stainless steel in a sodium environment at 700 K (and also at 811 K, not shown here) are approximately the same as those for an air environment at room temperature. This characteristic has also been exhibited by Type 316 steel at 873 K. Figure 4-29 also shows that crack growth rates are essentially

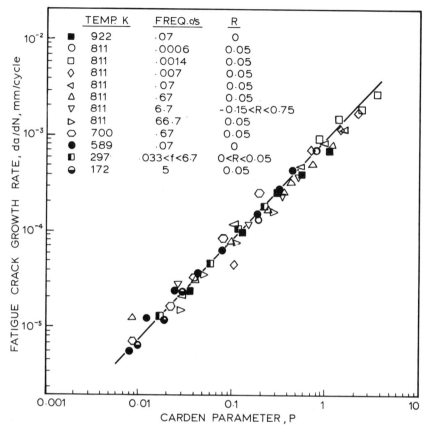

Fig. 4-28. Use of the Carden parameter P to correlate crack growth behavior of annealed Type 304 stainless steel tested in air over a wide range of temperatures, cyclic frequencies, and stress ratios (Ref. 82).

the same in vacuo and in sodium, thereby suggesting that liquid sodium is approximately as inert as a vacuum.

New data now show that fatigue crack growth can be accelerated in heavily carburizing and heavily decarburizing sodium environments. Results reviewed by Lloyd[71] indicate that, at 873 K, fatigue crack growth rates follow the trend: carburizing sodium > decarburizing sodium > air > low oxygen sodium. Typically crack growth rates in the carbon-bearing sodiums are increased by a factor of 5 over those observed for low-oxygen sodium or air. The reasons for this observation are not understood, however.

Temperature (and Thermal Aging). As might be expected from the earlier fatigue discussion, fatigue crack growth rates in austenitic steels and their weldments increase with increasing temperature, at least in an

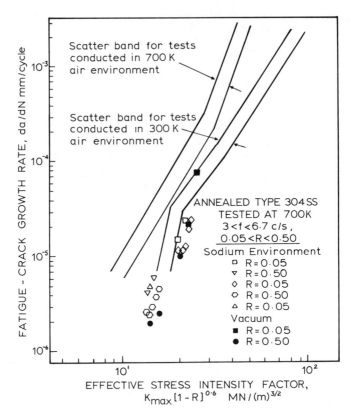

Fig. 4-29. Fatigue-crack propagation behavior of annealed Type 304 stainless steel in sodium and vacuum environments at 700 K (Ref. 80).

air environment. Measurements in air have been made from room temperature up to ~970 K. A simple Arrhenius rate-process equation cannot be used to correlate the data over such a wide temperature range, but, as shown in Fig. 4-28 earlier, a more complex rate process equation such as that due to Carden appears to apply.[82] The effect of temperature on fatigue crack growth in a sodium environment has only been measured over a narrow temperature range, but the crack growth rates measured at 700 and 811 K in Type 304 stainless steel and at 773 K in Type 316 stainless steel are similar, thus indicating an absence of thermal activation, at least at temperatures up to ~$0.5T_m$.[80]

The long-term effects of temperature, namely thermal aging, on crack growth behavior have been studied extensively.[83] To simulate 20–30 year conditions, aging times are usually accelerated by aging at temperatures higher than the intended service temperature. Such a procedure is only valid so long as the difference between service and test temperatures is not so great as to result in different types, sizes, or distributions of precipitates.

In the absence of hold times, thermal aging in austenitic stainless steels and weldments appears to have little or no influence on fatigue crack growth rates. In those cases where an effect is noted, however, the trend is toward slightly lower growth rates in the aged material.[83] The mode of crack extension under continuous cycling is generally predominantly transgranular. Since precipitation of second-phase particles is concentrated in the vicinity of the grain boundaries, it is therefore reasonable that tests involving transgranular cracks would not exhibit large thermal aging effects.

When hold times are employed in the testing, however, thermal aging effects do become apparent. As noted earlier in the discussion of Maiya and Majumdar's[69] work on fatigue life at elevated temperatures, as the hold time is increased (or cyclic frequency is decreased) a transition from predominantly transgranular to predominantly intergranular crack extension occurs. The hold time (or frequency) at which this transition takes place is a function of the alloy, temperature, and environment.

Michel and Smith[89] reported no apparent effect of hold time (or frequency) in unaged Type 316 stainless steel tested at 700 K, and somewhat mixed results in the aged material. Cracking was primarily transgranular for both types of material and crack growth rates were generally lower in the aged material. As Fig. 4-30 shows, however, at 866 K the hold-time effect was observed in unaged Type 316 (with a transition from transgranular to intergranular cracking occurring with increasing hold time) but not in the aged material (where cracking was primarily transgranular at all hold times). Michel and Smith[89] concluded that the second-phase precipitation along the grain boundaries in the aged material delayed the mode transition by suppressing grain boundary sliding. This is consistent with the explanation offered by Maiya and Majumdar[69] for the increased fatigue life of certain heats of Type 304 stainless steel.

Thus, in general, long-term thermal aging does not degrade fatigue-crack growth resistance and, in several instances, it has been shown to be beneficial. It must be cautioned, however, that once again these are all air environment data. In sodium, the long-term carburization effects could influence fatigue crack growth rates as they influenced fatigue life in Zeman and Smith's[72] study discussed earlier. This is an area where clearly more work is required.

Irradiation. Tests on irradiated austenitic steels have produced mixed results with respect to influence on fatigue crack growth. Some studies have shown higher growth rates in irradiated material, while others show just the reverse. As did Michel and Korth[74] in their review of irradiation

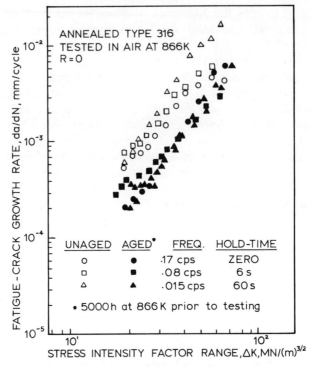

Fig. 4-30. Fatigue-crack growth behavior of unaged and aged annealed Type 316 stainless steel tested at 866 K with and without hold times (Ref. 83).

effects on fatigue life, James,[83] in his review of fatigue crack growth behavior, found that where differences (either higher or lower) between unirradiated and irradiated data are noted, a relatively large difference exists (e.g., greater than 50 K) between the irradiation and test temperature. For instance, neutron irradiation up to fluences of 5×10^{26} n/m² produced no significant effect on the crack propagation rate of annealed austenitic stainless steels in air when the temperature (700 K) is within ±50 K of the irradiation temperature. However, the increase in test temperature to 866 K produces a small increase in crack propagation rate when compared with unirradiated tests at 866 K. Likewise, for cold-worked austenitic steels, comparison with unirradiated results indicates that there is no significant effect of neutron irradiation at 700 K on crack propagation rate, but an increase in test temperature from 700 to 866 K results in a substantial increase in crack propagation rate which exceeds the increase produced in unirradiated material by the equivalent increase in test temperature.[74]

The inclusion of tensile hold times up to 60 s produces no significant effect on crack propagation rate in annealed and cold-worked neutron-irradiated and unirradiated Type 316 stainless steel at 700 K. However, at a test temperature of 866 K, the effect of hold time is to produce a small increase in crack propagation rate. In cold-worked, neutron-irradiated Type 316 steel, the effect of hold time at 866 K is to produce nearly an order of magnitude increase in crack propagation rate when compared with equivalent unirradiated test results.[74]

For austenitic stainless steel welds, the available results show that there is little effect of neutron irradiation to 1.2×10^{26} n/m^2 on crack propagation rate in Type 316 weld metal. Results for Type 308 weld metal indicate that an increase in fluence from 1.2 to 3.6×10^{26} n/m^2 produces an increase in crack propagation rate for submerged-arc welds at 866 K, as opposed to shielded-metal-arc welds where no significant effect on crack propagation rate is observed.[74,83]

Once again, all these data were collected from post-irradiation tests in an air environment. The synergistic effects of *in-situ* neutron irradiation and sodium therefore need to be evaluated. For pure cyclic fatigue it is reasonable to believe that the sodium environment would have the same effect on crack propagation performance as in unirradiated material, that is, a beneficial effect. However, with the incorporation of hold times, which more realistically simulate the expected duty cycle, the possibility of a complex creep–corrosion interaction exists, and insufficient data are available to fully characterize this more prototypic operating regime.

4.3.2.3. Closing Remarks

Reference to Fig. 1-8 in Chapter 1 shows that the structural integrity analysis (of generally thick sections) starts from the premise that a flaw exists in the material. Crack-propagation analysis methods must therefore be used to deal with the growth of pre-existing defects, and, for LMFBR vessel and piping materials, fatigue has been assumed to be the predominant growth mechanism. The fatigue crack growth data described in Section 4.3.2.2 indicate that (a) most of measurements have been made in air, which appears to be more aggressive than pure sodium; (b) thermal aging and irradiation have little effect on crack growth at least in post-exposure tests; and (c) the weld materials do not behave any better or worse than Types 304 and 316 stainless steel. With that as a basis, then, the LEFM-based parametric correlation due to Carden,[82] shown in Fig. 4-28, should conservatively describe fatigue crack growth in the austenitic stainless steels over the temperature range of interest in the LMFBR. The single greatest uncertainty, however, is whether a relation based on relatively short-time, high-frequency data can be valid for a 30- to 40-year time frame, with cycle frequencies of months. Since Carden's parameters lack a physical basis, it is difficult to answer the question of extrapolation to low frequencies. However, the initial success in correlating such a broad range of variables surely justifies continued work on this approach.

In thin sections, or where the probability is high that a material is free of significant defects (and that eventuality must be considered attainable) use of crack-growth analysis methods alone is impractical because credit for the life expended in the initiation stage is lost, and this stage can account for a large percentage of the total life. In the time-independent or subcreep regime, either the Code Case 1592 curve (Fig. 4-20) or the Coffin–Manson life-prediction equation[65] appear to appropriately describe low-cycle fatigue behavior. Again, air environment data are generally conservative and could be utilized to account for uncertainties due to the time-dependent sodium and temperature (aging) effects. In the time-dependent fatigue regime, either the frequency separation or strain range partitioning method can be extended to treat the prototypic low-strain ($<0.5\%$ over $\geqslant 30$ years) problem.[65] The former method is basically set up to treat low strains since it emphasizes environmental fatigue; the latter method is basically conceived to work with plastic strains, although there is no requirement that the strains be large. The damage law of Maiya and Majumdar[69] seems to be more appropriate at higher temperatures (and under irradiation) where creep effects, and in particular creep cavitation, become important. However, this is at the "upper end" of the normal LMFBR operating regime and moving into the CTR regime. In any event the primary need is for data under conditions that simulate the combined LMFBR environment, rather than separate effects properties.

4.4. Materials Improvements

This section will be devoted almost exclusively to improvements to LWR pressure boundary materials inasmuch as experience with the corresponding LMFBR materials is too limited to assess the direction(s) that improvements should take. Nevertheless, it should be evident that the trends evident in LWR pressure boundary technology are likely to continue in the advanced reactors, particularly the move to cleaner steels and the search for "more forgiving" materials. Further, the techniques to reduce residual stresses and to produce defect-free castings are directly applicable to LMFBR components.

4.4.1. Improvements to Vessel Steels

Recently, considerable improvements have been made in the properties of LWR pressure steels as a result of the increase in understanding of the role of residual elements.

It is generally realized that to obtain steel with a high fracture toughness where the failure mode is ductile–dimple fracture, the basic requirement is to combine a relatively low yield strength with the absence of any large number of inclusions, since these have been found to nucleate the microvoids in the crack tip plastic zone that link up for easy crack propagation.[12] The principal inclusions are oxides, slag, and sulfides, and, on a finer scale, carbides. Increased fracture toughness would result if these inclusions could be reduced in total quantity, and any that remain could be dispersed on a finer scale, preferably with uniform particle shape.

The approach to achieving these improved steels has involved reducing both the sulfur and carbon contents. With respect to the former, the trend is towards a range of 0.010 to 0.015 wt.%; in the case of the latter element, reduction of the upper permitted carbon level to below 0.22 wt.% appears to offer several advantages including reduction in the incidence of large chromium carbide particles, limiting the yield strength (to an upper limit of ~450 MPa) and reducing the incidence of under-clad cracks and cracks in the HAZs of the main vessel welds.[23,90,91] The reduction in chromium has also been found to be beneficial in reducing under-clad cracks and reheat cracks in vessel steel.

The process for making these steels involves the melting of carefully selected scrap (to achieve a low level of copper and other residual elements) in a basic electric furnace, followed by vacuum treatment to remove hydrogen and to allow some inclusions to float out. The proposal to adopt electroslag refining for the final stage is very attractive since some impurities would be reduced in concentration and any remaining second-phase particles would be very finely dispersed.[91] Steel of increased toughness should therefore become available if this process were adopted. One drawback of ESR, however, which will be discussed in Chapter 7 in connection with turbine rotor ingots, is the limitations on ingot size.

Heat treatment is also an important method to enhance the toughness properties of the ferritic steels. Austenitizing in the 1143 to 1193 K temperature range, associated with careful control of the grain-refining elements, such as aluminum, are effective in avoiding austenitic grain growth before quenching.[90] (Austenitic grain size acts as a significant factor in the location of the ductile–brittle transition temperature.)

Control of tempering and postweld heat treatment (PWHT) temperatures is a second factor of importance and preliminary results indicate that the tempering temperature and parameter specified by ASME for reduction of residual stresses may not be optimum[90] for toughness and strength.

4.4.2. Remedies for BWR Pipe Cracking

4.4.2.1. Modifications to Type 304 Stainless Steel

Of more immediate interest are the remedies emerging from the pipe cracking program that apply to existing plants for repairs and replacement of pipes and to plants under construction that are committed to the use of Type 304 stainless steel piping. Based on the data to date, near-term remedies have focused on the sensitization and stress contributors to IGSCC.[44,46,92] Four potentially practical remedies that are attractive from the standpoints of cost and ease of implementation have been identified.[44,46,92] These are solution heat treatment (SHT), corrosion-resistant cladding (CRC), induction heating stress improvement (IHSI), and heat-sink welding (HSW). Additional details are given below.

(*a*) *Solution heat treatment* of the weldment eliminates weld sensitization and also relieves the weld residual stresses and, therefore, is being adopted for most new BWR shop butt welds of Type 304 piping. This technique is difficult to control under field conditions, but nevertheless is also being developed by the Japanese for existing field welds. Piping subjected to solution heat treatments has been shown to be highly resistant to IGSCC under BWR operation conditions.

(*b*) *Corrosion-resistant cladding* is proposed for both shop and field welds. The approach is to isolate the HAZ in Type 304 stainless steel weldments from the reactor water by using corrosion-resistant cladding—duplex microstructure, consisting of austenite and ferrite—on the pipe's inside surface. Type 308L stainless steel, with a ferrite content of at least 8% after welding, appears to be a good cladding material for this application and is being evaluated in U.S. systems.

(*c*) *Induction heating stress improvement* is a technique that redistributes tensile residual stresses so that the inside surface of the pipe is left in a favorable state of axial compression. This is achieved by the use of induction heaters to heat the outer pipe wall, while the inside surface of the pipe is cooled with water. When the heat is applied, the outside surface undergoes thermal expansion causing the cooler inside surface to be stretched inelastically. Upon cooldown, the outside surface contracts and the inside surface is compressed.

This technique is being adopted widely in Japan.[44,93] It appears to provide adequate stress relief in smaller-diameter piping, but there is some question about its effectiveness in larger-diameter lines because the greater wall thicknesses provide an excellent heat sink that will compete with the water cooling on the inside of the piping.

Also, the use of IHSI on existing plant welds raises some questions, because the IHSI initially subjects the inside surface of the pipe to a high tensile stress. As a result, any existing cracks may be driven deeper, and the thermal-mechanical cycle may make the material more susceptible to low-temperature sensitization. Nevertheless a U.S. project is undertaking tests to determine how much IHSI improves the residual stress distribution in pipe welds of various diameters.

(d) *Heat-sink welding* for new welds has the same objective as IHSI, namely, to ensure a compressive axial residual stress on the pipe's inside surface in the vicinity of the weld. In HSW, after the root-weld pass, the pipe's inside surface is sprayed with cold water, or the pipe is filled with cold water. The heat in the outer portion of the weldment is provided by the welding process itself. Environmental pipe tests on heat-sink-welded 4-inch-diameter Type 304 stainless steel pipes have shown that heat-sink welding produces a significant improvement in resistance to IGSCC compared to reference welding procedures.[46] To date, heat-sink-welded pipe welds have been on test approximately five times longer than failed reference welds without showing any signs of cracking. This improvement factor is increasing as test time accumulates. However, it should be noted that the benefit due to heat-sink welding is reduced if grinding is performed after the heat-sink welding is complete. Grinding after the root pass or first several passes, but before the heat-sink welding, does not appear to cause any reduced benefit.

The benefit from heat-sink welding was originally thought to be due to reduced weld sensitization. While heat-sink welding does reduce the level of sensitization, Rybicki *et al.*[94] have recently calculated that heat-sink welding should also produce a residual stress distribution that would lower the overall stress level and thus reduce the tendency for IGSCC (Fig. 4-31). An interesting discovery by Rybicki and co-workers is that the favorable residual stress distribution may possibly be accomplished by applying inside cooling only during the final few welding passes, and that high heat input during these final passes enhances the favorable residual stress profile.

A statistical qualification test program is underway to demonstrate that these remedies could prevent IGSCC for the plant design lifetime.[92] This involves the testing of full-size welded pipes to provide results that could be scaled to reactor service. The criterion for qualification is improvement by a factor of 20 in the time to first failure of the improved pipe over the mean time to failure of the reference 304 stainless steel pipe. The factor of 20 translates to a conservative improvement that is

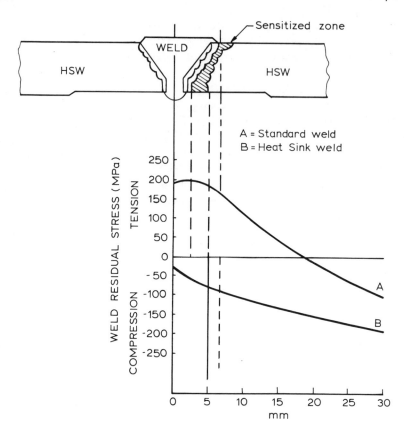

Fig. 4-31. Heat-sink welding-stress benefit in Type 304 piping components (courtesy R. Smith, EPRI).

equivalent to the plant design lifetime. The SHT and shop CRC pipe remedies have already been shown to far exceed (factors of improvement of 46 and 48 respectively) the test criterion, and, accordingly, they are currently being implemented in domestic and foreign BWRs, both in the repair of existing plants and in plants under construction.[92] Although the HSW and field CRC did not achieve the desired factor of 20 improvement, tests are continuing and, since they nevertheless did represent a considerable improvement over reference 304 stainless steel (factors of 9.3 and 6 respectively), they are being considered for limited applications in pipes that cannot be treated by the other remedies.[92]

4.4.2.2. Alternative Materials

As noted earlier in this chapter, more recent BWR designs have made increasing use of carbon steels, so that today the current BWR/6 standard plant has carbon steel in virtually all the primary pressure boundary piping systems. Also, replacement of 100-mm bypass lines with carbon steel piping has already been accomplished in some early BWRs.

While operating experience with carbon steel piping appears excellent—no incident of environmentally assisted cracking has been reported—laboratory studies do indicate a susceptibility to SCC.[95] Screening tests on welded pipe sections of SA106-GrB and SA333-Gr6 indicate that severe combinations of cyclic applied stress and high-temperature oxygenated water can result in environmentally enhanced cracking. No difference in cracking propensity was observed between SA106-GrB and SA333-Gr6, nor did postweld heat treatment show a beneficial effect. The cracks were transgranular, and the fracture surfaces showed evidence of cleavage or quasi-cleavage. Thus, the transgranular stress corrosion susceptibility of carbon steel does not appear to be associated with any specific metallurgical condition, such as the weld heat-affected zone. Rather, the transgranular stress corrosion of carbon steel occurs at geometrical stress concentrators, such as notches, the weld counterbore, and the weld fusion line. The potential impact of this lack of localized susceptibility to in-service inspection requirements has not yet been fully evaluated.

The carbon steels, however, appear to be more resistant to stress corrosion than sensitized Type 304 stainless steel especially at elevated temperatures. At 567 K, environments and stress rates that would produce stress-corrosion cracks in sensitized Type 304 stainless steel did not indicate stress corrosion for carbon steel. When stress corrosion did occur, high stresses and strains were required, indicating that stress corrosion of the carbon steels is a difficult process.

Preliminary results on SA333-Gr6 indicate that there is a region of tensile-ductility degradation (compared to results in dry air) at low temperatures and intermediate oxygen contents (~ 1.8 ppm O_2) in BWR quality water at strain rates of $\sim 10^{-5}$ min^{-1}. Note that a similar observation was previously made for sensitized Type 304 stainless steel (Fig. 4-18). However, in contrast to the sensitized stainless steel, where the degradation was due solely to intergranular stress-corrosion cracking, the degradation in carbon steel is attributed jointly to pitting and transgranular stress-corrosion cracking at the lower temperatures, while, at higher temperatures, transgranular stress-corrosion cracking is the main mode of ductility degradation. A *tentative* outline for this SCC susceptibility region for carbon steel is shown in Fig. 4-32, which is an O_2/T plot similar to Fig. 4-18. The data show that carbon steel (SA333-Gr6) might be susceptible to transgranular stress-corrosion cracking during start-up operations, especially with the oxygen/temperature combinations associated with the upper limit of observed BWR start-up conditions. Precisely what happens in terms of crack propagation rate when dynamic straining decreases at

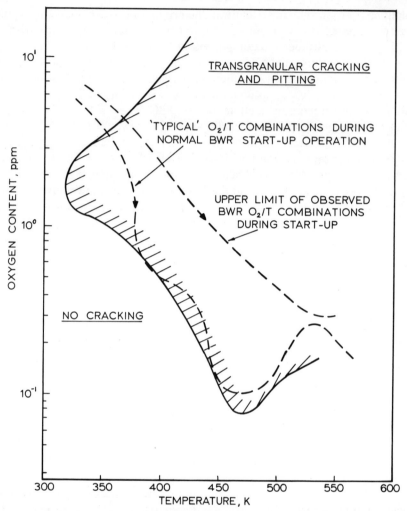

Fig. 4-32. Tentative boundary between cracking and no cracking for SA333-Gr6 carbon steel under dynamic straining at 2.6×10^{-7} s^{-1} as a function of oxygen content and temperature (Ref. 95).

the end of the start-up period and crack blockage ensues during long on-load periods is not known, nor is it known at this stage whether a particular crack initiated and grown in one start-up period necessarily continues to propagate on the second start-up period. Current research is investigating these points with the objective of developing realistic design stress rules for the carbon steels. In addition to controlled extension rate tests (CERT), crack growth rate tests, low-cycle fatigue studies, corrosion fatigue studies, and full-scale pipe tests on 100-mm-diameter pipe welds are being conducted.[44,46]

Other steels that are being actively evaluated as alternative piping materials include Types 316L, 304L, CF-3 (i.e., cast 304L) and 347 stainless steels.[44,46] Types 316L, 345, 304L, and CF-3 stainless steels have all shown over a factor of 20 improvement in IGSCC over reference pipe welds of Type 304 in environmental pipe testing. Based on work to date, Type 316L (K grade) is considered the prime replacement material with Type 304L (K grade) as the backup.[96]

4.4.3. Improvements to Castings

Cast valve and pump bodies and, in some plants, cast pipes comprise a significant part of the nuclear pressure boundary. The primary materials are the alloys CF8 and CF-8M, which are the cast counterparts of the wrought austenitic stainless steels, Types 304 and 316, respectively. Recently efforts have been initiated to improve the quality of castings through the use of hot isostatic processing (HIP).[97]

One of the major problems with cast materials is the high rejection rate and significant rework required to meet nuclear code specifications. To the Utility, this is manifested as high cost and unpredictable delivery schedules, both of high nuisance value. The major causes for rejection of cast stainless steel components are subsurface shrinkage and porosity, which are revealed by radiography, and surface imperfections, which are revealed by dye penetrant indications. Hot tearing is not as prevalent but has been encountered when constraint against shrinkage occurs in association with improper mold design. Center section shrinkage is by far the biggest problem and results in considerable scrap or appreciable rework and repair welding.

The hot isostatic processing or HIP route to improving castings is being evaluated as an alternative to the conventional weld repair approach, inasmuch as the latter introduces its own set of defects, with the result that repeated welding can result in a large casting containing up to one-third weld metal.

The first phase of a HIP evaluation program for the nuclear industry[97] produced worthwhile results. A HIP treatment for 3 h at 1376 K and 103 MPa was found to be effective in shrinking internal tears and gas pores until they become undetectable by radiographic means. However, this was only true for those defects which were not connected by a gas path to the surface of the casting. These stainless steel castings (plate and valve bodies) were characterized by a large amount of surface-connected porosity, which could only be eliminated by first sealing the component in a container. Obviously this is not an economic approach for a large casting, and so a method of surface sealing needs to be developed.

Of particular interest was the observation that HIP upgraded the defective materials to normal tensile mechanical property levels. For example, tensile elongation in defective regions ranged from nil to 20%. Similar regions after HIP were raised to a nominal 75% elongation. Associated increases in strength values also resulted. A statistical data base will be required, however, before ASME code acceptability of HIP is gained.

From these early results HIP promises to be an effective means of improving the quality of cast valve bodies. However, extension to larger castings such as pumps or piping will only proceed as the HIP equipment size progresses upwards.

References

1. USAEC—Advisory Committee on Reactor Safeguards, The Integrity of Reactor Vessels for Light-Water Power Reactors; WASH-1285, Washington (January 1974).
2. P. P. Zemanick, F. J. Witt, and R. F. Sacramo, Probabilistic Assessment of Primary Piping Integrity, 1976 ANS Annual Meeting, Toronto, Canada, June 14–18, 1976, *Trans. Am. Nucl. Soc.*, **23**, 382 (1976).
3. R. H. Koppe and E. A. Olson, Nuclear and Large Fossil Unit Operating Experience, Electric Power Research Institute, EPRI NP-1191 (1979).
4. R. L. Jones, T. U. Marston, S. T. Oldberg, and K. E. Stahlkopf, Pressure Boundary Technology Program: Progress 1974 through 1978. Electric Power Research Institute, NP 1103-SR (March 1979).
5. C. A. Rau, Jr., Quantitative Decisions Relative to Structural Integrity, paper presented at Conference on Structural Integrity Technology, Washington, D.C., May 9–11, 1979, sponsored by Materials Division, ASME.
5a. R. Snaider, BWR Feedwater Nozzle and Control Rod Drive Return Line Nozzle Cracking, NUREG-0619 for comment, U.S. Nuclear Regulatory Commission, Washington, D.C. (April 1980).
6. L. D. Blackburn and R. L. Knecht, Irradiation Effects and Design of LMFBR Permanent Reactor Structures, presented at the 5th International Conference on Reactor Shielding, Knoxville, Tennessee, April 18–23, 1977, *Atomkernenergie*, **30**, 278–281 (1977).
7. P. M. Besuner, L. M. Cohen, and J. L. McLean, The Effects of Location, Thermal Stress, and Residual Stress on Corner Cracks in Nozzles with Cladding, in: *Transactions of the 4th International Conference on Structural Mechanics in Reactor Technology*, 1977, T. A. Jaeger and B. A. Boley, eds., Paper G4/5.
8. L. A. James, Fatigue Crack Propagation Analysis of LMFBR Piping, in: *Coolant Boundary Integrity Consideration in Breeder Reactor Design*, Series PVP-PB-027, R. H. Mallet and B. R. Nair, eds., ASME, New York (1978).
9. T. U. Marston, The EPRI Ductile Fracture Research Program, in: *Proceedings of the Seminar on Fracture Mechanics*, ISPRA, Italy, (April 2–6, 1979).
10. J. C. Merkle, G. D. Whitman, and R. H. Bryan, An Evaluation of the HSST Program Intermediate Pressure Vessel Tests in Terms of Light Water Reactor Pressure Vessel Safety, ORNL-TM-5090, Oak Ridge National Laboratory, Oak Ridge, Tennessee (November 1975).

11. D. A. Canonico, The Heavy Section Steel Technology (HSST) Program, *Met. Prog.*, 32–40 (July 1979).
12. EPRI Ductile Fracture Research Review Document, prepared by Electric Power Research Institute, Palo Alto, California, T. U. Marston, ed., EPRI NP-701-SR (February 1978).
12a. C. E. Turner, Description of Stable and Unstable Crack Growth in the Elastic-Plastic Regime in Terms of J_r Curves, Proceedings of the 11th National Symposium on Fracture Mechanics, *ASTM STP 677*, Philadelphia, Pennsylvania.
13. P. C. Paris, H. Tada, A. Zahoor, and H. Ernst, The Theory of Instability of the Tearing Mode of Elastic-Plastic Crack Growth, U.S. NRC Report NUREG-0311 (1977).
14. J. R. Rice, Elastic-Plastic Models for Stable Crack Growth, in: *Mechanics and Mechanism of Crack Growth*, M. J. May, ed., Proc. of Conf. at Cambridge, England, April 1973, Physical Metallurgy Center Publication (1975), pp. 14–39.
15. J. W. Hutchinson and P. C. Paris, The Theory of Stability Analysis of *J*-Controlled Crack Growth, presented at ASTM Symposium on Elastic-Plastic Fracture, November 1977, ASTM-STP-668 (1979), pp. 37–64.
16. B. A. Szabo, G. G. Musicco, and M. P. Rossow, An Analysis of Ductile Crack Extension in BWR Feedwater Nozzles, EPRI-1311 (June 1979), Project RP1241-1, final report.
16a. H. Tada, P. Paris and G. Irwin, A Parametric Analysis of Tearing Instability of Nozzle Cracks in Pressure Vessels, Proceedings of the Specialist Meeting on Elastoplastic Fracture Mechanics, Daresbury, U.K., May 1978.
17. F. J. Loss, Dynamic Tear Test Investigations of the Fracture Toughness of Thick-Section Steel, NRL Report 7056, HSSTP-TR-7, Naval Research Laboratory, Washington, D.C., May 14, 1970.
18. Flaw Evaluation Procedures: ASME Section XI, prepared by American Society of Mechanical Engineers, New York, New York, T. U. Marston, ed., Electric Power Research Institute Report EPRI NP-719-SR (August 1978).
19. T. U. Marston, Statistical Estimates of Fracture Toughness, *Electr. Power Res. Inst. J.*, 49–50 (May 1979).
20. K. F. Stahlkopf and T. U. Marston, A Comprehensive Approach to Radiation Embrittlement Analysis, paper presented at IAEA Meeting on Irradiation Embrittlement, Thermal Annealing and Surveillance of Reactor Pressure Vessels, Vienna, Austria, February 26–28, 1979.
21. D. M. Norris, Jr., J. E. Reaugh, B. Moran, and D. F. Quinones, Computer Model for Ductile Fracture: Applications to the Charpy V-Notch Test, Electric Power Research Institute Report EPRI NP-961, Project 603, Phase I Report (January 1979).
22. R. A. Wullaert, W. L. Server, W. Oldfield, and K. E. Stahlkopf, Development of a Statistically-Based Lower Bound Fracture Toughness Curve (K_{IR} Curve), in: *Transactions of the 4th International Conference on Structural Mechanics in Reactor Technology*, 1977, T. A. Jaeger and B. A. Boley, eds., Vol. G, paper G6/5.
23. S. H. Bush, Structural Materials for Nuclear Power Plants, *J. Test. Eval.*, **2**, No. 6, 435–462 (1974).
24. G. R. Odette, W. L. Server, W. Oldfield, R. O. Ritchie, and R. A. Wullaert, Analysis of Radiation Embrittlement Reference Toughness Curves, EPRI Research Project RP886-1, Final Report NP1661 (January 1981).
25. K. Ohmae and T. O. Zeibold, The Influence of Impurity Content on The Radiation Sensitivity of Pressure Vessel Steels: Use of Electron Microprobe for Irregular Surfaces, *J. Nucl. Mater.*, **43**, 254–257 (1972).
26. F. A. Smidt, Jr., and H. E. Watson, Effect of Residual Elements on Radiation Strengthening in Iron Alloys, Pressure Vessel Steels, and Welds, *Metall. Trans.*, **3**, 2065–2073 (1972).
27. F. A. Smidt, Jr., and J. A. Sprague, A Parametric Study of Vacancy Trapping during

Irradiation, NRL Memo Report 2531, Naval Research Laboratory, Washington, D.C. (August-October 1972), p. 31.
28. F. A. Smidt, Jr., and J. A. Sprague, Suppression of Void Nucleation by a Vacancy Trapping Mechanism, *Scr. Metall.*, **7**, 495–502 (1973).
29. An Assessment of the Integrity of PWR Pressure Vessels, report by a study group chaired by Dr. W. Marshall, UKAEA (October 1976).
30. W. G. Clark, Effect of Temperature and Specimen Size on Fatigue Crack Growth in Pressure Vessel Steel, *J. Mater.*, **6**, 134–149 (1971).
31. P. C. Paris, R. J. Bucci, E. T. Wessel, W. G. Clark, Jr., and T. R. Mager, An Extensive Study of Low-Cycle Fatigue Crack growth Rates in A533B and A508 Steels, in: *Stress Analysis and Growth of Cracks*, Proceedings of the 1971 National Symposium on Fracture Mechanics, Part 1, ASTM STP-513 (1972), pp. 141–176.
32. F. Shahinian, H. E. Watson, and H. H. Smith, Fatigue Crack Growth in Selected Alloys for Reactor Applications, *J. Mater.*, **F(4)**, 527–535 (1972).
33. T. R. Mager and S. A. Legge, Effects of High-Temperature Primary Reactor Water on the Subcritical Crack Growth of Reactor Vessel Steel, HSST Prog. Rep. ORNL-4855, Oak Ridge National Laboratory (April 1973).
34. T. Kondo, Fatigue Crack Propagation Behavior of ASTM A533B and A302B Steel in High-Temperature Aqueous Environment, HSST 6th Annual Information Meeting, Paper No. 6 (April 1972).
35. N. E. Dowling, Geometry Effects and the *J*-Integral Approach to Elastic-Plastic Fatigue Crack Growth, in: *Cracks and Fracture*, ASTM Philadelphia (1976), ASTM STP-601, pp. 19–32.
36. W. H. Bamford, D. M. Moon, and L. J. Ceschini, Crack Growth Rate Testing in Reactor Pressure Vessel Steels, in: *Proceedings of the Fifth Water Reactor Safety Information Meeting*, Gaithersburg, Maryland, November 1977.
37. W. H. Bamford, The Effect of Pressurized Water Reactor Environment on Fatigue Crack Propagation of Pressure Vessel Steels, in: *The Influence of Environment on Fatigue*, Inst. Mech. Eng., London (1977).
38. D. A. Hale, J. Yuen, and T. Gerber, Fatigue Crack Growth in Piping and Reactor Pressure Vessel Steels in Simulated BWR Environment, GEAP-24098/NRC-5, General Electric Co. (January 1978).
39. W. H. Bamford, Application of Corrosion Fatigue Crack Growth Rate Data to Integrity Analyses of Nuclear Reactor Vessels, paper presented at 3rd ASME National Congress on Pressure Vessels and Piping, San Francisco, California, June 1979.
40. P. M. Scott, Corrosion Fatigue in Pressure Vessel Steels for Light Water Reactors, *Met. Sci.*, 396–401 (1979).
41. L. A. James, Fatigue-Crack Propagation in Neutron-Irradiated Ferritic Pressure Vessel Steels, *Nucl. Saf.*, **18**, No. 6, 791–801 (1977).
42. G. D. Whitman, Heavy Section Steel Technology Program Quarterly Progress Report for July–September 1978, ORNL-NUREG-TM-275 (January 1979).
43. K. E. Stahlkopf, R. E. Smith, and T. U. Marston, Nuclear Pressure Boundary Materials Problems and Proposed Solutions, *Nucl. Eng. Des.*, **46**, 65–79 (1978).
43a. P. P. Holz and S. W. Wismer, Half-Bead (Temper) Repair Welding for HSST Vessels, ORNL/NUREG/TM-177, Oak Ridge National Laboratory, Tennessee.
44. Technical Report: Investigation and Evaluation of Stress-Corrosion Cracking in Piping of Light Water Reactor Plants, U.S. Nuclear Regulatory Commission NUREG-0531 (January 1979).
45. Seminar on Countermeasures for BWR Pipe Cracking, January 22–24, 1980, Palo Alto, California; EPRI WS-79-174, Workshop Report (May 1980).
46. Alternate Alloy for BWR Pipe Applications, Quarterly Rep. NEDC-23750-2, EPRI Contract RP168.

47. M. Fox, An Overview of Intergranular Stress Corrosion Cracking in BWRs, paper presented at Seminar on Countermeasures for BWR Pipe Cracking, January 22-24, 1980, Palo Alto, California.
48. R. L. Cowan and C. S. Tedmon, Jr., Intergranular Corrosion of Iron–Nickel–Chromium Alloys, *Adv. Corros. Sci. Technol.*, **3**, 293-400 (1973).
49. C. S. Pande, M. Suenaga, B. Vyas, and H. S. Isaacs, Direct Evidence of Chromium Depletion near the Grain Boundaries in Stainless Steels, *Scr. Metall.* **11**, 681-684 (1977).
50. H. E. Chung and J. B. Lumsden, Grain Boundary Characterization, Ohio State University Report, FCC-7704 (1977).
51. A. J. Giannuzzi, Studies on AISI Type 304 Stainless Steel Piping Weldments for Use in BWR Application, Electric Power Research Institute Final Report NP 944 (December 1978).
52. M. J. Povich and P. Rao, Low-Temperature Sensitization of Welded Type 304 Stainless Steel, *Corrosion*, **34**, 269-275 (1978).
53. M. J. Povich, Low-Temperature Sensitization of Type 304 Stainless Steel, *Corrosion*, **34**, 60-65 (1978).
54. W. L. Clarke and V. M. Romero, Detection of Sensitization in Stainless Steel: II. EPR Method for Nondestructive Field Tests, General Electric Company Report No. GEAP-12697 (February 1978).
55. W. J. Shack, W. A. Ellingston, and L. E. Paris, The Measurement of Residual Stresses in Type 304 Stainless Piping Butt Weldments, Argonne National Laboratory Report (September 1978).
56. W. L. Clarke and G. M. Gordon, Investigation of Stress Corrosion Cracking Susceptibility of Fe–Ni–Cr Alloys in Reactor Water Environments, *Corrosion*, **29**, 1-12 (1973).
57. E. L. Burley, Technical Highlights, Alternate Water Chemistry Program, General Electric Co., San Jose, California, COO-2985-14 (September 1978).
58. W. L. Pearl, W. R. Kassen, and S. G. Sawochka, Oxygen Monitoring and Control in BWR Plants, *Nucl. Technol.*, **37**(2), 94-98 (1978).
59. F. P. Ford and M. J. Povich, The Effect of Oxygen/Temperature Combinations on the Stress-Corrosion Susceptibility of Sensitized 304 Stainless Steel in High-Purity Water, Paper No. 94, in: *Corrosion 79*, Atlanta, Georgia, March 1979.
60. R. E. Hanneman, Prakash Rao, and J. C. Danko, Intergranular Stress Corrosion Cracking in 304SS BWR Pipe Welds in High-Temperature Aqueous Environments, General Electric Company Report No. 78CRD-92 (October 1978).
61. D. O. Harris, The Influence of Crack Growth Kinetics and Inspection on the Integrity of Sensitized BWR Piping Welds, EPRI NP-1163, Project 1325-2, final report (September 1979).
62. *Nuclear Systems Materials Handbook,* Vols. 1 and 2, Handford Engineering Development Laboratory (HEDL), TID-2666 [distribution limited to U.S. DOE approved recipients].
63. Code Case 1592-7, ASME Boiler and Pressure Vessel Code.
64. B. van der Schaaf, M. I. deVres, and J. D. Elen, Effect of Irradiation on Creep-Fatigue Interaction of DIN1.4048 Stainless Steel Plate and Welds at 823°K, in: International Atomic Energy Agency Specialists Meeting on Properties of Primary Circuit Structural Materials Including Environmental Effects, Bergisch Gladbach, West Germany, October 17-21, 1977.
65. L. F. Coffin, Jr., S. S. Manson, A. E. Carden, L. K. Severud, and W. L. Greenstreet, Time-Dependent Fatigue of Structural Alloys: A General Assessment (1975), Oak Ridge National Lab. Rep. ORNL-5073, (January 1977).
66. J. D. Heald and E. Kiss, Low-Cycle Fatigue of Nuclear Pipe Components, *J. Pressure Vessel Technol.*, 171-176 (August 1974).
67. D. R. Diercks and D. T. Raske, Elevated-Temperature Strain-Controlled Fatigue Data

on Type 304 Stainless Steel: A Compilation, Multiple, Linear Regression Model, and Statistical Analysis, Argonne National Laboratory, ANL-76-95 (December 1976).
68. K. D. Sheffler and G. S. Doble, Thermal Fatigue Behavior of T-111 and Astar 811-C in Ultrahigh Vacuum, in: *Fatigue at Elevated Temperatures,* ASTM STP-520, American Society for Testing and Materials (1973), pp. 491–499.
69. S. Majumdar and P. S. Maiya, A Mechanistic Model for Time-Dependent Fatigue, *J. Eng. Mater. Technol.,* **102,** 159–167 (1980).
70. C. Y. Cheng and D. R. Dierks, Effects of Hold Time on Low-Cycle Fatigue Behavior of AISI Type 304 Stainless Steel at 593°C, *Metall. Trans.,* **4,** 615–617 (1973).
71. G. J. Lloyd, Mechanical Properties of Austenitic Stainless Steels in Sodium, *At. Energ. Rev.,* **16,** 155–208 (1978).
72. G. J. Zeman and D. L. Smith, Low-Cycle Fatigue Behavior of Types 304 and 316 Stainless Steel Tested in Sodium at 550°C, *Nucl. Technol.,* **42,** 82–89 (1979).
73. P. S. Maiya and S. Majumdar, Elevated-Temperature Low-Cycle Fatigue Behavior of Different Heats of Type 304 Stainless Steel, *Metall. Trans.,* **8A,** 1651–1660 (1977).
74. D. J. Michel and G. E. Korth, Effects of Irradiation on Fatigue and Crack Propagation in Austenitic Stainless Steels, in: *Proceedings of the International Conference on Radiation Effects in Breeder Reactor Structural Materials,* Scottsdale, Arizona, June 19–23, 1977, Met. Soc. AIME, pp. 117–137.
75. G. E. Korth and M. D. Harper, Fatigue and Creep-Fatigue of Irradiated and Unirradiated Type 304 and 316 Stainless Steel, in: *Semi-annual Progress Report for the Irradiation Effects of Reactor Structural Materials Program for the Period Ending February 1975,* T. T. Claudson, ed., HEDL-TME 75-23, pp. ANC-1 to ANC-9 (March 1975).
76. C. R. Brinkman, G. E. Korth, and J. M. Beeston, Fatigue and Creep-Fatigue Behavior of Irradiated Stainless Steels—Available Data, Simple Correlations, and Recommendations for Additional Work in Support of LMFBR Design, ANCR-1096 (February 1973).
77. G. D. Korth and M. D. Harper, Fatigue and Creep Fatigue Behavior of Irradiated and Unirradiated Type 308 Stainless Steel Weld Metal at Elevated Temperatures, ASTM-STP-570 (1975), pp. 172–190.
78. C. R. Brinkman, G. E. Korth, and J. M. Beeston, Influence of Irradiation on Creep-Fatigue Behavior of Several Austenitic Stainless Steels and Incoloy 800 at 700°C, ASTM-STP-529, pp. 473–492 (1973).
79. C. R. Brinkman, G. E. Korth, and R. R. Hobbins, Estimates of Creep-Fatigue Interaction in Irradiated and Unirradiated Stainless Steels, *Nucl. Technol.,* **16,** 297–307 (1972).
80. L. A. James, Fatigue Crack Propagation in Austenitic Stainless Steels, *At. Energ. Rev.,* **14,** 37–86 (1976).
81. M. Reich and E. P. Esztergar, Compilations of References, Data Sources, and Analysis Methods for LMFBR Primary Piping System Components, BNL-NUREG-50650 (March 1977).
82. A. E. Carden, Parametric Analysis of Fatigue Crack Growth, in: *International Conference on Creep and Fatigue in Elevated Temperature Applications,* Philadelphia (1973), Paper C324/73.
83. L. A. James, Effects of Irradiation and Thermal Aging upon Fatigue-Crack Growth Behavior of Reactor Pressure Boundary Materials, paper presented at IAEA Technical Meeting, November 20–21, 1978, Innsbruck, Austria.
84. L. A. James, Some Questions Regarding the Interaction of Creep and Fatigue, *J. Eng. Mater. Technol.,* **98,** 235–243 (1976).
85. L. A. James, The Effect of Frequency upon Fatigue Crack Growth of Type 304 Stainless Steel at 1000F, in: *Stress Analysis and Growth of Cracks,* Proceedings of the 1971 National Symposium on Fracture Mechanics, Part I, ASTM STP-513 (1972), 218–229.

86. L. A. James, Hold-Time Effects on the Elevated-Temperature Fatigue-Crack Propagation of Type 304 Stainless Steel, *Nucl. Technol.*, **16**, 521–530 (1972).
87. L. A. James, Frequency Effects in the Elevated-Temperature Crack Growth Behavior of Austenitic Stainless Steel—A Design Approach, *J. Pressure Vessel Technol.*, **101**, 171–176 (1979).
88. L. A. James, The Effect of Stress Ratio on the Elevated-Temperature Fatigue Crack Propagation of Type 304 Stainless Steel, *Nucl. Technol.*, **14**, 163–170 (1972).
89. D. J. Michel and H. H. Smith, Effect of Hold Time and Thermal Aging on Elevated-Temperature Fatigue Crack Propagation in Austenitic Stainless Steels, Memorandum Report 3627, Naval Research Laboratory (1977).
90. B. Houssin and G. Slama, Metallurgical Practice and Assessment of Toughness for Components in SA 508 Cl3, in: European Nuclear Conference, Hamburg, Germany, May 6–11, 1979; abstract in: *Trans. Am. Nucl. Soc.*, **31**, 579–580 (1979).
91. J. H. Gross, Pressure Vessel Steels: Promise and Problem, *J. Pressure Vessel Technol.*, 9–14 (February 1974).
92. J. Danko, BWR Piping Remedies, *Electr. Power Res. Inst. J.*, 47–48 (June 1979).
93. Residual Stress Improvement by Means of Induction Heating, *Ishikawajima-Harima Eng. Rev.*, **18**, No. 1 (1978).
94. E. F. Rybicki, P. M. McGuire, and R. B. Stonesifer, The Improvement of Residual Stresses in Girth-Butt Welded Pipes through an Internal Heat Sink, presentation in Division F10, No. 8, 5th International SMIRT Conference (August 1979).
95. BWR Environmental Cracking Margins for Carbon Steel Piping: First Semi-annual Progress Report, July 1978–December 1978, EPRI Contract RP1248-1, General Electric Report No. NEDC-24625 (January 1979).
96. J. E. Alexander, Alternative Alloy for BWR Pipe Applications: Third Semi-annual Progress Report, Oct. 1978–March 1979, General Electric Report NEDC-23750-5 on EPRI Contract RP-968 (May 1979).
97. J. J. Mueller and Mohamed Behravesh, Improvement of Nuclear Castings by Application of Hot Isostatic Pressing (HIP), Electric Power Research Institute, final report on Project 1249, EPRI NP-1213 (November 1979).

5

Fusion First-Wall/Blanket Materials

At first sight it might appear somewhat incongruous that a discussion on CTR materials should be included in what is admittedly a fission-reactor-dominated work. However, the intent here is not to embark on a detailed discussion of CTR materials development (which could fill a book in itself) but, rather, to continue the theme developed in preceding chapters by examining the structural integrity aspects of the emerging CTR concepts. Thus discussion is limited to first-wall/blanket designs and material considerations, with particular emphasis being placed on differentiating between those requirements that can be met by existing fission reactor technology and those that are unique to the fusion environment.

The first-wall/blanket structure is the primary "pressure boundary" region of a CTR.[1,2] Its functions range from a vacuum chamber to contain the plasma to an energy converter which converts the kinetic energy of the fusion reaction into heat (refer to Figs. 1-5 and 1-6). While performing these functions, the first wall will be exposed to intense neutron radiation from the plasma, which can penetrate well into the blanket. In addition, about 20% of the fusion energy impinges onto the first wall in the form of ionized and neutral particles and electromagnetic radiation. The primary effects of neutron irradiation are swelling and changes in mechanical properties, which must be accommodated by design (refer to Figs. 1-18 to 1-23). The portion of the fusion energy deposited on the first-wall surface leads to vaporization, sputtering, and, perhaps, blistering (refer to Fig. 1-24). These processes, the so-called plasma–materials interactions, not only result in erosion of the first wall, but lead to the introduction of impurities of high atomic number into the plasma. Plasma contamination alters the fuel ion density and decreases the temperature through radiation losses. Therefore, more energy is required to achieve and maintain ignition, and a buildup of impurities during the plasma operation can limit the overall burntime.

Even though irradiation effects are crucially important for the fusion reactor first wall, there are a number of other considerations that cannot be ignored. Since net electrical energy is to be extracted from the blanket, the temperature must be high enough for a reasonable thermodynamic efficiency. In addition, the first wall is exposed to high heat fluxes of 0.2 to 0.8 MW/m^2 so that an effective coolant must be employed. Candidate coolants include Li, He, fused salts, and even water. The structure must be compatible with the coolant at the operating temperature. Finally, for the deuterium–tritium (DT) system, the blanket must contain the breeding material, lithium, or a compound such as LiAl, Li–Pb, or Li$_2$O. The breeding material must also be cooled and should be compatible with both the structure and the coolant.

It is immediately evident from the foregoing that the conditions in a CTR vessel will likely combine the worst features of the environments of LWR and LMFBR vessels; and requirements for compatibility among coolant, structure, and breeding material severely reduce the number of material choices that are available to the designer. First-wall loadings (largely from pulsed thermal stresses) and neutron flux will both be higher than experienced by LWR vessels, and wall temperatures will be at least as high, if not some 100 K higher than the LMFBR vessel. Accordingly, the materials selection and the design approach are very different from that of the fission reactor pressure boundary, favoring thin-walled metal tubular or spherical modular structures that are more reminiscent of fission reactor fuel rod designs. In fact, LMFBR fuel rod cladding technology has provided a basis for many CTR first-wall design studies. There is one important difference, however, which has far-reaching ramifications, and that is: In contrast to fission reactors where development of a few leaking fuel rods does not severely affect operation, the absolute integrity of the CTR first wall must be maintained since loss of vacuum and in-leakage of coolant will quench the plasma and terminate the fusion process. Thus, along with improved materials performance, there is a need to develop highly reliable first-wall designs that can operate with a low failure probability for the estimated life of the component. This latter requirement mandates the use of an improved structural integrity (SI) analysis of the type being developed for the fission reactor pressure boundary.[3]

The area of SI analysis is where the major technology transfer effort from fission to fusion should be undertaken since, clearly, current fission reactor materials will either be unacceptable or, at best, marginal for operation in the fusion environment. With this in mind the following sections will explore progress being made on two basic design approaches—metal first walls and ceramic first walls—for the two leading fusion concepts.

5.1. First-Wall/Blanket Designs

The concepts described below are general in nature and will serve as illustrations for establishing design and materials requirements.

5.1.1. Magnetic Fusion

One of the leading designs for a magnetically confined CTR first-wall/blanket structure is that of the UWMAK† tokamak.[4] As shown in Fig. 5-1, the first wall is of modular construction, comprised of metal U-bend heat transfer cells. These first-wall cells are 200 mm in depth and have a structural thickness of 2.5 mm. The U-bend radius is 75 mm. Lithium coolant flows through each set of four U-bends before returning to the headers. These cells are supported by locking studs on the blanket or heater structure. A graphite block reflector, and/or ceramic coating on the plasma side of the wall, might be employed to reduce radiation damage and erosion.[5]

† University of Wisconsin Tokamak.

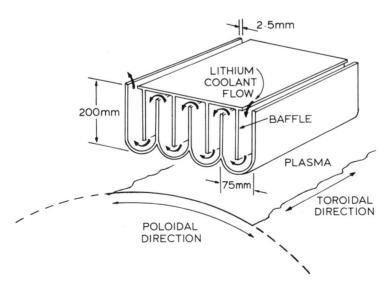

Fig. 5-1. U-bend heat transfer cell first-wall configuration for UWMAK tokamak.

As will be discussed in more detail later, there are definite advantages to using ceramic materials entirely in the construction of the CTR first-wall blanket,[6, 7] although their development for this purpose must overcome some severe problems. Two general ceramic first-wall/blanket design concepts—panels and modules—are illustrated in Fig. 5-2.[7] The panels or modules will be of the order of 1 m square or 1 m by 0.5 m diameter, respectively, with as many as 2500 being required for a tokamak first wall. The panel design (Fig. 5-2a) consists of a plate-type first wall welded with a joint bonding material and supported by a compliant insulator material. The blanket region would be located between the ceramic plates and a metallic vacuum wall. The modular design (Fig. 5-2b) consists of a ceramic shell forming the first wall and enclosing the blanket volume. Many of these blankets pods could be attached to the metal support structure.

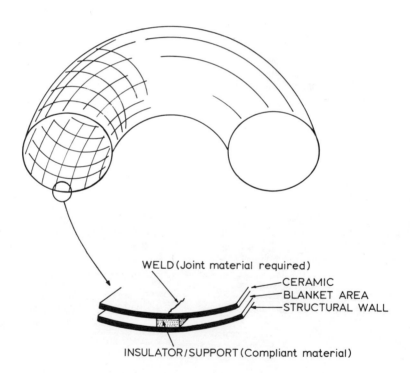

Fig. 5-2a. Ceramic panel for tokamak CTR (Ref. 7).

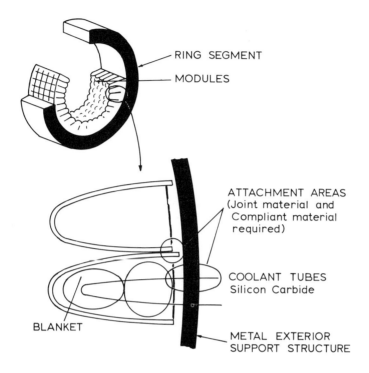

Fig. 5-2b. Ceramic module conceptual designs (Ref. 7).

5.1.2. Inertial Confinement Fusion

The combination of cyclic stresses with high-energy neutrons, and charged particle and x-ray pulses in ICF systems provides a much more severe environment for first-wall structures than do magnetic fusion devices. Conn[5] showed that a bare, unprotected ICF reactor chamber will not survive reasonable microexplosions at economical values of neutron wall loading. Some form of wall protection will therefore be required. Maniscalco et al.[8] recently reviewed five designs that are primarily distinguished by the method employed to protect the first wall from the short-range debris and x-rays. All but one concept use liquid lithium for blanket cooling and tritium breeding. Two designs, termed the wetted wall and the fluid wall concepts, also use liquid lithium in the vacuum chamber for first-wall protection. Another design uses a buffer gas of xenon to stop x-ray and charged particle debris. Two others employ graphite as either sacrificial liner or deflector.

The lithium wetted-wall concept used as the (laser) ICF example in Fig. 1-6 earlier has been widely studied.[9] This concept utilizes a cavity wall which is formed by a porous refractory metal through which coolant lithium flows to form a protective coating for the first-wall surface. Typically the coating will be about 1.2 mm thick, of which about 0.1 mm will be evaporated and ablated following each pulse.

Ceramics have also been considered as primary-wall/blanket structures in the ICF concept.[10] In a novel design, shown in Fig. 5-3, the first wall is envisioned as a freely falling layer of small ceramic spheres with the primary function of absorbing the x-ray, debris, and alpha particle output of the pellet. The blanket region is taken as a simple bed of larger ceramic spheres which are moving slower at a rate sufficient only for heat removal of the energy deposited by the pellet neutron output. Both the blanket and first wall would be recirculated and heat would be removed in the external portion of the loop; thus the spheres would be in the chamber atmosphere and the vacuum barrier and vessel wall would be well away from the high neutron damage regions. Sphere size ranges could be typically 1 mm diameter for the first wall and 0.1 m diameter for the blanket.

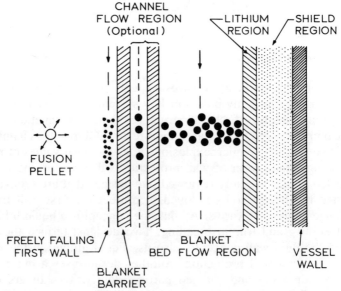

Fig. 5-3. Ceramic pellet and ball inertial fusion reactor design (Ref. 10).

5.2. Materials and Structural Integrity Considerations

At this early stage in the development of fusion power a number of materials have been or are currently being actively considered for first-wall/blanket structures because of different property attributes (Table 1-1). The properties of concern, as for the fission reactor pressure boundary, determine short-term and long-term structural integrity. A priority list of criteria for selecting first-wall materials is given in Table 5-1, and a recent IAEA workshop summary,[11] Table 5-2, lists the possible effects of the unique fusion environment on the important properties. Most have been discussed earlier in relation to either the fission reactor pressure boundary or the fuel assembly. CTR material selection processes start from the basis of this fission reactor experience.

Table 5-1. Criteria for Selecting First Wall Materials[a] in Fusion Reactors in General Priority Order[b]

Criteria	Favored materials	Less favored
1. Radiation damage and lifetime		
a. Swelling (dim, stability)	Ti, V, Mo, FS	SS, Nb, Al, C
b. Embrittlement	C, Nb, V, Ti, FS	SS, Mo, Al
c. Surface properties	V, Ti, Al, C, SiC	FS, SS, Nb, Mo
2. Compatibility with coolants and tritium		
a. Lithium	Ti, V, Nb, Mo, SS, FS	(Al, C)[c]
b. Helium	SS, FS, Ti, Mo, Al, C, SiC	(Nb, V)[c]
c. Water	SS, FS, Al, Ti	(C)[c]
d. Tritium	Mo, Al, FS, SS	Ti, V, Nb, C
3. Mechanical and thermal properties (irradiated)		
a. Yield strength	Mo, Nb, V, Ti, FS, SS	Al, C
b. Fracture toughness	SS, FS, Ti, Al	V, Nb, Mo, C
c. Creep strength	Mo, V, Ti, FS, SS, SiC	C, Al, Nb
d. Thermal stress parameter $M \equiv 2\sigma_y k(1 - v)/\alpha E$	Mo, Al, Nb, V, FS, C	Ti, SS, SiC
4. Fabricability and joining	SS, FS, Al, Ti	Nb, V, Mo, C, SiC
5. Industrial capability and data base	SS, FS, Al, Ti, C, SiC	Mo, Nb, V
6. Cost	C, Al, FS, SS, Ti, SiC	Mo, Nb, V
7. Long-lived induced radioactivity	V, C, SiC, Ti, FS, Al	SS, Nb, Mo
8. Resource availability (U.S.A.)	C, SiC, Ti, Mo, Al, FS, SS	Nb, V

[a] Alloys, Ti–6Al–4V, V–20Ti, TZM, Nb–1Zr, 316 SS, ferritic stainless steel (FS) [9–12 wt.% Cr], Al-6061. This is an illustrative list.
[b] Adapted from Conn, Ref. 4.
[c] Materials in parentheses are unacceptable with stated coolant.

Table 5-2. Effect of Fusion Environment on Materials Properties[a]

Effect of	Fracture toughness	Fatigue	Thermal creep rupture	Swelling incubation/growth	Irradiation creep
Gases and transmutants	Major	Major[d]	Major	Major {Major[d] / Minor[e]}	Minor
Compatibility with coolant	Major	Major	Major	Major	Minor
Dose rate	Minor[c]	Minor[c]	Minor[c]	Minor[c]	Minor
Difference between post-irradiation and in-situ testing	Major[b]	?	Major[b]	In-situ testing indispensable	
Time-varying irradiation	Minor	Minor	Minor	Minor	Minor
Time-varying temperature $T_{op} > T_{down}$	Major	Major	Minor	Minor	Minor

[a] Ref. 11.
[b] Post-irradiation results may give worst case.
[c] For stable alloys.
[d] For $T > 0.4T_m$ (T_m = melting temperature).
[e] For $T < 0.4T_m$.

As can be seen from Table 5-1 there is today no single class of candidate materials that obviously will be able to meet the requirements of a commercial CTR where the goal is a service life of 40 MWyr/m² for structural materials. For this reason research and development is following a parallel path approach with several metal alloy classes. Development programs are active in the U.S., Germany, U.K., and Japan. Using the U.S. *Fusion Reactor Materials Program Plan*[12] as an example, we list the following paths: Path A includes the austenitic alloys that have been characterized so well for LMFBR application in near-term machines; Path B comprises the Fe–Ni–Cr superalloys, under study for second-generation LMFBR core components (refer to Chapter 3); Path C is unique to the fusion concepts, that is, the use of reactive (e.g., titanium) and refractory metals (e.g., vanadium, niobium) and alloys; Path D is the so-called new concepts, and the ferritic stainless steels (e.g., HT-9 and the 9Cr–1Mo alloys) have recently been included in this path following a successful evaluation analysis[13] and encouraging results from the LMFBR program (refer to Chapter 3).

The exposure goal of 40 MWyr/m² makes irradiation effects—mechanical property degradation due to helium embrittlement and void swelling—of greatest concern to integrity (Table 5-2). These problems are relatively insensitive to confinement system or details of reactor design and are potential problems for all possible material choices. All irradiation phenomena experienced in fission environments are expected to be greatly exaggerated in the fusion environment due to the higher neutron energies.[14]

Although bulk radiation damage and its effect on first-wall integrity is clearly of major engineering importance, the primary consequence of plasma–first-wall interactions is likely to be a physics problem, namely the degradation of the plasma through the introduction of impurities from the wall material. Thus, there is a great incentive to protect the plasma by low-Z materials at the plasma–materials interface, either as coatings or liners, diverters, or as the first-wall structure itself. A new class of materials, the ceramics (e.g., graphite and silicon carbide), therefore, enter the competition for candidate first-wall structural materials, and are also included in Path D of the U.S. Program.[12]

As one can imagine, in a developing field of this type in which the system design is not yet frozen, let alone the materials, the literature abounds with design concepts and material combinations that are invariably based on very little hard materials data. The fact is, at this point in the development of CTRs, the data base for materials properties in the anticipated fusion environment [e.g., wall temperatures $\geqslant 873$ K, neutron flux of $\sim 3 \times 10^{18}$ n/m^2 sec (20% of neutrons have $E > 10$ MeV), lithium, helium, or water coolants, and cyclic stresses up to 200 MPa with frequencies up to 20 cps] is minimal. As will be seen in the following subsections, only the austenitic stainless steels (Types 304 and 316) are sufficiently well characterized for detailed structural analyses to be conducted and, unfortunately, results to date indicate that they will be marginal for this application. Thus, the discussion here will be primarily on what data are required for the potential replacements for the austenitics, rather than what data are in hand. Inasmuch as the property requirements for ceramics are somewhat different (e.g., emphasis on thermal shock and surface erosion resistance) the status of these materials will be reported in a separate subsection.

5.2.1. Candidate Metals and Alloys

The closest conditions to those expected in the thin-walled, tubular (Tokamak, Fig. 5-1) or spherical (ICF, Fig. 1-6) first-wall structures are encountered in the core of the LMFBR (Chapter 3). Therefore it is not surprising that both the LMFBR cladding and duct materials' development experience and their test reactors have been used for screening CTR alloys. The LMFBR data, however, must be used with some care in predicting the effects of the fusion reactor environment. The difficulty is with the different neutron spectra in fission and fusion reactors; the average energy of fission source neutrons ranges from 0.1 to 1.0 MeV, while the energy of source neutrons from the D–T reaction is ~14 MeV.[14] This difference is especially important for (n,p) and (n,α) transmutation reactions as these reactions in many metals have thresholds near 10 MeV. Since it is known that helium has a strong effect on both the mechanical properties and swelling characteristics of Types 304 and 316 stainless steel under LMFBR conditions (refer to Chapter 3), this is a major area of attention in CTR research. Facilities such as the high-flux isotope reactor (HFIR) come close to reproducing the dpa/helium production ratio expected in some CTR first-wall materials,[14] and results of material property measurements after irradiation to CTR damage levels are now becoming available. Unfortunately, we find that the differences in the neutron spectra do render radiation effects in fusion reactors more deleterious to reactor operation than in a fission reactor.[15,16]

The following discussion will be restricted to bulk radiation effects in the alloys, since, as a result of metal surface damage studies such as those summarized in Chapter 1, it is expected that low-Z materials will bear the brunt of the surface damage produced by the plasma. Also, only three property areas will be emphasized, viz. (a) high-temperature embrittlement due to helium, (b) void swelling, and (c) thermal fatigue, because a number of design studies[15, 17–19] have indicated that phenomenon (a) will probably place an upper temperature limit on first-wall life, phenomenon (b) will likely contribute most to "internal" stresses and strains, and phenomenon (c) is the most probable failure mechanism.

5.2.1.1. Helium Embrittlement

The phenomenon of helium embrittlement from fast neutron irradiation was introduced in Chapter 1 (Fig. 1-18) and reference to its impact on LMFBR fuel rod integrity was made in Chapter 3 (Fig. 3-10). Both figures were constructed from post-irradiation test data on austenitic stainless steels after irradiation in EBR-II to fast neutron fluences $\sim 5 \times 10^{26}$ n/m² $(E > 0.1$ MeV).[20] The normally observed grain boundary weakness above $0.5T_m$ is exacerbated by the nucleation and growth of helium bubbles on the grain boundaries. Intergranular fracture is further enhanced by the lattice hardening that accompanies irradiation.

One observes from Figs. 1-18 and 3-10 that helium embrittlement worsens as the temperature is increased due to increased migration of helium to grain boundaries. By the same token, it should be expected that the effect will be greater for the higher helium-production-to-damage ratios expected in CTR first walls (e.g., the He/dpa ratio is calculated at ~25, compared to 0.4 in EBR-II). Recent data from post-irradiation mechanical property tests of HFIR-irradiated stainless steels justify this concern. The combined effect of grain matrix hardening and grain boundary helium is a marked reduction in high-temperature ductility that is reflected in both short-term (tensile) and long-term (creep rupture) mechanical properties.[14-16]

The major changes to tensile properties are shown in Fig. 5-4.[15] Figure 5-4a is a data plot for 20% CW Type 316 stainless steel after irradiation in HFIR at temperatures between 622 and 973 K to 50 dpa and 4000 at. ppm helium. Irradiation has significantly reduced both the yield and ultimate tensile stress. Of even greater importance, however,

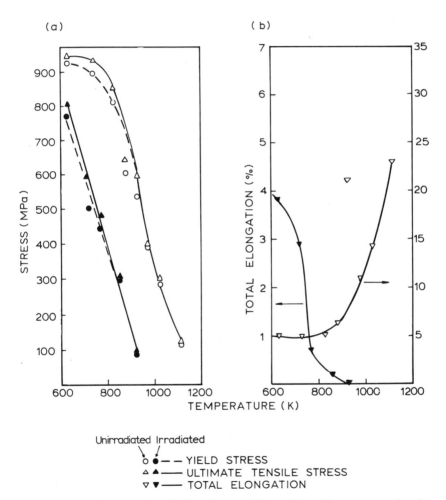

Fig. 5.4. The tensile properties of 20% cold-worked Type 316 stainless steel as a function of test temperature comparing unirradiated and irradiated data (Ref. 15). (a) Stress vs. temperature; (b) total elongation vs. temperature.

is the loss of ductility, shown in Fig. 5-4b. For the lowest irradiation and test temperature, 623 K, the ductility of the irradiated sample was only slightly less than that of the control. At increasing temperatures the tensile elongation continually decreases, dropping to zero at 923 K. At temperatures above 848 K this loss of ductility is achieved at relatively low damage levels (a fluence of less than 10^{26} n/m^2 and 100 at. ppm helium).

In creep-rupture tests, the effects of helium embrittlement are manifested as reductions in rupture life. At the highest damage level (43 dpa and 2990 at. ppm helium) the rupture life was about 10^{-5} of the unirradiated value.[15] There do not appear to be any data on helium effects on fatigue properties of stainless steels, but, if one uses the approach outlined in Section 4.3.2.1 and Eq. (4-9), Chapter 4, the observed reduced UTS and ductility would translate into reductions in both high- and low-cycle fatigue life. An increase in fatigue crack propagation rate and an intergranular fatigue fracture should also be expected.

Thus, to avoid helium embrittlement effects and maintain a 1% strain criterion, CW Type 316 stainless steel wall temperatures would have to be ~770 K or lower, which would severely compromise thermal efficiency using a lithium coolant.

Improvements in the post-irradiation tensile strength and ductility properties of Type 316 stainless steel can be achieved by the addition of 0.23 wt.% Ti.[22] The data on annealed material obtained over the temperature range 842 to 1023 K, after neutron fluences producing 1850 to 4000 appm helium and 30 to 60 dpa, are shown in Fig. 5-5. It might be mentioned that this type of alloy is of high interest to LMFBR cladding developers (e.g., D9 alloy, Table 3-1, Chapter 3).

Detailed microscopy studies of titanium-modified material have shown that the property improvement correlates well with the formation of a fine intragranular TiC precipitate, which accommodates helium in small cavities at its surfaces, and with smaller grain boundary cavities than in unmodified material. As Fig. 5-5 indicates, a dramatic improvement in properties results at 873 K, which is associated with the replacement of large grain boundary cavities by intergranular $M_{23}C_6$, in addition to the intragranular TiC. The fracture mode is also changed from brittle intergranular to ductile transgranular by the titanium additions. Coarsening of the TiC particles at higher temperatures reduces their effectiveness in trapping helium, with resultant loss in ductility. However, using the 1% strain criterion, one sees that the Ti-modified alloy could operate at wall temperatures up to 100 K higher than unmodified Type 316.

Of the alternative alloys, the higher-nickel superalloys are also helium embrittlement limited,[16] but the ferritic stainless (9–12 wt.% chromium, zero nickel) class appears to be immune to helium effects, at least under LMFBR conditions. Historically, the concern with the latter alloys has

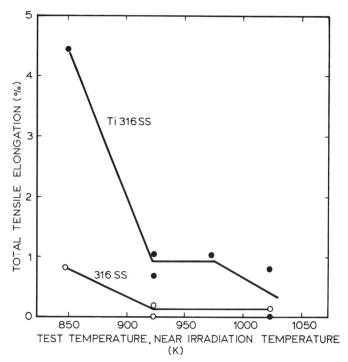

Fig. 5-5. Total tensile elongation of annealed Type 316 stainless steel with and without 0.23 wt.% Ti, as a function of test temperature. Tensile tests conducted at or near the irradiation temperature after irradiation in HFIR to fluences producing 1850 to 4000 at. ppm He and 30 to 60 dpa (Ref. 22).

been their marginal strength above ~750 K. However, a new developmental 9Cr1MoNb stabilized ferritic stainless steel appears to possess high-temperature strength equivalent to the Types 304 and 316 austenitic steels, at least in the unirradiated condition (refer to Fig. 6-16, Chapter 6). This alloy will be evaluated in the U.S. Fusion Program, together with the commercial ferritic, HT-9. The dispersion-hardened ferritic stainless steels (Chapter 3, Section 3.3.2) also exhibit improved high-temperature mechanical properties, which are retained after irradiation; but to the author's knowledge this alloy class is not included in any fusion materials research program.

The titanium alloys are candidate first-wall materials because they are low-activation alloys and possess favorable mechanical properties, particularly fatigue strength, at least in the unirradiated condition. The effect of helium on their tensile properties has recently been studied by Jones et al.[23] using a tritium charging technique; room temperature and 723 K tensile properties were measured. Helium had little effect on the yield and ultimate tensile strengths and uniform elongation of commercial purity titanium (Ti-70A) at 723 K for helium concentrations up to 430 appm. The major effect of helium bubbles in Ti-70A is on the total elongation and reduction in area, both of which decrease with increasing helium content. The largest helium effect was found for the Ti–6Al–4V (Ti64) alloy tested at 723 K where both the strength and ductility were strongly dependent on helium concentration, as shown in Fig. 5-6. The reduction in ductility is highly dependent on outgassing temperature, which, in turn, appears to determine the morphology of the β phase at the grain boundaries; according to the authors, a continuous intergranular β network could easily account for the lower failure strains recorded after outgassing at 873 K.

The limited data available on radiation embrittlement of the refractory metal alloys (at least those alloys of vanadium and niobium examined to date) indicate a trend similar to other alloys: that is, with high concentrations of helium, the high-temperature ductility is reduced and the fracture mode becomes intergranular.[16] The advantage of the refractory alloys, however, is that these effects appear at higher temperature (but approximately the same homologous temperature of 0.5) and, therefore, they offer the possibility of operating at higher wall temperatures. For example, helium embrittlement of vanadium alloys (e.g., V–Cr–Ti) is observed at temperatures above ~1050 K, where the uniform elongation in samples containing 25 to 30 appm helium was reduced to ~3%;[24] this

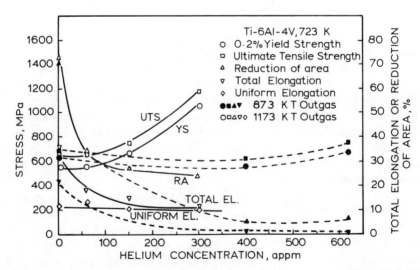

Fig. 5-6. Tensile properties of Ti–6Al–4V at 723 K versus helium concentration (Ref. 23).

embrittlement temperature limit is at least 200 K higher than either the titanium alloys or the stainless steels.

5.2.1.2. Swelling Behavior

The swelling characteristics and, in particular, the creep-swelling interaction of stainless steels were discussed in Chapter 3 where it was shown that this phenomenon is life limiting for Type 316 stainless steel fuel assemblies. The adverse swelling properties of the 18–8 austenitic steels was of course the basis for the current intense alloy development effort in the LMFBR program, out of which grew the similar parallel path effort for the CTR program noted earlier.

Unlike the LMFBR, however, a CTR will not experience coolant flow restriction if excessive swelling occurs in structural materials. The primary concern is that shape distortion due to swelling might lead to excessive stresses, especially at points of support or constraint, which would exceed the already low ductility limit of the material. Swelling must, therefore, be known and should be minimized or, at least, allowed for; typically, designs are based on a 3 vol.% swelling limit or less.

Since it is known from the LMFBR studies (Chapter 3) that helium has a major effect on swelling, the CTR-oriented swelling studies have concentrated on simulating the simultaneously high displacement damage rate and high helium production rate in stainless steels in the high-flux isotope reactor (HFIR).

As expected, early results point to greater swelling of both annealed and cold-worked Type 316 stainless steel under CTR conditions. Swelling as high as 1.6 vol.% at a fluence of 7×10^{26} n/m² ($E > 0.1$ MeV) was found in 20% cold-worked Type 316 material irradiated at 653 K.[25] This contrasts with a slight densification in material irradiated in EBR-II. The swelling is high above 673 K and continues up to at least 1000 K (Fig. 5-7). However, the addition of 0.23 wt.% Ti reduces the swelling of both the annealed and cold-worked Type 316 steels, but only in the 873 to 973 K range; the trend at higher temperatures is to the same swelling levels as unmodified Type 316 steels (Fig. 5-7).[26] This is explained by the reduced effectiveness of the TiC precipitates as a helium sink as the particles coarsen.

This study, and the previously mentioned study of mechanical properties,[22] led Maziasz and Bloom to conclude that alloy optimization in terms of composition and pre-irradiation microstructure should include development of a fine distribution of TiC to maximize the surface area for a given volume of precipitate. This should also be combined with the precipitation of grain boundary $M_{23}C_6$ to prevent formation of large grain boundary cavities. The problem here, however, as it was for the LMFBR

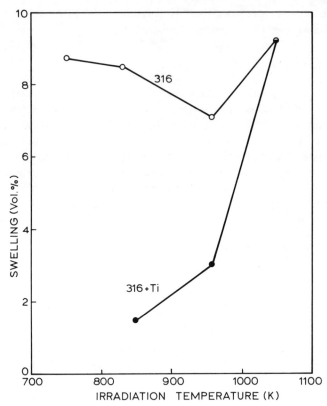

Fig. 5-7. Swelling, as determined by cavity volume fraction, as a function of irradiation temperature for Type 316 stainless steel with and without 0.23 wt.% Ti, irradiated to fluences producing 3000 to 4200 at. ppm He and 40 to 60 dpa (Ref. 26).

cladding, is long-term microstructural stability at high temperatures. Consequently, in this writer's opinion, it would be better to select a material that inherently produced less helium and fewer voids under irradiation.

The ferritic stainless class of steels fit this category, and early LMFBR-related data are promising with respect to both swelling and helium embrittlement (Section 3.3.2, Chapter 3). Vanadium- and titanium-based alloys also are quite resistant to swelling. With respect to the former, the binary V–Ti alloys containing more than 3% Ti, the binary V–Cr alloys, and the ternary V–Ti–Cr alloys are all quite resistant to swelling for irradiation temperatures up to 870 K. The maximum neutron-induced void swelling yet observed in these alloys is only ~1% (for V–10Cr irradiated at 973 to 1073 K).[27] The studies of titanium alloys mentioned earlier[23] concluded that both Ti-70A and Ti-64 were resistant to void formation under heavy ion irradiation (Ni^{++} + He^+ ions) up to about 920 K. Above 920 K, void swelling is detected in Ti-64. This temperature corresponds roughly to 770 K in a neutron environment where the damage rate is slower. Since 770 K is the likely upper service temperature limit for titanium alloys, it is therefore possible that void swelling will not be a significant problem.[28] This conclusion must be substantiated by further research, however, and in particular by high-fluence neutron irradiations in the 620 to 770 K temperature range.

It is important to emphasize that the aforementioned swelling studies were conducted under steady-state irradiation conditions. According to Ghoniem and Kulcinski,[29] the pulsed nature of CTR irradiation will produce a different point-defect behavior in materials and hence different swelling characteristics from steady-state irradiations. Under pulsed irradiation, void growth behavior is a strong function of both pulse repetition rate and the irradiation temperature. Their calculations showed that voids in Type 316 stainless steel exhibit slower growth rates at high temperatures when there are long times between successive pulses. Applying their analysis to actual CTR first walls, specifically the ICF reactor, Ghoniem and Kulcinski[29] conclude that the amount of swelling in the first wall can be reduced by (a) using higher-yield pellets, which will result in higher mutual point-defect recombination rates and also longer annealing times between microexplosions, and (b) operating at the highest temperatures allowed by other design factors (e.g., helium embrittlement). However, the "beneficial effect" of pulsed irradiation on void swelling needs to be verified experimentally before credit can be taken in design analyses.

5.2.1.3. Fatigue Behavior

Fatigue is likely to be the predominant failure mode in both the tokamak and the inertial confinement fusion (ICF) systems because of the pulsed nature of operation. The first wall will be subjected to cyclic loading from thermal stresses, in addition to the steady pressure due to the coolant, with the combined effect creating a significant creep-fatigue interaction that could range into the high-cycle regime (Refer to Chapter 1, Section 1.3.4.3, for basic details).

For the tokamak the cycle loading frequency may be as low as 0.0001 cps, but for the ICF concepts cyclic loading frequencies will range from 1 to 20 cps; peak stresses could be over 200 MPa in both types of fusion reactor. These conditions are much more severe than those experienced by any existing nuclear system and the early quantification of the resulting mechanical damage is essential for guiding fusion alloy development. It is of interest to note that Hafele *et al.*,[1] in a recent report for the International Institute for Applied Systems Analysis (IIASA), called fatigue the "Achilles heel" of the pulsed fusion reactor.

Inasmuch as the fatigue characteristics of the austenitic stainless steels have already been described (refer to Chapter 4, Section 4.3.2), and no data under actual CTR conditions (e.g., 14-MeV neutrons, lithium, etc.) exist, it is perhaps appropriate to review the use to which the existing, primarily LMFBR data, have been put in establishing performance limits for Type 316 stainless steel first-wall structures. This approach will also facilitate the discussion of data needs for the alternative materials.

A number of recent studies have questioned the adequacy of the fatigue and flaw growth properties of Type 316 stainless steel for both tokamak[5, 15-19] and ICF first walls.[5, 8] A recent detailed study of the thermal fatigue characteristics of the U-bend modular tokamak design, Fig. 5-1 earlier, undertaken jointly by groups at Argonne National Laboratory and McDonnell Douglas,[17] will be used for illustration purposes. In this "SI analysis" a rather modest wall lifetime goal of 10 MWyr/m^2 was selected.

Comprehensive thermal-hydraulic calculations were used as a basis for defining the stress-time history to be expected at the first wall for different reactor operating conditions. Wall loadings are sensitive to stress and temperature limits of the material. This study indicated that for stainless steel, if maximum structural temperatures are to be limited to less than 773 K, wall loadings must be kept less than ~2 MW/m^2, while nearly 4 MW/m^2 is allowed if the maximum temperature can reach 873 K (by using a diverter for example). Likewise, if thermal stresses in the stainless steel wall are limited to the yield strength (117 MPa), then wall loading must be limited to 1.7 MW/m^2, whereas they can reach 3.3 MW/m^2 if design limits are relaxed and thermal stresses are allowed to exceed the yield stress. The creep-fatigue analysis was made with these stress and temperature limits applied as boundary conditions.

Fatigue stresses are created mainly by the fluctuating thermal stresses that are set up between the burn and downtime. The creep stresses are mainly due to the pressure of the coolant. Damage due to creep-fatigue interaction was evaluated using the (Code Case 1592) linear-time-fraction damage rule, Eq. (1-11), Chapter 1, and the unirradiated fatigue data used in the analysis were representative of that described earlier in Chapter 4, Section 4.3.2.1.

During the early part of the operation, the total stress during the burn time is the sum of both the pressure and thermal stresses. These high stresses, if unrelaxed, could result in considerable creep damage in the first wall. In reality, however, these stresses are expected to relax due to creep deformation, and, although the stress relaxation per cycle is small, their cumulative effect over a large number of cycles is significant.

The following scoping results[17] illustrate the importance of the stress relaxation: If stresses are not allowed to relax, for a burn cycle of 95 s with 60-s burn time the life at the edge of the first wall towards the plasma side is computed to be 3200 cycles, with the creep damage accounting for almost 99% of the total damage. If, on the other hand, the stresses are allowed to relax with cycles because of *thermal* creep then the same location has a predicted life of 1.19×10^6 cycles, with the creep damage accounting for only 3% of the total damage; and, finally, if *radiation-enhanced creep* is assumed, the predicted cycles to failure increase to 1.24×10^6 and the creep damage is negligible. This latter situation is assumed for the wall lifetime calculations.

The effects of wall loadings and maximum metal temperature on the life of the stainless steel first wall as predicted by this study[17] are shown in Fig. 5-8. It is evident that the lifetime goal of 10 MWyr/m² is barely

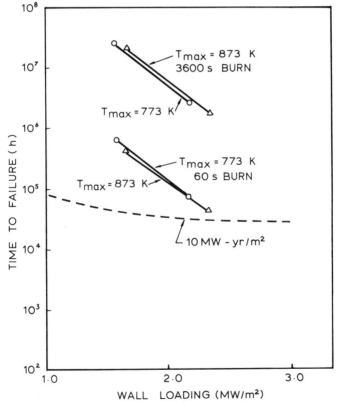

Fig. 5-8. Variation of fatigue life for a clamped stainless steel first wall with wall loading, assuming stress relaxation to occur during burn time (Ref. 17).

met at a wall loading of 2.35 MW/m² and a burn time of 60 s (95-s burn cycle time). If a scatter of 4 is incorporated to account for uncertainties in both data and analysis, then wall loadings are limited to 2.0 MW/m² or less for these operating conditions. It is also important to note that, if the property data used in this analysis were limited to the conservative design curves in Code Case 1592, the situation would be even worse for stainless steel. The lifetime is independent of temperature because creep damage is minimal and the fatigue life has been assumed to be approximately the same for all temperatures. Since the life in this case is limited by fatigue, the number of cycles to failure is the controlling factor, and consequently the time to failure increases with burn time.

An indication of the magnitude of the improvements in fatigue properties required to achieve the 10 MWyr/m² lifetime at economic wall loadings (2 to 4 MW/m²) can be obtained from Figs. 5-9 and 5-10. These figures were constructed in another part of this same study,[17] using slightly different criteria than those employed in the development of Fig. 5-8, for example; nevertheless the outcome is the same, as will be shown below.

With respect to fatigue strength improvements on the Type 316 stainless steel properties, Fig. 5-9 compares the calculated stress range values

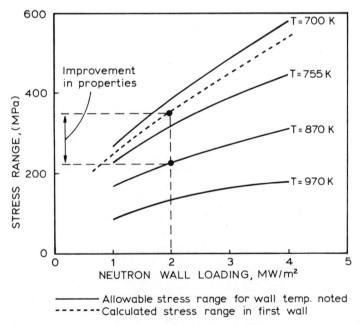

Fig. 5-9. Fatigue property requirements for stainless steel assuming scatter factor of 4.0 (Adapted from Ref. 17).

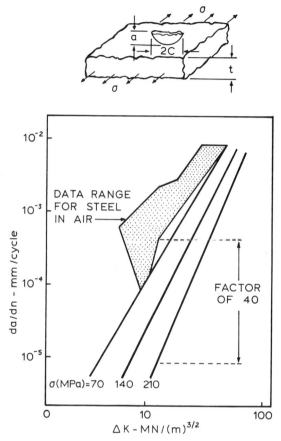

Fig. 5-10. Required flaw growth rate properties for stainless steel depending on stress levels (Ref. 17).

expected in a first wall operating at neutron wall loadings from 1.0 to 4.0 MW/m² with fatigue strength data for irradiated Type 316 stainless steel (for a fluence of $\sim 10^{23}$ n/m² and scatter factor of 4) over the temperature range equivalent to the wall loadings (i.e., as wall loading increases, wall temperature increases for constant coolant outlet and inlet temperatures). At each wall loading, a different number of cycles is required to meet the 10 MWyr/m² requirement. Therefore, the allowable stress range increases with increasing wall loading, as indicated in the figure. Figure 5-9 indicates that only the fatigue strength at 700 K can survive the calculated wall stresses, and this wall temperature translates to a wall loading of ≲1 MW/m², which is not acceptable; to survive a wall loading of 2 MW/m² (~870 K wall temperature) for the 10 MW yr/m² goal lifetime, a 50% improvement in allowable stress range for irradiated Type 316 stainless steel (i.e., from ~220 to ~340 mPa) is required. As wall loading, and hence wall temperature, is increased, the corresponding improvement level increases (e.g., to almost a factor of 3 at the 4 MW/m^2, 970 K wall condition).

Turning now to the fatigue crack growth situation, we show in Fig. 5-10 the effect of stress level (normal to the crack) on crack growth, using conventional LEFM procedures. The three crack growth lines shown correspond to those properties which would be required at operating stresses of 70, 140, and 210 MPa using a scatter factor of 4. Conditions assumed in establishing these curves are a wall life of 10 MWyr/m^2, a structural thickness of 2.5 mm, an initial flaw depth of 1.24 mm, and a flaw aspect ratio of 0.5. Also shown in the figure is the shaded area corresponding to the available crack growth properties of austenitic steel tested in air at a typical first wall loading frequency of ~9×10^{-5} cps. Inasmuch as a typical wall loading of 2 MW/m^2 would produce a stress level of 210 MPa in a 25-mm-thick structure, it can be noted from Fig. 5-10 that an improvement of approximately 40-fold would be required in the steel properties based on tests in air.

While fatigue properties of austenitic steels are improved in either sodium or in vacuo (e.g., crack growth rates are reduced by ~7; refer to Fig. 4-29), the intergranular corrosion and penetration observed when these steels are exposed to liquid lithium[30] is likely to reduce fatigue life (by eliminating the crack initiation stage) and might even increase fatigue crack growth rates. Earlier discussion in Chapter 4 also showed that, below helium embrittlement temperatures, irradiation does not have much effect on fatigue crack growth. Above ~770 K, however, the helium bubbles on the grain boundaries should greatly increase fatigue crack growth rates. Some extension of the helium embrittlement temperature limits could probably be achieved by using the Ti-modified Type 316 stainless steel, but clearly the results of this study[17] provide a strong incentive for developing or qualifying materials that possess improved fatigue properties.

SI analyses such as the one described above are useful, not only in establishing operational limits on first-generation materials and designs, but also in pointing to avenues via which improvements can be made. For instance, in the referenced study[17] it was noted that alloys that possessed improved thermal properties (e.g., higher M values in Table 5-1) could achieve the goal lifetime, even without improved fatigue properties. Vanadium alloys were used as the example, and it was shown that thermal stresses might be reduced by a factor of 2, principally because of almost a factor of 2 reduction in thermal expansion. Unfortunately, the fatigue properties of these alloys are unknown; in fact only one study of this class exists in the literature, that of vanadium at room temperature.

The fatigue life of vanadium at room temperature is comparable to that of stainless steel tested at higher temperatures. Although one would expect fatigue life to be reduced somewhat at higher temperatures, any loss should be offset by improvement in fatigue properties by alloying. Hence, it can be assumed that the fatigue characteristics of unirradiated vanadium alloys are at least as good as those of Type 316 stainless steel. Under irradiation, they should be significantly better at the same temperature since helium embrittlement is delayed and void formation is minimal. Using just such an argument the SI analysis concluded that creep-fatigue of vandium alloys is not a limiting factor in first-wall operations[17] even at first-wall temperatures of 1023 K and wall loadings as high as ~5 MW/m^2, assuming a scatter factor of 4 (Fig. 5-11). Obviously though, data are

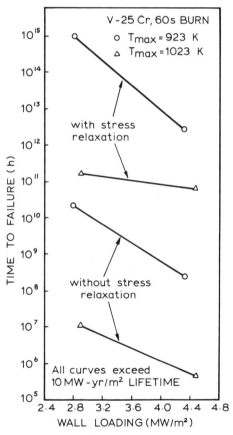

Fig. 5-11. Variation of fatigue life for a clamped vanadium alloy first wall with wall loading (Ref. 17).

badly needed on the in-pile creep-fatigue and lithium compatibility behavior of this class of materials.

Similar arguments can be forwarded for both the ferritic stainless steels and the titanium-based alloys. In addition to favorable irradiation properties, the developing stabilized ferritic stainless steels (9–12% Cr, 1–2% Mo, Nb, V) can have creep rupture and fatigue strengths better or comparable to the austenitic stainless steels up to temperatures of ~870 K (refer, e.g., to Fig. 6-16, Chapter 6) and they develop about half the thermal stress for a given surface heat flux than do the austenitics.[13] Based on the vanadium alloy discussion, the 10 MWyr/m^2 goal lifetime should easily be achieved, assuming no unusual degradation in lithium. This latter point is of concern because one would expect lithium to attack ferritic steels in a similar manner to the austenitics, and this effect may determine the wall temperature limit. Clearly, qualification of this class of steel for fusion first-wall use must emphasize lithium compatibility studies early in the program.

For titanium alloys, the lower Young's modulus and thermal expansion coefficient (compared to austenitic steels) counter the low thermal conductivity, such that the thermal stresses generated are less than in steel. In addition, fatigue properties are better than stainless steel, at least in the unirradiated state.[28] Accordingly, the alloy Ti–6Al–4V, for example, is considered able to achieve 10 MWyr/m^2 lifetime at wall loadings as high as ~4 MW/m^2.[31] Based on a relatively small data base in unirradiated environments,[23] titanium appears to be compatible with liquid lithium or helium at temperatures below 770 K. Corrosion in water is similar to that of Zircaloy (see Chapter 2) although hydriding rates are faster. Compatibility with molten salts is unknown. Therefore, this is another instance of a material class showing promise in design studies but lacking sufficient data for detailed evaluations to be made.

5.2.1.4. Closing Remarks

First-generation CTR first walls that are constructed from Type 316 stainless steel are likely to be severely limited with respect to both wall loading and wall lifetime. The combination of helium embrittlement and fatigue strength considerations will limit neutron wall loadings to uneconomic values of 1 MW/m^2 or less (i.e., wall temperatures of $\leq$700 K), while either or both swelling strains and fatigue crack growth rates will make the modest goal of a 10 MWyr/m^2 wall lifetime barely achievable. This marginal situation is improved somewhat by substitution of Ti-modified austenitic steel, but the data base on this alloy is currently insufficient to allow that move. In particular, the long-term aging effects on properties need to be better characterized.

It is, therefore, highly unlikely that austenitic stainless steels will feature in designs for second-generation, "commercial" CTRs. Although the same phenomena that cause the elimination of the austenitics are also likely to establish operating limits for their replacements, the goal of the development programs is to push those limits well into the economic regime. Based on the data to date it is probable that the combination of helium embrittlement (which determines the maximum allowable wall temperature) and creep-fatigue (which determines the wall lifetime) will limit the alloy selection for second-generation CTRs to ferritic stainless steels (e.g., 9Cr–1Mo,Nb), vanadium alloys (e.g., V–20Ti and V–15Cr–5Ti have both been used in analyses), and titanium alloys (e.g., Ti–6Al–4V), most probably in that order of preference.

Ferritic stainless steels are considered to be a leading candidate first-wall material, despite their late entry into the fusion program, because of the encouraging data base emerging from the LMFBR program. For instance, the Combustion Engineering developmental 9Cr–1MoNb steel is likely to be accepted in the high-temperature design Code Case 1592 in the near future as a result of efforts in the LMFBR steam generator development program (refer to Chapter 6, Section 6.3.1).

It is evident from the foregoing, however, that the CTR environment is quite unique and that major efforts must be expended to develop a data base on (a) bulk damage effects under pulsed 14-MeV neutron irradiation conditions, in particular the effects on fracture toughness and DBTT, (b) *in-situ* irradiation (rather than post-irradiation) creep-fatigue behavior, and (c) long-term compatibility in liquid lithium, inasmuch as several leading designs involve lithium. Items (b) and (c) will no doubt have to be combined in some manner. Ferrous alloys are susceptible to intergranular attack in lithium,[30] and this phenomenon would significantly affect fatigue and fracture properties. The available data indicate that the rate of attack depends on stress level, the composition (C, N) of the lithium, and the composition and microstructure of the material.[30] Thus, in principle, attack could be minimized or even eliminated through control of the lithium purity, and design to reduce wall stresses. Nevertheless, the lithium effect on creep-fatigue could impose a temperature limit on a ferritic steel first wall that is lower than the helium embrittlement limit.

Of the more reactive metals, vanadium and titanium (alloys) both possess attractive properties for CTR application. However, based on past experience elsewhere, fabrication and joining of these materials will be difficult, and will require special techniques to prevent (gaseous) impurity contamination and resulting embrittlement (see for example, Chapter 7, Section 7.3.2). This aspect of the problem has barely been addressed.

In the area of material properties, the problem with vanadium alloys is the complete absence of fatigue data. Clearly, fatigue life and fatigue crack growth rate data constitute the current most critical properties needed; both unirradiated and irradiated data are needed, with the latter again emphasizing *in-situ* effects. Needless to say, also, no vanadium alloys are approved in Code Case 1592, and, if such acceptance is necessary, a major effort will be necessary.

Next to fatigue, the most critical need is for additional information and data on the compatibility of vanadium-base alloys with lithium. Interstitial mass transfer is particularly important[27]; lithium getters oxygen from vanadium, for example, but vanadium getters carbon and nitrogen from lithium, and possible embrittlement due to excessive pickup of these interstitials poses serious concern. To complete the program, more data on bulk irradiation damage effects under CTR conditions is also required.

The titanium alloys are much better characterized with respect to mechanical properties (unirradiated, in air) because of their use in the aircraft industry. Still, no titanium alloy is in Code Case 1592. From a strength viewpoint the Ti64 alloy offers no advantages over other candidates, being limited to wall temperatures of 700 K or less. Therefore, for titanium alloys to be worthwhile, the high-strength Ti–6Al–2Sn–4Zr–6Mo (Ti–6246) alloy or the high-temperature creep-resistant Ti–6Al–2Sn–4Sr–2Mo (Ti–6242) alloy (or alloys of similar properties) would have to be qualified. These materials have been well characterized for aircraft turbine service conditions but little if any data exist on their behavior under CTR conditions.

The major problem for titanium alloys, and the one that may eliminate them from the CTR development program, is hydrogen (tritium) absorption. The effect of hydriding on the key properties controlling integrity is, of course, of concern, but, perhaps more of a problem is the large amount of tritium that might be stored in a titanium alloy first wall, and the associated health hazard. This concern clearly requires resolution before an in-depth qualification effort can proceed. The attractive properties of titanium alloys are unlikely to override the tritium inventory problem if it proves to be significant.

5.2.2. Candidate Ceramics

Ceramics as insulators, coatings, and liners have been included in almost all CTR conceptual designs from the early days.[32,33] The more recent consideration of ceramics for first-wall structures in CTRs, however, was motivated by the desire to use low Z (low atomic number) materials in order to avoid plasma energy loss by impurity ions.[6] Several additional advantages of low Z ceramics over refractory metals and stainless steels emerged as evaluation proceeded[6]; these include: (a) neutron interactions produce only very small amounts of long half-life radioactive isotopes, which would greatly reduce radioactive waste and disposal problems, and access for maintenance purposes; (b) short half-life radioactivities are generally small, and the total energy associated with their decay is small enough that after-heat problems would be of no concern; (c) raw materials are abundant; and (d) they are capable of withstanding very high temperatures, $\geqslant 1773$ K, while still retaining desirable mechanical properties. Thus, employment of these high-temperature materials would allow for utilization of fusion energy, both in high-efficiency thermodynamic energy conversion cycles for power generation and high-temperature-process heat applications.

Design analyses to date[6,7,10,34,35] have served as an initial selection process for further in-depth, development work. They have demonstrated the sensitivity of ceramic performance to the thermal and mechanical properties, which in turn are controlled initially by processing history and, later, during operations, by irradiation and/or plasma damage effects. Thus, regardless of whether the ceramic is intended as an insulating wall liner or as the wall itself, three significant problem areas are apparent in the further development of these materials for economic fusion reactor applications:

(1) design for brittle materials whose properties are tied to fabrication processing and material composition (e.g., brittle fracture, thermal shock resistance, etc.)
(2) plasma–particle first-wall interaction (to limit energy loss from the plasma)
(3) neutron irradiation damage including transmutation products at appropriate high (operating) temperatures.

An additional area that must be addressed is the bonding or joining of ceramics to achieve the large-size structures necessary for commercial systems. In some applications a vacuum-tight seal that is as strong as the primary material must be obtained.[36]

Several low Z ceramics have been evaluated and active research is ongoing on the most likely candidate materials. A recent review article summarized the state-of-the-art with respect to the range of applicability of a series of ceramics: carbon, SiC, Be_2C, B_2C, TiC, BN, Si_3N_4, Al_2O_3, and BaO.[6] Carbon (graphite) and silicon carbide (SiC) were selected as the most promising. In addition, they have a significant production and use history, including reactor applications, and are available in fiber or composite forms. These two materials will provide the basis for the discussion of the state-of-the-art with respect to properties relevant to CTR environments.

Carbon (graphite) and silicon carbide (SiC) are available in several forms and their properties differ widely. As background to the following discussion of properties, a short description of ceramic forms is therefore necessary. The form of carbon used must be impermeable. Chemical-vapor-deposited (CVD) carbon can be made impermeable but the material is anisotropic, which adversely affects heat transfer through the first-wall structure. Isotropic graphites are therefore preferred. Of the many grades available, two have been selected for further study: near-isotropic nuclear grade graphite and Poco graphite. CVD carbon may be useful in thin coatings, combined with a suitable structural material such as SiC or a refractory metal.

Silicon carbide is available in self-bonded, hot pressed, recrystallized, and CVD forms, all of which are isotropic. The self-bonded variety generally provides the highest thermal stress resistance and near-zero porosity and permeability. However, its free silicon content (8–15%) limits maximum operating temperature to about 100 K less than for pure SiC. Hot-pressed SiC has slightly inferior properties but no free silicon. Various hot-pressing aids (typically Al_2O_3 and B_4C) are added and other impurities are also generally present at higher concentrations. The properties of CVD SiC are similar to those for a hot-pressed material, but it has far fewer impurities.

5.2.2.1. Brittle Fracture Characteristics

The strength of ceramics depends on the stress required to propagate small inherent flaws which are distributed throughout the material. As is the case for metals, the fracture strength σ_f is related to the flaw size a by the standard fracture mechanics relation, Eq. (1-7) of Chapter 1, where K_{Ic} is the fracture toughness. The principal advantage of carbon and SiC is the retention of strength to very high temperatures (Fig. 5-12). Tensile fracture strengths for SiC are an order of magnitude greater than for some graphite materials. However, it must be noted that differences in strength (or, for that matter, any property value) can result from different fabrication procedures; variations of a factor of 2 or more are not uncommon.

Two major differences exist between metals and ceramics, with respect to strength. First, for ceramics there are a wide range of strength-controlling flaw sizes and thus a wide variation in the fracture strength from one specimen to another (hence the scatter bands in Fig. 5-12); and, second, there is an effect of specimen size on ceramic fracture strength due to the higher probability of encountering a larger critical flaw with increasing volume of stressed material. This variation in fracture stress can be characterized by a "weakest link" model due to Weibull,[37] where the risk of rupture is defined as

$$R = \int [\sigma_f/\sigma_0]^m \, dV \tag{5-1}$$

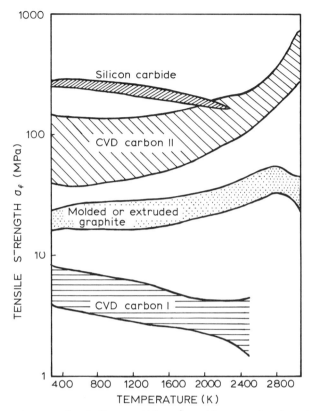

Fig. 5-12. Tensile strengths of silicon carbide and graphites as a function of temperature (Ref. 40).

and the probability of failure is

$$P = 1 - e^{-R} = 1 - \exp[-(\sigma_f/\sigma_0(V))^m \cdot V] \quad (5\text{-}2)$$

where V is the volume (surface area could also be used for fracture controlled by surface flaws), σ_0 is a normalizing constant termed the characteristic stress, and m is the so-called Weibull modulus. The Weibull modulus is a measure of the scatter in the strength distribution and the size effect, a small m value indicating a large amount of scatter and a large size effect. Brittle materials are characterized by a low value of m. For example, typical m values for SiC and graphite are 5 to 10,[6,38] which can be compared to a value of about 50 for steel at room temperature. Thus, in addition to the average fracture strength (Fig. 5-12), the unique material property that is required is the Weibull modulus.

At least two additional brittle-fracture-related properties must be considered in first-wall design; these are fatigue strength and thermal shock. With respect to the first, subcritical crack growth (i.e., when $K < K_c$) in ceramics at high temperatures can be described by Eq. (1-8), Chapter 1. Normally, for ceramics the time to failure t_f is assumed to consist entirely of the time required for the subcritical crack growth of a pre-existing flaw; hence the crack velocity can be integrated for constant stress (σ_f) and constant stress rate cyclic ($0 \rightarrow \sigma_f$) loading. For constant loading $\sigma_f^n t_f = C_1$, a constant; for cyclic loading $\sigma_f^n t_f = C_1(n + 1)$, where t_f is the number of cycles multiplied by the period of the cycle. Thus, for strength degradation due to subcritical crack growth, n, the strength degradation parameter, is the key material property. For a self-bonded silicon carbide at 1473 K, n is about 70, and the strength for 1000 h of constant stress would be about 80% of the room-temperature strength. Fatigue strengths (after 10^6 cycles) of 50% of the instantaneous fracture strength are reported.[6]

Inasmuch as the first wall will be subjected to many thermal cycles, and a temperature gradient will exist through the wall even under steady-state conditions, its resistance to thermal shock is a critical design parameter. Actual thermal shock data on graphite and silicon carbide are very limited. The M parameter (refer to Table 5-1) can be calculated from the available data to indicate relative thermal shock resistance, and it is interesting to note that graphite, despite its low strength, has a high resistance to thermal shock (i.e., high M), while the very-high-strength SiC has only a moderately high value.[6]

The implications of these basic brittle fracture properties to ceramic first-wall/blanket designs have been studied by Rovner, Hopkins, and co-workers[6,10,34] and by Trantina.[7,35] In particular, Trantina[7,35] has developed a complete structural integrity analysis in which a Weibull statistical analysis is coupled with finite-element thermal and stress analysis to predict failure probabilities of ceramic first-wall/blanket design concepts such as those in Figs. 5-2 and 5-3. The usefulness of this approach was demonstrated in one study[7] in which Trantina optimized the geometry of these structures to produce the minimum probability of failure. Regardless of the sophistication of the design analysis, however, the general conclusions of these studies are similar and are as follows:

1. As a consequence of the scatter in the fracture strength of ceramics and the size effect, large ceramic first-wall/blanket structures should be modular.

The results in Fig. 5-13 illustrate the size effect. The fracture strength of the material determined by a laboratory bend specimen is divided by the maximum stress in the structure and is plotted as a function of the volume. A first-wall module such as those shown in Fig. 5-2 is an order of magnitude larger than a "large" ceramic gas turbine component, although well within the range of potential capability for one single com-

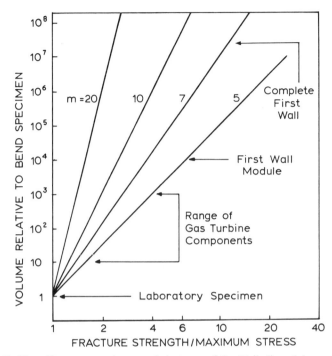

Fig. 5-13. Size effect on ceramic strength in terms of the Weibull modulus m (Ref. 7).

ponent. Even so, more than 2000 modules are required for a demonstration-size tokamak. Figure 5-13[7] indicates that for a Weibull modulus of 7, the laboratory fracture strength would have to be nearly four times the maximum stress in the structure for such a size module. This figure also shows that if the module size is increased to reduce the number required for the wall, either or both the Weibull modulus and fracture strength would have to be increased. Maximization of both the fracture strength and the Weibull modulus is, therefore, a key factor in the use of ceramics, and any in-service degradation of these properties is to be avoided (see Section 5.2.2.3 later).

2. Based on the large numbers of modules and, hence, volumes of ceramics that are likely to be involved in these first-wall designs, extensive proof testing of each module may have to be made.

Figure 5-14[7] indicates the probability of failure of a first-wall module and the entire first wall. For a probability of failure of the first wall of 10^{-4}, Fig. 5-14 shows that the fracture strength of a small bend specimen must be 40 times the maximum stress, for a Weibull modulus of 7. Clearly this would be nearly impossible to achieve. However, if each first-wall module were proof tested where 1 in 10 failures would be allowed, the

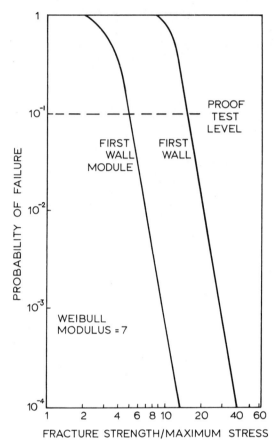

Fig. 5-14. Failure probabilities and proof test considerations for ceramics (Ref. 7).

fracture strength would need to be only 5 times the maximum stress in the structure, which is achievable with state-of-the-art technology.

3. Strength degradation due to subcritical crack growth would tend to shift the curves in Fig. 5-13 to higher values of *required* fracture strength to maximum stress ratio (i.e., larger safety factors) or, in Fig. 5-14, to higher probabilities of failure. The magnitude of this shift would depend on the strength degradation exponent n and the load history. A large amount of degradation caused by a large number of thermal cycles, (i.e., a small n) can have a significant effect, which again would have to be accounted for by proof testing under simulated service conditions. The level of stress in the proof test would be of such a level to assure a certain given lifetime; however, use of a material with a high n would help mitigate this concern.

5.2.2.2. Erosion Characteristics

The second area of concern is the influence of the ceramics on wall erosional loss rates from sputtering by energetic particles. Even with the selection of low Z materials, the influence on the plasma energy balance due to wall erosion remains of importance.

Sputtering data are now available on graphite and SiC for a range of incident atoms or ions. Table 5-3 is adapted from the report by Rovner et al.[34] and contains sputtering coefficients (atoms removed per incident atom or ion) derived from available data, by extrapolation or interpolation, assuming a maximum in the sputtering yield occurs in the range of 1–5 keV particle energy. Rovner et al.[34] point out that for the D–T, thermalized α-particles, and C or Si, the sputtering coefficients should decrease for energies that are either smaller or larger than the indicated nominal value 1–5 keV.

Reference to the table shows that sputtering coefficients for He^+ are 5 to 10 times higher than for H^+ in the low-energy range. For carbon, the experimental evidence with H^+ ions supports a chemical sputtering mechanism, with the formation of volatile hydrocarbons.[39] The situation is less clear for SiC; the absence of a temperature dependence up to 880 K supports a physical sputtering mechanism, but at higher temperatures the increase in erosion that occurs is a result of chemical reactions.[39]

The mechanism of erosion changes dramatically under high-energy (MeV) He^+ bombardment.[34] Both carbon and SiC were observed to undergo surface uplifting at low doses. For SiC, larger doses produced flaking of layers from the surface, with the thickness of the flakes being comparable to the implantation depth (Fig. 5-15). This behavior was not observed in the graphite experiments.

Rovner et al.[34] provide a qualitative explanation for the different

Table 5-3. Sputtering Coefficients for Carbon and Silicon Carbide[a]

Projectile	Energy (keV)	Sputtering coefficient	
		Carbon	Silicon carbide
D, T	1–5	0.03	0.015
He	1–5	0.15	0.15
He	3,500	≤3	100
n	≤14,100	10^{-4}	10^{-4}
C	1–5	0.6	—
Si, C	1–5	—	0.3

[a] Adapted from Ref. 34.

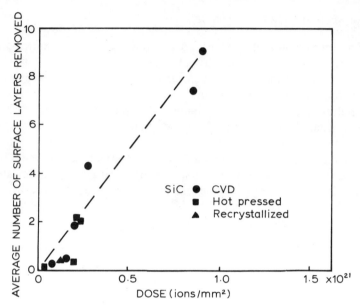

Fig. 5-15. Average removal rate of surface layers from silicon carbide bombarded with 3.5-MeV He$^+$ (Ref. 34).

responses to MeV He$^+$ in terms of the models for blistering established for metals[39] (Section 1.3.4.5, Chapter 1). These models deal with the effect of the implanted helium before it finds an escape path to the surface. For high-energy helium, the implant depth is large and, if there are no easy release paths, the helium accumulates below the surface until reemission occurs by blister rupture with attendant large effective sputtering ratios. This appears to be the case for the SiC samples, which were either of theoretical density or large grain size. At the other extreme, the high porosity and relatively small grain size of graphite facilitates release of the implanted helium and, in fact, *in-situ* measurement of helium release shows it to be reemitted with a continuously increasing rate from the outset.

The experimental results of Rovner *et al.*[34] only constitute an initial investigation of the results of bombarding various types of carbon and SiC with helium ions in the MeV range. However, it has been clearly demonstrated that very large effective erosion rates can result (as is normal in metals), but that there is a strong dependence on the type of material and energy of the bombarding ion. Several points remain unclear, and basic unresolved questions relate to formulating a more quantitative understanding of the processes involved in the helium implantation and release, and consequent surface deformation and erosion. There are several elements required to achieve this goal.

An open question is whether the inherently lower strength of carbon is playing a major role in leading to its smaller effective sputtering ratios, or whether structural effects dominate. Resolution of this question has importance as to whether SiC with appropriate surface structural features can be fabricated to greatly reduce its effective sputtering ratio. Rovner et al.[34] calculated that reduction in the SiC sputtering ratio by a factor of 10 would more than halve the wall erosion rate (from 2.5 to 0.9 mm/yr). It is important to note that other calculations in this same study, of the plasma impurity concentration buildup from wall erosion, indicated that such improvements might be essential even with these low Z materials.

Other important parameters that have been neglected, and which are important for predicting performance of a fusion reactor first wall, are angle of incidence of the ion beam, sample temperature, dose rate, and a more complete energy dependence. For metals, there is a strong dependence of surface deformation on temperature during bombardment due to an increasing helium mobility and decreasing yield strength with increasing temperatures.[39] For ceramics, the yield strength variation is essentially absent, but the temperature dependence of helium mobility may have an influence on effective erosion rates. Overall, however, there does seem to be better prospects for minimizing erosion and plasma poisoning through optimization of ceramic properties than by metal alloy deployment.

5.2.2.3. Irradiation Effects

The third important area of concern is the effect of neutron irradiation on the properties and performance of these ceramics. As is the case with metals, caution must be exercised when using available fission irradiation-induced property changes because both the neutron spectrum and temperatures are so different in a fusion system. The differences caused by transmutation products need to be investigated, due to the high threshold energy for these reactions and their relatively large cross sections above threshold. Influences on irradiation-induced strain and efficiency of annealing procedures need to be evaluated. As for displacement damage, however, the displacement cross section is relatively insensitive to the neutron spectrum, and so fission effects can be translated to fusion effects, particularly for carbon.

There are extensive data for carbon and SiC derived principally from the high-temperature gas reactor (HTGR) development program,[6,40] but irradiation effects at very high temperatures have not been adequately explored. Most emphasis has been placed on the measurement of irradiation-induced strain. SiC is relatively stable under neutron irradiation, and in several respects behaves in a predictable isotropic manner. The absolute magnitude of the strains are small, and there is no dependence on the sample microstructure or fabrication technique. For temperatures below ~1170 K the fractional expansion saturates at about 3×10^{24} n/m^2 and no further swelling is observed for fluences out to 4×10^{26} n/m^2. The data for temperatures above 1273 K are limited, but indicate two successive stages of apparent self-annealing.[6]

By contrast, the irradiation strain behavior of carbon is strongly anisotropic, reflecting the peculiar layer structure of the carbon lattice.[41,42] In some measure, a near isotropic behavior can be obtained in "isotropic" carbons, in which the swelling appears to be an average of strains observed in the perpendicular and parallel directions for the anisotropic graphite. The fabrication process therefore has an important influence on the irradiation behavior. For highly oriented pyrolytic graphite, irradiation (swelling) strains of as much as 45% at a fluence of 2.5×10^{26} n/m^2 have been reported. In contrast, the low-density amorphous, glassy carbons have been irradiated to moderate fluences at 873 K and in the range 1273 to 1373 K, and in all cases a contraction has been observed which possibly saturates at 6 to 9% linear strain.

Other property changes that result from neutron irradiation have been reported.[43,44] Neutron-induced damage results in decreased thermal conductivities of both SiC and carbon. No change in coefficient of thermal expansion of SiC has been reported to ~2.5×10^{25} n/m^2 fluence, but, for carbon, the changes are structure, temperature, and fluence dependent. If the material is isotropic, a decrease is generally found amounting to as much as a factor of 3.

The effect of neutron irradiation on mechanical properties is of particular importance. In a recent study,[45] the effect of neutron fluence (at ~403 K) on post-irradiation fracture strength, Weibull modulus, and strength degradation exponent was measured at 298 and 1373 K or 1473 K.

The data of Matheny et al.[45] are summarized in Figs. 5-16 and 5-17. For graphite at 298 K the mean fracture strength increases with irradiation and then remains constant. The scatter in strength data give rise to a lower Weibull modulus as the fluence is increased. On the other hand, for graphite tested at 1373 K the mean fracture strength remained constant as the neutron fluence was increased, indicating that some recovery of the strength had occurred. Significant strength degradation, indicated by the rather low strength degradation exponents n of 16.5 and 21.2, was observed for the irradiated material.

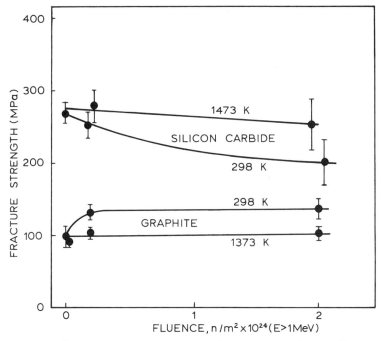

Fig. 5-16. Mean fracture strength versus neutron fluence for graphite and silicon carbide (adapted from Ref. 45).

The results of the silicon carbide tests are somewhat similar. At 298 K both the mean fracture strength and Weibull modulus decrease. At 1473 K, however, no loss of strength is observed for samples exposed to low fluence (2×10^{23} n/m^2), while there is an approximate 10% decrease in fracture strength after 2×10^{24} n/m^2 fluence. At 1473 K the Weibull modulus for the unirradiated and low-fluence SiC is essentially the same, but a large decrease occurs in the high-fluence tests.

The increased fracture strengths at 1473 K for SiC for both fluences, compared with the 298 K data, indicate that substantial annealing of defects has occurred in the ~1273–1473 K temperature range, which restores some of the mechanical strength degraded by the fast neutrons. The high values of strength degradation exponents n of 217 and 242 for SiC indicate that loading rates do not significantly influence the fracture strength.

Of the observations made in this study, the reduction in Weibull modulus (Fig. 5-16) is perhaps of most concern. The design studies mentioned earlier[6,7,10,34,35] concluded that m should be as high as possible throughout life; the irradiation degradation measured in this preliminary

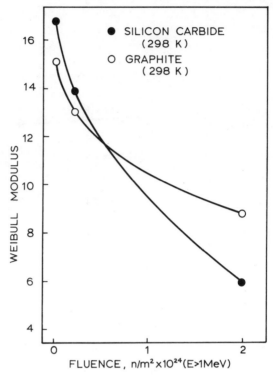

Fig. 5-17. Weibull modulus versus neutron fluence for graphite and silicon carbide at room temperature (Ref. 45).

study would make the use of either SiC or graphite difficult. However, the data are too preliminary to base a final judgment.

5.2.2.4. Closing Remarks

From the foregoing, it appears that sufficient flexibility exists in C and SiC microstructures that can be fabricated by current technology, so that erosion and irradiation damage can be minimized. Thus, these materials will be prime candidates for sacrificial coatings or liners. However, their inherent brittleness and difficulties of joining represent great disadvantages for their extensive use for the first-wall and blanket structure. If bulk graphite or SiC are to be used for first-wall/blanket primary structures, major advances would be required in brittle materials design, manufacturing processes for improved mechanical property reproducibility, and mechanical property characterization to establish design allowables and inspection procedures. Good advances are being made in the former by Trantina[7,35] and by Rovner, Hopkins, and co-workers.[6,10,34] Even if these improvements are accomplished to the maximum degree possible, however, there are some[46] who still feel that bulk ceramics would be acceptable primary structures only in smaller-size reactors (or small-size-reactor components) and under reduced operational requirements.

The general view is that there is a much better chance of success if these materials are used in fiber-reinforced composite forms.[46] For example, ceramic composite structures have been more successful in aerospace applications than bulk ceramics. This includes graphite fiber-reinforced/graphite matrix composites for nose tips, rocket motor parts, and Space Shuttle leading edges. These composites are less brittle and stronger than bulk ceramics and can be more readily manufactured into hardware. However, their state of development is currently very limited and there are many unknowns regarding radiation damage and interaction with energetic hydrogen. For example, considering the significant dimensional change associated with highly oriented graphite, the fibers may undergo rapid shrinkage because they have a high degree of preferred crystallographic orientation and are below theoretical density. In any event, there is sufficient incentive to pursue the optimal use of such materials in fusion reactors based on current state-of-the-art, and the prospects for technology transfer from other areas that have experience with the use of carbon and SiC in high-temperature structural applications are clear.

References

1. W. Hafele, J. P. Haldren, G. Kessler, and C. L. Kulcinski, *Fusion and Fast Breeder Reactors*, International Institute for Applied System Analysis (IIASA) A-2361 Laxenburg, Austria, RR-77-8 (November 1976; Revised July 1977), Chapter IX, pp. 351–436.
2. Fusion Power: Status and Options; Report to Electric Power Research Institute, EPRI ER-510-SR (June 1977), prepared by McDonnell Douglas Astronautics Co.
3. R. L. Jones, T. U. Marston, S. T. Oldberg, and K. E. Stahlkopf, Pressure Boundary Technology Program: Progress 1974 through 1978. Electric Power Research Institute, NP 1103-SR (March 1979).
4. R. W. Conn, Magnetic Fusion Reactors, to be published in *Fusion*, E. Teller, ed., Academic Press, New York (1980).
5. R. W. Conn, First Wall and Diverter Plate Material Selection in Fusion Reactors, *J. Nucl. Mater.*, **76** and **78**, 103–111 (1978).
6. L. M. Rovner and G. R. Hopkins, Ceramic Materials for Fusion, *Nucl. Technol.*, **29**, 274–302 (1976).
7. G. G. Trantina, Design Techniques for Ceramics in Fusion Reactors, *Nucl. Eng. Des.*, **54**, 67–77 (1979).
8. J. A. Maniscalco, D. H. Berwald, and W. R. Meier, The Material Implications of Design and System Studies for Inertial Confinement Fusion Systems, *J. Nucl. Mater.*, **85** and **86**, 37–46 (1979).

9. L. A. Booth, Central Station Power Generation by Laser-Driven Fusion, LA-4858-175, Los Alamos Scientific Laboratory (February 1972).
10. G. R. Hopkins, Design with Ceramics for Fusion Reactors, *J. Nucl. Mater.*, **85 and 86**, 409–414 (1979).
11. R. W. Conn, T. G. Frank, R. Hancox, G. L. Kulcinski, K. H. Schmitter, and W. M. Stacy, Jr., Fusion Reactor Design II. Report on the 2nd IAEA Technical Committee Meeting and Workshop, Madison, Wisconsin, October 10–21, 1977, *Nucl. Fusion* **18**, 985–1016 (1978).
12. The Fusion Reactor Materials Program Plan, DOE/ET 0032, Vols. 1 to 4 (July 1978).
13. S. N. Rosenwasser, P. Miller, J. A. Dalessandro, J. M. Rawes, W. E. Toffolo, and W. Chen, The Application of Martensitic Stainless Steels in Long Lifetime Fusion First-Wall/Blankets, *J. Nucl. Mater.*, **85 and 86**, 177–182 (1979).
14. F. W. Wiffen, Radiation Effects in Structural Materials for Fusion Reactors, in: *Critical Problems in Energy Production*, Charles Stein, ed., Academic Press, New York, (1976), pp. 164–188.
15. E. E. Bloom, F. W. Wiffen, P. J. Maziasz, and J. O. Stiegler, Temperature and Fluence Limits for a Type 316 Stainless Steel Controlled Thermonuclear Reactor First Wall, *Nucl. Technol.*, **31**, 115–122 (1976).
16. E. E. Bloom, Mechanical Properties of Materials in Fusion Reactor First-Wall And Blanket Systems, *J. Nucl. Mater.*, **85 and 86**, 795–804 (1979).
17. M. Abdou, S. D. Harkness, S. Majumdar, V. Maroni, B. Misra, B. Cramer, J. Davis, D. DeGreece, and D. Kummer, The Establishment of Alloy Development Goals Important to the Commercialization of Tokamak-Based Fusion Reactors, ANL/FPP Technical Memorandum, Number 99, MDCE-1743 (November 1977).
18. R. F. Mattas and D. L. Smith, Model for Life-Limiting Properties of Fusion Reactor Structural Materials, *Nucl. Technol.*, **39**, 186–198 (1978).
19. R. W. Conn, Tokamak Reactors and Structural Materials, *J. Nucl. Mater.*, **85 and 86**, 9–16, (1979).
20. J. J. Holmes and J. L. Straalsund, Effects of Fast Reactor Exposure on the Mechanical Properties of Stainless Steels, in: *International Conference on Radiation Effects in Breeder Reactor Structural Materials*, M. L. Bleiburg and J. W. Bennett, eds., AIME (1977), pp. 53–63.
21. D. J. Michel and G. E. Korth, Effects of Irradiation on Fatigue and Crack Propagation in Austenitic Stainless Steels, in: *Proceedings of the International Conference on Radiation Effects in Breeder Reactor Structural Materials*, Scottsdale, Arizona, June 19–23, 1977, Met. Soc. AIME, pp. 117–137.
22. P. J. Maziasz and E. E. Bloom, Alloy Development for Irradiation Performance, Quarterly Progress Report January–March 1978, DOE/ET-0058/1 (August 1978), pp. 54–81.
23. R. H. Jones, B. R. Leonard, Jr., and A. B. Johnson, Jr., An Assessment of Titanium Alloys for Fusion Reactor First-Wall and Blanket Applications, EPRI-RP1045-3 Project Final Report (1979), Electric Power Research Institute.
24. R. F. Mattas, H. Wiedersich, D. G. Atteridge, A. B. Johnson, and J. F. Remark, Elevated Temperature Tensile Properties of V-15Cr-5Ti Containing Helium Introduced by Ion Bombardment and Tritium Decay, in: *Proceedings of the Second ANS Topical Meeting on Technology of Controlled Thermonuclear Fusion*, CONF-760935, U.S. Energy Research and Development Administration (September 1976).
25. M. L. Grossbeck and P. J. Maziasz, in: Alloy Development for Irradiation Performance, Quarterly Progress Report, January–March 1978, DOE/ET-0058/1 (August 1978), pp. 82–85.
26. P. J. Maziasz and E. E. Bloom in: Alloy Development for Irradiation Performance, Quarterly Progress Report, January–March 1978, DOE/ET-0058/1 (August 1978), pp. 40–53.

27. R. E. Gold, D. L. Harrod, R. L. Ammon, R. W. Buckman, Jr., and R. C. Svedberg, Technical Assessment of Vanadium-Base Alloys for Fusion Reactor, in: Alloy Development for Irradiation Performance, Quarterly Progress Report, January–March 1978, DOE/ET-0058/1 (August 1978), pp. 119–127.
28. NUWMAK: A Tokamak Reactor Design Study, University of Wisconsin, Madison, Wisconsin (March 1979), UWFEDM-330, Chapter IX.
29. N. M. Ghoniem and G. L. Kulcinski, The Effect of Pulsed Irradiation on the Swelling of 316 Stainless Steel in Fusion Reactors, *Nucl. Eng. Des.*, **52**, 111–125 (1979).
30. J. H. DeVan, J. E. Selle, and A. E. Morris, Review of Lithium Iron-Base Alloy Corrosion Studies, ORNL/TM-4927, Oak Ridge National Laboratory (1976).
31. R. W. Conn, G. L. Kulcinski, and C. W. Maynard, NUWMAK: An Attractive Medium Field, Medium Size, Conceptual Tokamak Reactor, in: *Proceedings of the 3rd Topical Meeting on the Technology of Controlled Thermonuclear Fusion*, American Nuclear Society, Sante Fe, New Mexico, May 9–11, 1978.
32. F. W. Clinard, Electrical Insulators for Magnetically Confined Fusion Reactors, in: *Critical Problems in Energy Production*, Charles Stein, ed., Academic Press, New York (1976), pp. 142–163.
33. Special-Purpose Materials for the Fusion Reactor Environment: A Technical Assessment (February 1978) U.S. DOE/ET-0015.
34. L. H. Rovner, R. F. Bourque, and K. Y. Chen, Applications of Low Atomic Number Ceramic Materials to Fusion Reactor First Walls, Electric Power Research Institute, EPRI ER-216, final report on Project 115-2 (August 1976).
35. G. G. Trantina, Ceramics for Fusion Reactors—Design Methodology and Analysis, *J. Nucl. Mater.*, **85** and **86**, 415–420 (1979).
36. W. E. Hauth, R. D. Blake, H. L. Rutz and J. M. Dickinson, Fabrication of the 320-cm OD All-Ceramic ZT40 Torus, *J. Nucl. Mater.*, **85** and **86**, 433–438 (1979).
37. A. Weibull, A Statistical Approach to Engineering Design in Ceramics, *Proceedings of the British Ceramic Society*, Vol. 22, pp. 429–452 (1973).
38. R. L. Jones and D. J. Rowcliffe, Tensile-Strength Distributions for Silicon Nitride and Silicon Carbide Ceramics, *Caram. Soc. Bull.* **58**, 836–839 (1979).
39. J. Gittus, *Irradiation Effects in Crystalline Solids*, Applied Science Publishers, London (1978), Chapter 10, pp. 475–476.
40. J. T. A. Roberts, Ceramic Utilization in the Nuclear Industry, *Powder Metall. Int.*, **10**, 212 (1978); **11**, 24–29 (1979); **11**, 72–82 (1979).
41. G. B. Engle and W. P. Eatherly, Irradiation Behavior of Graphite at High Temperature, *High Temp.-High Pressures*, **4**, 119–158 (1972).
42. G. B. Engle, Irradiation Behavior of Nuclear Graphites at Elevated Temperatures, *Carbon*, **9**, 539–554 (1971).
43. R. J. Price, Thermal Conductivity of Neutron-Irradiated Reactor Graphites, General Atomic Publication No. GA-A13157 (1974).
44. R. J. Price, Thermal Conductivity of Neutron-Irradiated Pyrolytic Silicon Carbide, *J. Nucl. Mater.*, **46**, 268–272 (1973).
45. R. A. Matheny, J. C. Corelli, and G. G. Trantina, Radiation Damage in Silicon Carbide and Graphite for Fusion Reactor First-Wall Applications, *J. Nucl. Mater.*, **83**, 313–321 (1979).
46. Conference Proceedings: Low-Activation Materials Assessment for Fusion Reactors, Electric Power Research Institute Publication No. EPRI ER-328-SR (1977).

6

Heat Exchanger Materials

As the name implies, the heat exchanger serves to transfer heat between coolant circuits in a power plant. The steam generator, intermediate heat exchanger (IHX), and condenser were introduced in Chapter 1 as the components employed for this purpose in nuclear systems. To reiterate, the PWR requires a steam generator, the LMFBR (and most probably the CTR) requires both a steam generator and IHX, and all reactors require a condenser.

Inasmuch as these components operate in aggressive environments (air, water, sodium, etc.), all problems associated with heat exchangers are either directly or indirectly a consequence of corrosion. As was already noted in Chapter 4, water is a particular "bad actor," and waterside corrosion phenomena will therefore be a major topic in this chapter. The intent here is: to review recent research into corrosion mechanisms and potential remedies for PWR steam generators (Section 6.2); to summarize the key aspects of the data base assembled in support of the reference LMFBR steam generator and IHX designs, review early field experience, and discuss development of alternative or advanced LMFBR heat exchanger materials (Section 6.3); and to review corrosion problems in condensers (exclusively LWR) and the search for improved tubing materials (Section 6.4). By way of introduction, Section 6.1 describes typical designs and materials of construction for these heat exchanger components. The modifications to conventional designs necessitated by the emerging CTR concepts are also included here.

6.1. Design and Materials of Construction

Although the specific design varies, all heat exchangers consist essentially of a series of tubes within a shell. In order to minimize vibration due to coolant flow over the tubes and to direct flow over the tubes so that maximum efficiency of heat transfer results, tube support plates and/or baffles are provided within the shell. Also, tube sheets are required to separate or form header or plenum regions within the shell for the entering and exiting coolant that flows through the tubes. Tube sheets also seal the coolant flowing through the tubes from the coolant on the shell side.

Table 1-1, earlier, reveals a broad range of materials used for heat exchanger construction in the attempt to minimize corrosion damage. Component-specific details will be presented in the following subsections.

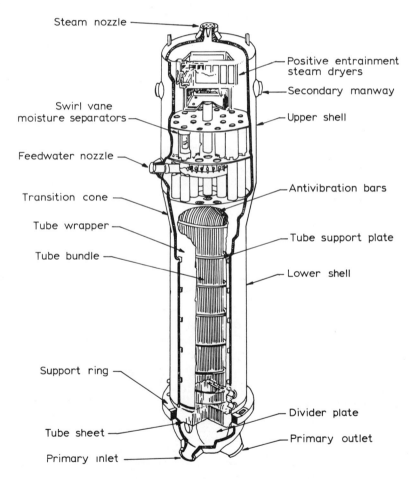

Fig. 6-1a. Schematic of a recirculating D&S steam generator (Westinghouse Model F is the example).

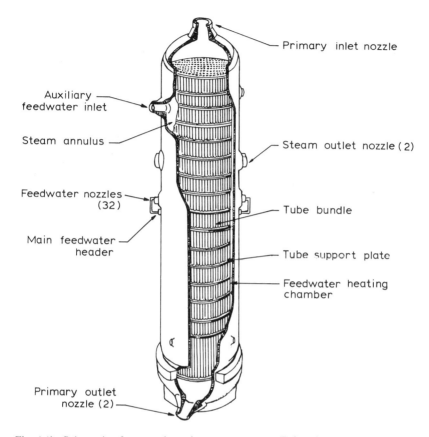

Fig. 6-1b. Schematic of a once-through steam generator (Babcock and Wilcox model).

6.1.1. PWR Steam Generator

There are two basic commercial PWR designs in operation: a recirculation system which produces dry steam at the saturation temperature (D&S). A vertical U-tube D&S unit is shown in Fig. 6-1a; and a once-through steam generator (OTSG), illustrated in Fig. 6-1b, in which all the water entering the generator is converted to steam and superheated at the outlet in a single pass. Westinghouse, Combustion Engineering (CE), and Kraftwerk Union (KWU) produce variants of the inverted U-tube design, but there are important differences of detail between vendors and between the older and current designs. Babcock and Wilcox (B&W) produce the OTSG units. Typically, steam generators are ~20 m high by ~4 m in diameter, and weigh about 310,000 kg.

Steam generator materials in current use in PWRs are summarized in Table 1-1, Chapter 1. A variety of tube materials have been used, including 1Cr–½Mo steel, 300 stainless steels, Alloy 800, and Inconel 600; experience with them has been reviewed by Berge and Donati.[1] As will be discussed later in Section 6.2, materials evaluation programs are continuing throughout the world in an attempt to maximize tube lifetime.

The tube-sheet and tube-support materials include mild steel (which has been shown to be unsatisfactory and which will therefore be replaced by 12Cr material in all new designs from Westinghouse and CE, and by 13Cr materials in some French units) and 300 series stainless steel in KWU units.

The design details of the steam generator such as temperature, heat flux, fluid flow, and methods of fabrication influence vulnerability to corrosion attack just as much as the materials of construction. However, a detailed discussion of the effects of design differences on thermal hydraulics, although important, is considered outside the scope of this book. Two aspects of steam generator design that are worthy of note here are tube to tube-sheet joints and tube supports (refer to Fig. 6-1 for locations). These areas will feature strongly in the discussion of corrosion damage in Section 6.2. In older designs the method of joining the tube to the tube sheet produces a crevice between tube and plate on the secondary side in which aggressive solutions can be generated. In more recent D&S designs the tubes are expanded into the tube sheet, thereby virtually eliminating the crevice. The method of expanding the tube into the tube sheet can also influence corrosion vulnerability, however, because the extent of residual stress and cold work produced in the tube will affect SCC susceptibility. Mechanical rolling can produce high residual stresses, whereas expansion by explosion or static hydraulic pressure has been shown to be far superior and is now standard practice for new designs.[2]

With respect to the method of tube support, the mild steel, rigid plate, drilled hole design illustrated in Fig. 6-2a is the type which has caused problems. Figures 6-2b–d show the broached and "egg-crate" designs that are either in operation or will be used in new designs, and which are much more open and more flexible than drilled plates.

6.1.2. LMFBR Heat Exchangers

In the LMFBR, the heat generated in the reactor core is first transferred from the primary sodium to a second sodium coolant circuit in the intermediate heat exchanger (IHX). The IHX is included to prevent possible contamination of the steam with the radioactive sodium of the primary coolant circuit. The IHX and steam generator differ in both design and materials of construction due to their different operating environments.

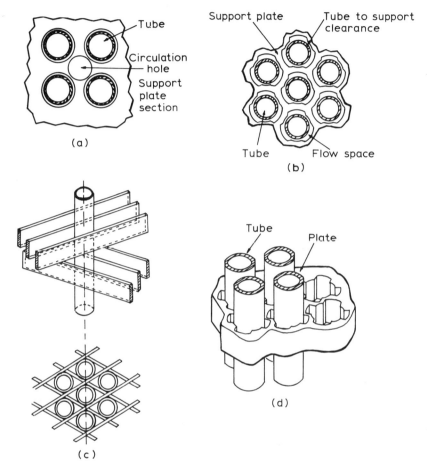

Fig. 6-2. Tube support plate designs: (a) drilled hole; (b) B&W Trifoil plate for OTSG; (c) KWU "Egg-crate"; (d) Westinghouse Quatrafoil for Model F units (Ref. 2).

6.1.2.1. Steam Generator

The LMFBR steam generator can have a design similar to that for the PWR (the so-called "saturated cycle") or can incorporate a superheater that heats the wet steam to above its saturation temperature ("superheat cycle"). The evaporator and superheater may consist of individual units or be combined in a single unit; the tube bundle can be either recirculating or once through. The superheat cycle appears to have the greater potential for higher temperatures and accompanying improvements in thermal efficiency and, therefore, has been adopted for several prototype LMFBRs around the world.[3] On the other hand, the U.S.

utilities favor the lower-temperature saturated cycle for the first commercial LMFBR in order to eliminate material reliability problems.[4] A general Electric/Bechtel design has a maximum sodium outlet temperature of 740 K and a saturated steam cycle approach based on a straight tube, double-walled, recirculating steam generator to deliver steam at 7.1 MPa and 560 K.[5]

The current trends in steam generator designs can be illustrated by the U.S. prototype CRBRP (Fig. 6-3a) and the French demonstration

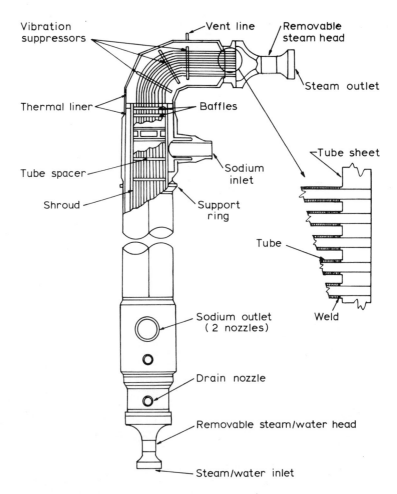

Fig. 6-3a. Schematic of recirculating steam generator for CRBRP.

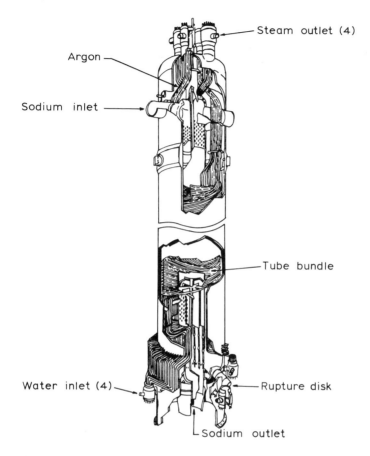

Fig. 6-3b. Schematic of once-through steam generator for Super Phenix.

reactor Super Phenix (Fig. 6-3b). In the former, the steam generator module consists of two evaporators and one superheater (three steam generators will be used in the plant).[6] The overall height is 2.0 m and diameter is 1.3 m. The evaporators and superheaters are vertically oriented shell and recirculating tube heat exchangers arranged in a "hocky stick" configuration with a 90° bend in the shell and tubes to provide for differential thermal expansion between the tube bundle and shell. The tubes (757) are supported along their length by 19 tube spacers in the active heat transfer region, and the shell connects to upper and lower tube sheets. The CRBRP steam generator should produce superheated steam at 10 MPa and 755 K, while cooling the sodium to ~620 K.[6]

In contrast, the steam generator selected for the Super Phenix LMFBR is a true once-through unit with three heat transfer zones (economizer, evaporator, and superheater) located along the same tube, whose total length is 22 m and whose diameter is ~3 m.[7] The bundle consists of 357 tubes, 19.8 mm I.D., which are coiled within the shell and allowed to move axially and also radially as a result of thermal expansion. This design is expected to deliver superheated steam at 18.5 MPa and 760 K, while cooling the sodium to ~620 K.

As noted in Table 1-1 earlier, both ferritic and austenitic materials have been chosen for steam generator construction, and it is evident that opinions differ on the optimum material.[3] The U.S. selected $2\frac{1}{4}$Cr–1Mo steel (Croloy) in the annealed condition for use in CRBRP for its metallurgical stability and good service record in boiler tubes and main steam piping systems in fossil-fired steam generating plants.[6] Further, $2\frac{1}{4}$Cr–1Mo steel is qualified to meet ASME Boiler and Pressure Vessel Code requirements, and it has also been used as the steam generator material in EBR-II, which has operated virtually trouble-free since 1965.

The French chose both $2\frac{1}{4}$Cr–1Mo steel and a niobium-stabilized version of this alloy for use in the evaporators and Type 321 stainless in the superheaters of the earlier Phenix reactor, but switched to Type 316 L stainless steel for the shell and internal structures and Alloy 800 for the tube bundle for Super Phenix. The reasons given for selection of Alloy 800 are that it is much more tolerant to SCC on the waterside than austenitic steels, and provides greater protection from wastage due to the sodium–water reaction, in the case of small leaks, than other austenitic or ferritic steels.[7]

Other demonstration reactors have or will employ both ferritic and austenitic materials. For instance, the U.K. CDFR evaporators and superheaters will utilize ferritic 9Cr–1Mo in the normalized-and-tempered condition, and the German SNR-300 is using a niobium-stabilized version of $2\frac{1}{4}$Cr–1Mo steel. A very informative data sheet on LMFBR steam generator designs and materials has been provided in the review paper by Brinkman and Katcher.[3]

6.1.2.2. Intermediate Heat Exchanger

The IHX is a relatively simple component that functions on a counterflow principle: the primary sodium circulates down the outside of the tubes, whereas the secondary sodium flows up inside the tubes.

The diagram in Fig. 6-4 is of the IHXs for the French Phenix reactor series.[8] The Super Phenix IHX is essentially a scaled-up version of the Phenix IHX. The tube bundle (1) is held between an inner cylindrical shell (2) and an outer cylindrical shell (3), which encloses the primary

Heat Exchanger Materials

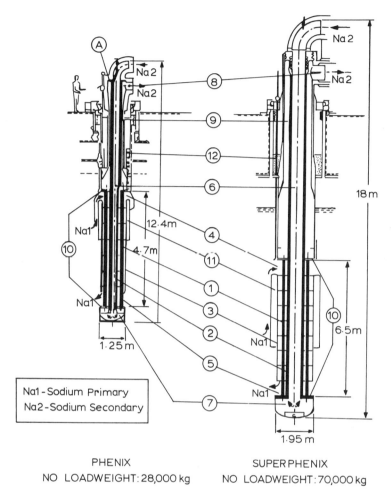

Fig. 6-4. Schematic of typical LMFBR intermediate heat exchangers (French Phenix and Super Phenix designs are the examples) (Ref. 8).

sodium circulating round the tubes. The primary sodium inlet (4) is at the top and the outlet nozzles (5) are at the bottom. The secondary sodium inlet pipe (6) axially penetrates the center of the tube bundle. The cold sodium flows down this pipe, turns through 180° (7), and flows up inside the tubes to the outlet header (9), from which it is conveyed by the secondary coolant loop pipes (8) to the steam generators. The tube bundle is comprised of straight tubes welded at each end to two annular tube plates. The tube bundle and shell are fabricated out of austenitic stainless steel, Types 304 and 316. It is common practice to choose the same material for the IHX as for other primary components.

6.1.3. CTR Steam Generator

Until recently very little attention had been given to the "balance of plant" for conceptual CTR designs, the general view being that first-generation CTRs would utilize the LMFBR balance of plant wholly (refer to Chapter 1). It is now realized, however, that (a) the need to reduce tritium permeation losses and (b) the attractiveness of heat transport media other than liquid metals calls for different heat exchanger designs. It is of interest to compare the modifications made to conventional steam generators and IHXs to incorporate the unique requirements of the CTR environment. As might be expected, while the designs differ, the materials remain essentially within the fission reactor technology state of the art. Therefore, the reader is referred to Sections 6.2 and 6.3 for appropriate materials data and experience.

The "conventional" heat transport system, as exemplified by the UWMAK-I tokamak,[9] utilizes a lithium primary coolant which transfers its heat in an IHX to a sodium secondary fluid. The sodium in turn transfers its heat to a steam generator. The steam generator for UWMAK-I was developed from B&W OTSG technology, with the main difference being that *double-walled* tubes are employed to minimize tritium diffusion into the steam. The helium in the gas gap between the walls contains a trace amount of oxygen to getter tritium diffusing through the walls. Inconel 600 tubes and $2\frac{1}{4}$Cr–1Mo steel platens comprise the tube bundle, which is contained within a 13-mm-thick carbon steel vessel with a 6 m outside diameter and an overall length of 14.6 m. It is expected that the unit would be mounted horizontally.

Other means of preventing tritium leakage in heat exchangers employ barrier metals (e.g., copper, tungsten), barrier oxides (e.g., oxidation films, BeO), and metal getters (e.g., titanium, cerium), and all might be placed into laminated (triplex) tubing.[10] The virtue of oxide barriers is that they are far less permeable than the least permeable of metals (tungsten). However, they suffer from the disadvantage of being difficult to fabricate. This is especially true if pore-free thin films are required as the intermediate layer.

In the UWMAK-II tokamak design,[11] helium is the primary coolant, transferring its heat, as before, to a sodium secondary fluid in the IHX. Since a gas is involved, the IHXs are of the same design as for LMFBRs but larger (21 m high and 3 m in diameter). Conventional materials are proposed, namely Incoloy 800 for tubes and Type 316 stainless steel for the shell. These materials are compatible with helium and sodium and the system operating temperatures. Likewise, the conceptual steam generator utilizes LMFBR design practice. A modular concept is proposed, with three modules [preheater ($\leq$670K), superheater ($>$670 K), and resuperheater (>700 K)] for each loop. The preheater modules are $2\frac{1}{4}$Cr–1Mo ferritic steel while the superheater and resuperheater modules will use Type 321 stainless steel. Such steam generators are expected to deliver superheated steam at 25.5 MPa and 780 K to the turbines.

The SOLASE ICF concept[12] (refer to Chapter 5) utilizes lithium oxide as the primary heat transport medium and operates on the Rankine steam power cycle. A special steam generator design is therefore necessary, but an intermediate coolant loop is not required because of the expected low tritium leakage rate.

The steam generator design selected has a C-shaped tube configuration, (Fig. 6-5) to ease thermal stress caused by differential expansion between tubes and the shell (compare to the "hockey stick" CRBRP design, Fig. 6-3a). The steam generator is oriented vertically so that the lithium oxide particles would flow by gravity through the passages between the tubes. Support bars provide vertical support for the tube bank.

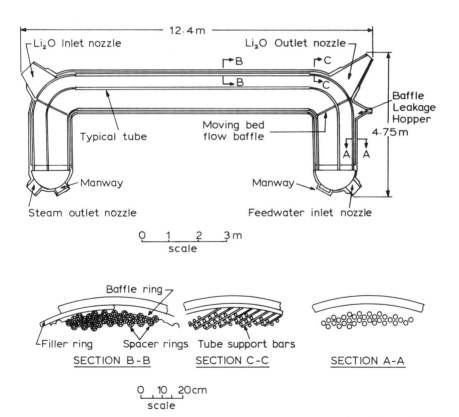

Fig. 6-5. Schematic of steam generator concept for the SOLASE CTR. (Adapted from Ref. 12.)

Each unit is expected to have over 10,000 2¼Cr–1Mo tubes, with the boiler length being ~2 m long and the superheater section, ~3 m long. The steam delivered is expected to be at 16.6 MPa pressure and at a temperature of 810 K.

6.1.4. Condenser

The major function of the condenser is to transfer the heat rejected from the steam cycle to a receiving body of water or to the atmosphere when a cooling tower is employed (refer to Figs. 1-1 to 1-3, Chapter 1).

Condensers for LWR and LMFBR (and fossil-fueled power plants) have the same configuration. A typical design[13] is shown in Fig. 6-6, which is an isometric cutaway view of a single-pass (of the cooling water) condenser with a divided water box. This particular design has two bundles of tubes, each with its own air removal section. Cooling water flows in parallel through the two bundles. It is conventional to introduce the exhaust steam from the low-pressure (LP) turbine into the top of the condenser.

The condenser tubes are affixed by rolling and welding to the tube sheets, and are supported along their length by support plates, just like the other heat exchangers. The key materials consideration in condenser design is the selection of tubing material (Table 1-1, Chapter 1). Different tubing materials might be used for the three sections, air removal, [8] impingement, [11] and condensing [16] (refer to Fig. 6-6 for locations). For instance, in most condensers with admiralty brass tubes in the condensing section, tubes of a harder material such as Type 304 or 316 stainless steel or copper–nickel (90–10 or 70–30) are used in the impingement area (i.e., outer few rows). In addition to these materials, a recent review of condenser experience[13] also identified the following tube materials in use in condensers: aluminum brass, aluminum bronze, and titanium.

6.2. PWR Steam Generator Experience

Steam generators have been consistently troublesome for PWR units, and a large variety of problems have contributed to the plant capacity losses due to this component.[14a–d] A recent analysis[14d] of steam generator problems concluded that tube-related problems are the principal cause of component downtime. Capacity factor losses were 4.2% for the early years (up to July 1976) and have been 3.5% recently (July 1976 to June 1978). This trend to smaller losses will no doubt continue as operating experience is gained, but until remedies to the key problems are found,

Heat Exchanger Materials 333

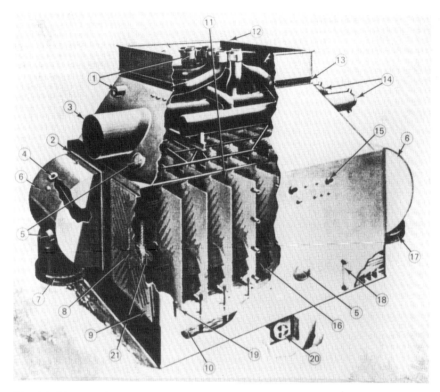

Fig. 6-6. Diagram of a single-pass surface condenser (Westinghouse design). (1) Extraction connection to external heater; (2) diaphragm connection between tube sheet and shell; (3) feed pump turbine drive exhaust; (4) noncondensable removal connection; (5) manhole; (6) water box; (7) cooling water connection (in); (8) tube bundle (noncondensable removal section); (9) tube sheet; (10) hot well; (11) tube bundle (impingement); (12) turbine exhaust hood connection; (13) expansion joint to turbine exhaust hood; (14) F.W. Heaters in condenser steam dome; (15) dumps and drains; (16) tube bundle (condensing); (17) cooling water connection (out); (18) hot well level; (19) tube-support sheets; (20) condensate pump connection; (21) cold-end baffle in noncondensable section.

1 to 2% capacity loss per year can be expected due to the requirement for regular inspections and repairs that primarily involve tube plugging. More importantly, there is a significant probability of steam generator replacement prior to the end of a 40-year power plant lifetime. Estimates of steam generator replacement cost and replacement power costs during the outage have ranged from $100 million to more than $300 million per plant.[15] Thus, research to solve steam generator malfunctions has recently been accelerated.

The major experience with steam generator performance has been gained with the U-tube D&S design (Fig. 6-1a). Some Westinghouse and CE designs have experienced severe corrosion damage which falls into three main categories[2,14a,b,15]: (a) caustic cracking of the tube within the tube sheet or just above the tube sheet in regions submerged in crud deposits; (b) tube wastage (phosphate thinning) in the region just above the tube sheet; and (c) "denting" at tube/tube-support intersections on some plants which operate with all-volatile (alkali) treatment (AVT). The distortion caused by "denting" has in some cases led to stress corrosion cracking of the tubing from the *inside* of the tube. Figure 6-7 schematically indicates the type and location of corrosion damage, which is primarily restricted to the secondary side of the system.

With the exception of the Obrigheim reactor, the KWU plants have not experienced caustic cracking, but very thick crud deposits have accumulated on the tube sheet, and progressive tube thinning has occurred in this region on some plants. None of these has experienced denting, however.[2]

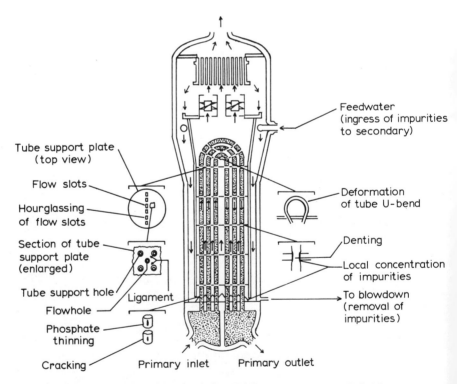

Fig. 6-7. Corrosion problems in recirculating (D&S) steam generators (Ref. 15).

The B&W OTSG units (Fig. 6-1b) have not had any of the above problems but have experienced mechanical problems, including high-cycle fatigue failure, fretting, pitting, and erosion/corrosion of tubes, all of which again originate on the secondary side of the system.[14c]

Before proceeding to discuss the corrosion and mechanical damage phenomena in some detail, it is important to note that not all the problems listed are encountered on one design of plant and for any one design the severity of a particular problem can range from very severe to undetectable. Secondary circuit corrosion damage, as on any steam generator, is a complex function of boiler design, operating conditions, boiler water chemistry, and the extent of ingress of air and cooling water contaminants.

6.2.1. Corrosion Damage Processes

Most of the D&S units in current operation have used phosphate to control the pH of the water, and a large fraction have experienced some secondary-side corrosion damage (pitting, cracking, and wastage) to the tubing.[2,15] As a result, most of those steam generators that were initially operated with phosphate were converted to all-volatile treatment (AVT), utilizing hydrazine or morpholine. The conversion from phosphate to AVT water treatment (including new units started with AVT) resulted in a decrease in tube wastage and pitting but introduced a new damage process, termed "denting," which is considered to be much more serious.[2,15,16]

"Denting" is a circumferential inward movement, or dent, of steam generator tubing at support plates and at tube-sheet locations having open crevices (see Fig. 6-7). Distortion of the tubes is caused by the volume expansion of oxides that form when accelerated corrosion of carbon steel occurs in the drilled holes of support plates. The oxide volume expansion also distorts the support plates, and this in turn can have serious side effects, such as deformation and sometimes cracking of inner-radius tubing of U-bends and cracking of the support-plate ligaments as indicated in Fig. 6-7. In a few instances, tube denting has caused leaks to develop in tubes in the vicinity of the support plate. The worst denting is in plants that use salt water for condenser cooling (i.e., resulting from chloride intrusion, see Section 6.4), that have drilled holes in carbon steel support plates and tube sheets (i.e., crevices available), and that are operating on AVT water chemistry.

New forms of damage, some of which appear to have been caused or aggravated by secondary-side corrosion-related mechanisms, have occurred in once-through steam generators (OTSG) in the past two years.[2,15] The observed types of tube damage include tube fractures, wear from combined erosion/corrosion, corrosion-cavitation, and pitting. Flow-induced vibration of steam generator tubes is believed to be a major contributing factor to the wear and tube fracture damage (see Section 6.2.2). Since the data from once-through steam generators is limited, due to the limited number in operation, it is difficult to prognosticate the extent of damage that may occur in this type of steam generator.

Local corrosion rather than general corrosion is therefore the cause of damage in PWR steam generators, and various terms are used in an attempt to "capture" the physical picture of the corrosion processes, e.g., pitting, wastage, intergranular stress corrosion cracking (IGSCC), and denting. For the most part, local corrosion tends to be restricted to certain zones on the secondary side of the steam generator, viz. (1) tube-sheet crevices, (2) areas just above the tube sheet and possibly just underneath the top of the tube-sheet sludge piles, (3) at and near support plate crevices, and (4) at the upper U-bend regions, particularly at tube-support devices (refer to Fig. 6-7). The common feature of all of these zones is that concentration of aggressive chemicals is favored there, and this observed correlation is the basis for the belief that *chemical concentration* is the probable cause of all corrosion damage. Thus, there are three essential steps in the corrosion damage process: (1) the ingress of impurities into the secondary system, (2) the concentration of impurities in the steam generator, and (3) corrosion attack of steam generator materials by the concentrated impurities. A schematic of these steps and the factors that are pertinent to them are shown in Fig. 6-8.[17]

The sources of impurities that enter steam generators via the feedwater are various and include condenser cooling water inleakage, makeup water, demineralizer regenerator chemicals, various soluble and insoluble corrosion and erosion products, some water treatment chemicals such as phosphates, and miscellaneous contaminants such as air, oil, preservatives, packing, etc., which enter when systems are under construction or are opened up for repair and inspection.

Factors that act to concentrate chemical inpurities in steam generators are also shown in Fig. 6-8. The dominant driving force for concentration is the boiling or evaporation process, which produces relatively pure steam and a chemical residue in the liquid phase. This tendency is counteracted by good hydraulic conditions that tend to redilute the local boildown areas and restore them to the bulk chemistry conditions. However, the large "open" volume on the secondary side, in combination with thermal-hydraulic asymmetries, makes it very difficult to assure fluid flow conditions that provide adequate wetting of heat transfer surfaces on a local scale approximating the size of a single tube. Furthermore, sludge and occluding conditions, such as scale and crevices, are important and complex factors affecting impurity concentration.

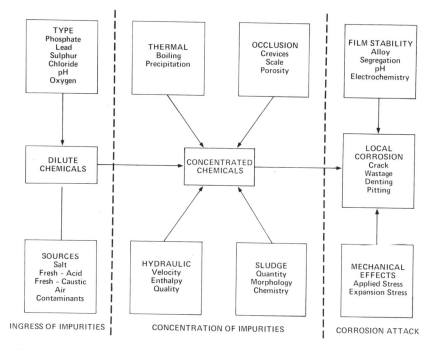

Fig. 6-8. Local corrosion process in the secondary side of steam generators (Ref. 17).

6.2.1.1. Denting

The cause of denting is best explained by a Potter–Mann type linear accelerated corrosion in which a nonprotective oxide layer is formed as the corrosion progresses.[2,15,16] Potter and Mann[18] observed that mild steel can be oxidized to magnetite (Fe_3O_4) in aqueous conditions at 570 to 670 K and that the oxide growth is nonprotective in the presence of ferrous, nickel, (e.g., in the vicinity of FeNiCr alloys), and chloride ions (Fig. 6-9). The nickel is believed to act as an efficient cathode in an electrochemical metallic oxidation reaction under cathodic control.

The dominant events in the growth of nonprotective magnetite are conceived as follows.[16] When the aqueous environment contains a sufficient concentration of dissolved iron, the outward movement of iron cations through an incipient magnetite film is virtually suppressed so that the film grows almost entirely at the metal/oxide interface. The resultant stresses are relieved by cohesive and adhesive failures in the magnetite, which is rendered so porous as to provide no obstruction to its own continued growth. The kinetic control of the oxidation can then reside elsewhere than in the film and, in particular, will pass to the cathode if

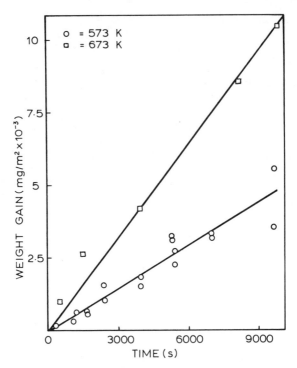

Fig. 6-9. Oxidation of mild steel in 0.1 M nickel chloride (initially) at 573 and 673 K (Ref. 18).

a more efficient surface than magnetite is available. Nickel metal (for example) on the magnetite surface is able to provide the needed cathodic stimulation of the oxidation, as long as there are chloride ions available to keep iron cations in solution.

Accordingly, research into denting is concentrated on the role of the coolant chemistry in the corrosion process.[2,15,16] Sample intersections of tubes and support plates removed from dented steam generators have shown local chloride concentrations of over 4000 ppm in the dented region.[15] Denting has been successfully reproduced in the laboratory with acid chlorides; in laboratory pot boiler tests, fast linear growth (nonprotective) of magnetite was generated in less than 56 days when operating with volatile chemistry and 1 ppm chloride (from seawater) in the bulk. In these pot boiler tests, a local chloride concentration measured in the dented region exceeded 10,000 ppm after the tests in some cases, and denting increased with increased chloride concentration.[15] The denting was, however, arrested by the addition of a neutralizer, a base. These results indicate that denting is an acid reaction and, therefore, one way to control denting is by addition of a base. To date, no denting has been obtained with sodium chloride, a neutral chloride.[15,16]

In laboratory testing, copper as well as nickel ions have acted as accelerators of the denting corrosion process, supporting the Potter–Mann mechanism.[16] In samples removed from actual steam generators, small beads of copper have been found in the dented region, and similar observations (of nickel also) have been made in laboratory model boiler dents.[15,16]

Van Rooyen and Weeks[16] consider three possible roles of an external oxidizing ion such as Cu^{2+} (or Ni^{2+}, which is less oxidizing) in accelerating the corrosion of carbon steel:

a. At reasonably high concentrations Cu^{2+} can be catalytic and, in addition, it can depolarize the cathodic reaction and produce elevated Cl^- together with increased acidity in a local crevice region.
b. At lower concentrations it may be merely a catalyst in speeding up the ferrous hydroxide conversion to magnetite, without much change in the oxidation potential.
c. It could be plated out, leaving a metallic surface with low H_2 overpotential.

The first two appear most feasible based on laboratory results to date.

At the present time, there are no laboratory data that show that oxygen is an aggravating factor for denting. However, there are concerns that oxygen can affect denting because it is known that oxygen, in combination with occluding or crevice conditions, can promote the generation of acids, which would be an obvious concern to denting.

Minimization of copper (and nickel) and oxygen ingress to steam generators is, therefore, considered desirable. However, this should not be considered the cure-all, since other chemical factors such as Fe^{3+} and SO_4^{2-} ions, and power level and local thermal/hydraulic effects also appear to be involved in the denting process.[15,16]

In addition to tube denting, there is some evidence that tube-sheet denting can be occurring (Fig. 6-7). This tube-sheet denting has so far been recorded only in plants with open crevices, i.e., where the tubes were not rolled or expanded to close the crevices. This phenomena has been observed in plants with AVT chemistry. Tube-sheet denting most likely occurs because of the combination of the following three factors: chemical impurities in the boiler water, chemicals from the sludge, and the concentrating mechanism resulting from local thermal/hydraulic conditions. Tube-sheet denting has been obtained in laboratory boiler tests that contain acid chloride and/or similar degrading resin fines.[15]

The chemical factors that cause tube-sheet denting are probably similar to those that cause tube denting in the support plates, and therefore similar remedies may apply.

6.2.1.2. Stress-Corrosion Cracking

In U-tube steam generators, intergranular cracking of the Inconel 600 tubing has occurred at various locations (e.g., within the tube sheets, just above tube sheets, in dents at support plates, and in tight-radius U-bends).[2,15] Cracks have also occurred in the ligament regions of the carbon steel support plate (Fig. 6-7). Interestingly, in contrast there are no reports of cracking of Alloy 800 tubes in the KWU steam generators.[2]

While the mechanisms causing the Inconel 600 cracks at these various locations are not well understood, their characteristics and the prevailing environment support the view that stress corrosion is dominant.[2,15] First, concentrated caustic environments are caused by dryout of boiler water containing in-leakage from condensers cooled by certain fresh water sources; second, although normal operational stresses for the tubes are far below the yield stress, analyses have shown that once the clearance annulus is closed by the support plate corrosion product, additional corrosion can readily cause yielding of the tube. Although yielding in the support plate is expected to precede yielding of the tube, large tube deformations are expected (and observed) to precede gross yielding of the plate.[19]

Examination of a tube pulled from an OTSG also revealed stress corrosion cracking.[2,15] Sulfur was present and may have been involved since it has long been recognized that sulfur can induce cracking of Inconel 600. Possible sources of sulfur are resin beads, resin fines, demineralizer regeneration with sulfuric acid, and cooling tower water treatment with sulfuric acid.

The mechanical, metallurgical, and environmental factors affecting the stress corrosion cracking of Inconel 600 were reviewed by Was[20] in 1978. He reported that several inconsistencies existed in the data obtained both in pure water and in caustic solutions, with the responsible parameters being material heat treatment, carbon content, and oxygen and contaminant concentrations in the water.

Sensitization (i.e., grain-boundary chromium depletion) and the presence of oxygen, either separately or in combination, do appear to increase the intergranular attack (IGSCC) of Inconel 600 in "pure" water.[20] However, mill-annealed Inconel 600 can also be cracked intergranularly in high-temperature, deaerated water when strained at low rates, as evidenced by the recent work of Bulischeck et al.[21] This latter study is in the preliminary stages and, although a microstructural effect appears probable, it has yet to be confirmed.

In recent years much work has been done on the SCC of Inconel 600 in caustic environments,[1,20,22-24] because of the caustic concentrating effects in the steam generator noted above. Again, due to the numerous parameters involved, the resistance of this alloy to SCC (particularly relative to other alloys such as the austenitic stainless steels and Alloy 800) is the subject of much controversy.[1]

The effect of the stress level on the rate of cracking in the alloys in a 100 g/liter solution of NaOH at 620 K is shown in Fig. 6-10a.[1] A stress level can be defined below which cracking does not occur in a short time for Type 316 stainless steel (~140 MPa) and Alloy 800 (~210 MPa). However, the situation is not so clear-cut for Inconel 600, although the threshold is clearly well below that for Alloy 800.

The relative susceptibilities to SCC are affected by caustic concentration and by oxygen.[1,20] With regard to the former, Fig. 6-10b shows crossover points, such that at high concentrations of 500 g/liter of NaOH Alloy 800 behaves less well than does Inconel 600 and shows very rapid cracking if the stress exceeds about 120 MPa. Inconel 600 therefore appears to be most susceptible to SCC in weak concentrations (4 g/liter). Based on this, the hypothesis has been put forward[25] that the phenomenon of SCC in Inconel 600 in pure water is of the same nature as that in a caustic medium. In deoxygenated caustic solutions Inconel 600 is more resistant to SCC than alloys of lower nickel content such as Alloy 800 or Type 316 stainless, but in oxygenated caustic solutions the situation is reversed.[23]

Since the attack is almost always intergranular, the role of grain-boundary carbides and the related chromium-depleted layer in the process has been extensively studied.[20] However, no universally acceptable model is available to explain the SCC behavior of Inconel 600, although grain-boundary carbides and the segregation of impurities (P and S) to the grain boundaries clearly play an important role in the process (see also Section 6.2.3.3a). For instance, in water and caustic solutions, the presence of grain-boundary carbides inhibits SCC, while in acid solutions cracking is accelerated. If one accepts the oxide film rupture model as the relevant model, then the presence of a grain-boundary carbide could influence either the grain-boundary deformation mechanism or the oxide film composition. Similarly, in the case of an oxide film dissolution mechanism the carbides could again affect the oxide film composition. In addition, there are differences in electrochemical behavior between the carbides and the matrix which will affect the SCC susceptibility. This influence of grain-boundary composition and microstructure on SCC sus-

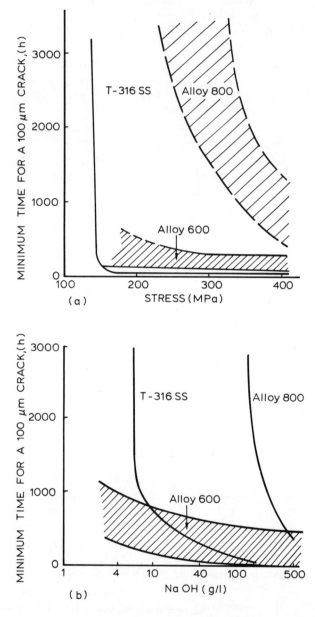

Fig. 6-10. Comparison of the resistance to stress-corrosion cracking of Inconel 600, Alloy 800, and Type 316 stainless steel in deaerated caustic solutions at 623 K: (a) effect of stress (NaOH = 100 g/liter); (b) effect of NaOH concentration ($\sigma \approx 0.8\sigma_y$) (Ref. 23).

ceptibility has been utilized in approaches to improve Inconel 600's resistance to caustic SCC, as will be discussed later in Section 6.2.3.3a.

6.2.1.3. "Phosphate Thinning"

Tube-wall thinning or wastage has been observed in the region above the tube-sheet (refer to Fig. 6-7) when phosphate chemistry has been used—hence the term "phosphate thinning".[2] Since the industry shifted to AVT chemistry the progression of tube-wall thinning appears to have diminished to the point where it is not a serious industry-wide problem.[15] However, thinning continues to be observed in plants that have remained on phosphate chemistry.

The mechanism of phosphate thinning is still a subject for discussion, but a viable mechanism, albeit qualitative, was recently proposed by Garnsey.[2] He bases his model on the sodium phosphate phase diagram, which indicates that, above 546 K, solutions with an Na/PO_4 ratio of between about 2.8 and 2, when sufficiently concentrated, will precipitate an immiscible liquid phase. This phosphate-rich liquid phase, which is a molten sodium hydroxyphosphate, is extremely corrosive to both ferritic steels and high-nickel alloys. It is the precipitation of this phosphate-rich liquid phase, which occurs as a result of concentrating phosphate solutions within the porous crud deposits, that is postulated to be the cause of phosphate thinning. Autoclave tests have shown that corrosion of Inconel and ferritic steels is very much greater when the second phase has been allowed to precipitate than in the aqueous phase at a concentration close to that at which precipitation occurs.

Garnsey[2] admits that some uncertainties in the model need to be resolved. For example, the corrosivity of the phosphate-rich mixture will depend, among other things, on whether the concentrated phosphate phase or the dilute aqueous phase "wets" the tube material. There is also insufficient information on how the presence of iron and nickel oxides influence the phase diagram in this region. The corrosivity of both liquids will also depend on the pH, or the Na-to-PO_4 ratio, but this ratio also affects the concentration at which the second liquid phase appears. Decreasing the Na/PO_4 ratio will make the dilute aqueous phase more corrosive to nickel alloys, but doing so also increases the concentration which must be achieved before the phosphate-rich phase separates. Thus, if the appearance of the phosphate-rich liquid phase is necessary for rapid thinning, then lower Na/PO_4 ratio phosphate solutions might be less aggressive than higher-ratio solutions provided that the concentration efficiency within the porous deposit is sufficiently low to prevent the appearance of the aggressive phase at that lower Na/PO_4 ratio.

With this model, Garnsey[2] can now reconcile the varied industry experience with phosphate water treatments, and he predicts a phosphate regime that would result in successful operation of PWR steam generators once the locations which give high salt concentration efficiencies (dryout in porous deposits and crevices) have been eliminated.

6.2.2. Vibration and Mechanical Problems

Mechanical damage has to date been restricted to OTSGs, but conditions are present in the D&S steam generators which could also result in fretting, wear, and fatigue over the long term once the short-term corrosion problems are eliminated. Again this type of damage process originates on the secondary side of the system, although effects of the primary side may contribute to its initiation.

6.2.2.1. Fretting and Wear

Vibration, of the tube within the tube-support plate clearance, is the cause of tube fretting and wear. Clearly, this damage mechanism is possible in both types of steam generator design, but has only been observed in OTSGs to date.[2,15]

Erosion-type metal loss and cracking has occurred in OTSGs where the tube interfaces the tube-support plate at the highest support plate along an inspection lane. Fretting has also been observed at this location. There was also some superficial cavitation near the upper tube-sheet along an inspection lane.

Another type of metal loss, from impingement, was located during one examination, remote from the tube inspection lane at the second highest tube-support plate. Preferential pitting under the top tube-support plate was also seen in regions remote from the inspection lane.

In addition to the detrimental effects of tube-metal loss, a major concern is that fretted regions are highly sensitive to fatigue cracks. Under fretting conditions, fatigue cracks can be initiated close to the material surface at very low stresses, well below the fatigue limit of nonfretted tube test material.[26]

6.2.2.2. Fatigue Cracking

High-cycle fatigue (refer to Chapter 1, Section 1.3.4.3) conditions exist in steam generators. In D&S units, the most likely regions for vibration and possible fatigue failure of the Inconel tubes occur just above the tube sheet and in the U-bend region where the flow velocities of the primary side are highest and cross-flow from the secondary side is significant. While tube cracking has not yet been observed in these regions, the steam generators have not been operated for long enough times. Inasmuch as cracking has been observed in OTSGs, it is planned to obtain data to predict whether or not fatigue cracks in these regions would occur over the expected 40-year lifetime of the D&S units.[15]

Circumferential tube cracking that has occurred in OTSGs exhibits characteristics of a fatigue mechanism.[15] Examination of the crack striations produced indicated that the propagation phase of the crack involved a large number of cycles. However, the calculated stress level of ~85 MPa necessary for crack propagation is much lower than the usually reported fatigue strength of Inconel 600[26]; therefore, initiating mechanisms such as fretting fatigue, localized corrosion, or corrosion fatigue must be considered as possible contributing factors. Nevertheless, environmental fatigue testing, completed for up to 10^7 cycles at 560 K in air, water, and steam, showed no difference in fatigue properties.[26] Also, there was no reduction in fatigue strength with cavitation-induced damage, but there was a 30% reduction in fatigue strength in a notched specimen, in which the notch was produced electrochemically. It has therefore been inferred that the process of crack initiation and propagation could be due to cyclic loading (e.g., from flow-induced vibration) of the tube in conjunction with an initial mechanical surface defect.[26]

6.2.3. Avoidance or Mitigation of Steam Generator Damage

Corrosion damage in PWR steam generators can be avoided by eliminating either the ingress of chemicals or their concentration, i.e., control of the *local* chemistry. Mechanical damage can be minimized by changes in design and fabrication procedures that reduce, respectively, the possibility of high residual stresses in the materials and flow-induced vibration and fretting. Improved materials also play an important role in ensuring adequate steam generator lifetime; however, the first two remedies are of most near-term interest.

6.2.3.1. Chemical Control

The presence of impurities in boiler water, in combination with certain adverse thermal/hydraulic factors, can cause severe local chemistry concentrations. Severely reduced flow, sometimes termed "flow starvation," which is affected by many factors such as overall and local hydraulic design, crevices, and sludge accumulation can combine with high heat transfer to result in adverse thermal hydraulics. Thus, the explicit factors that affect corrosion damage via chemical concentration are: (1) the presence of impurities, (2) hydraulic design, (3) thermal design, and (4) the presence of concentrators such as crevices or sludge.[2,15]

There are a variety of ways to control these factors, but they are subject to certain constraints. In an operating unit, thermal and hydraulic designs and the presence of crevice concentrators are not easily alterable. On the other hand, the presence of impurities and sludge concentrators are more easily controlled. In new units, the thermal/hydraulic design and the presence of crevices are all controllable. In addition, modification of system designs external to the steam generator can control the presence of impurities as well as sludge concentrators. Examples of impurity eliminators include use of cleanup systems to remove impurities from boiler water, use of leak-tight double tube sheets, titanium tube condenser designs (refer to Section 6.4), elimination of copper components, use of demineralizers, and/or use of deaerators. Sludge can be reduced through the use of more corrosion-resistant secondary system materials, use of less corrosive secondary system environments, and use of condensate and feedwater filter systems.

Another control parameter available to both operating and new units is the use of chemical additives to reduce impurity aggressiveness. This has been accomplished in the past by using chemicals such as sodium phosphates, ammonia, and morpholine. However, in the future, it may also be desirable to employ other additives in a controlled manner during shutdown periods (i.e., "off-line" use) to neutralize the acid chlorides that are one of the primary causes of denting. Candidate neutralizing additives are boric acid, calcium hydroxide, sodium hydroxide, lithium hydroxide, and sodium phosphate.[15,16] Model boiler tests indicate that phosphate chemistry has an inhibiting effect on denting when chlorides are present, and consequently an accelerated phosphate neutralization/water-soak program is measuring the benefits and the potential deleterious side effects of *off-line* use of phosphates as a near-term remedy for denting.[15] Nevertheless, extensive experience under prototypic conditions will be necessary before the industry will be convinced to try this approach, because the earlier *on-line* use of phosphate water treatment led to the phosphate wastage problems described in Section 6.2.1.3.

Basic research into the chemistry of corrosion-producing salts in water is, for the first time, enabling engineers to predict the most likely corrosion consequences of any present or proposed chemical environment in the system.

Figure 6-11 is a part of a Pourbaix diagram for iron/water which covers conditions of interest to PWRs. The standard Pourbaix parameters of oxidizing potential and pH have been correlated to parameters that are measured by the plant chemist, namely oxygen and ammonium hydroxide concentrations, respectively. The dashed lines separate ionic species in solution (at the 10^{-6} molar concentration level), whereas the solid lines separate solids such as Fe_3O_4 from both other solids, such as Fe_2O_3, and

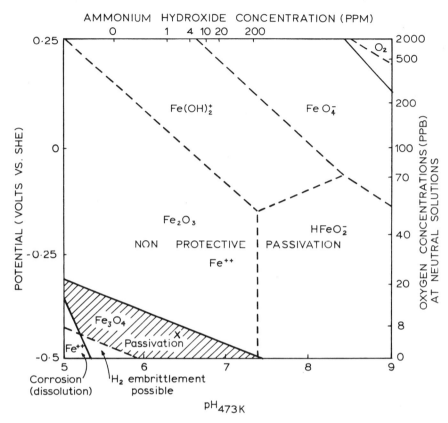

Fig. 6-11. Pourbaix diagram for predicting corrosion of iron in water at 473 K in terms of in-plant measured quantities. Fe_2O_3, a non protective passivating film, will cover metal in most of the area shown (adapted from Ref. 27).

dissolved ions. The point X denotes normal PWR operating conditions, which, according to the diagram, should yield a passivating oxide film. However, note also that small changes in chemical composition can easily move the system into a nonprotective regime. Through a new computer program, Pourbaix diagrams that show regions of corrosion vulnerability are now available for conditions up to 570 K, for iron and nickel alloys in water containing the common impurities (e.g., chloride and sulfate ions) and operating variables (e.g., oxygen and ammonia).[27] Using such diagrams one can specify the pH and oxidizing potential regime for which corrosion will be minimized; that is, where a stable oxide film will form. Regular on-line measurements of these parameters can be used to check that the "safe conditions" are being sustained during plant operation. Translating this system to regular practice, however, will be a long-term project, since the influence of all the other chemical species in the coolant must be included in the diagrams.

In the near term, a novel method for reducing impurity (i.e., chloride, sulfate, silica, calcium, magnesium, etc.) concentration in crevices is through periodically operating the steam generator at reduced power and performing periodic water soaks.[15] At some reduced power, the alternate wetting and drying in the crevices may be sufficiently reduced to leave the crevice wet so that impurities will diffuse out of the crevices into the bulk fluid and be removed from the steam generator by blowdown. The power at which impurities diffuse out of crevices is not known; it varies from plant to plant and probably from crevice to crevice within the same steam generator. Hence, an accelerated laboratory test program is underway to quantify (and optimize) the conditions under which water soaks are most effective in reducing crevice impurity concentrations and therefore denting. Preliminary capsule test results indicate that repeated water flushes reduce, and in some cases stifle, the denting progression.

6.2.3.2. Design and Fabrication Changes

Tube to tube-sheet joints and tube-support designs have received a great deal of attention in D&S units as a result of the denting phenomenon. The industry trend is to a more "open" support structure with respect to the latter (Fig. 6-2b–d), which, although minimizing the likelihood of tube pinching, increases the risk of vibration fatigue and fretting problems. A redesign of antivibration bars at the U-bends should help solve this, however. Inasmuch as vibration-related problems have been observed in OTSGs, the newer designs eliminate the vacant tube lanes and stiffen the support structure in order to minimize fatigue and wear damage.[2] Only in-service experience will determine whether these various design modifications are in the right direction.

In the fabrication area, the major issue is residual stresses (compare the BWR pipe crack situation described in Chapter 4). While the primary pressure stresses and the thermal stresses can be calculated quite easily (they represent about 120 MPa), this is not the case for residual stresses due to tube fabrication.[1] Measurements have shown that peak tensile stresses of more than 150 MPa can exist in the U-bends, perhaps exceeding 200 MPa on both inner and outer surfaces of the bending radii.[28] These stresses are very high and are likely to produce stress corrosion cracks under unfavorable chemical conditions. Therefore, reduction of residual tensile stresses is an important factor in minimizing steam generator damage.

A radical approach to this problem, practised on OTSGs in the U.S., is to heat treat the entire steam generator after assembly. Although desirable from the point of view of treating the tubes and the tube-to-plate joints in the same operation, there are obviously some problems with this approach. First and foremost, the steam generator must be specially designed to withstand this operation without distortion of the tube bundle; this could be a problem with D&S units. Second, the treatment temperature is limited to about 870 K because of the presence of the carbon and low-alloy steels, and consequently the stress relief in the bundle is minimized, leaving ~100 MPa residual stresses.[1] Third, the tubes can be sensitized in this temperature regime at long enough times. Consequently, the general practice is to separately heat treat different parts of the bundle at the appropriate stages of fabrication.

The more conventional approach to the reduction of residual stresses is to achieve it during tube fabrication. A number of methods are described by Berge and Donati in their review.[1] These include (a) avoiding tube distortion during fabrication, (b) "mechanical" stress relief of the straight tubes by applying tension or internal pressure, (c) introducing residual compressive stresses at the outer surface, by shot peening for example, and (d) heat treatment at 970 K or above (the final stress level after 16 h treatment at this temperature is ~35 MPa for Inconel 600).

6.2.3.3. Modified or Alternative Materials

From the foregoing, it is evident that a highly corrosive and variable environment potentially exists in a steam generator, and, consequently, radical changes in both design and operation will be essential in order to assure a 40-year steam generator lifetime. Material optimization plays a small but important role in the long-range plans. As might be expected from the foregoing discussion, the emphasis for tubing is to develop a material that is resistant to environmental cracking, whether stress corrosion or corrosion fatigue, while retaining high overall resistance to local attack or wastage. With respect to tube-supports or tube-sheets, the primary need is for a material that does not exhibit the nonprotective corrosion mechanism. Several materials and material processes are being evaluated for replacement tubes (nickel-base alloys such as Inconel 600 and 690, and Alloy 800) and support plates (ferritic stainless steels such as Types 405 and 409) using standard laboratory stress corrosion and general corrosion tests and long-term model boiler tests under aggressive fault chemistry conditions.[26] The following subsections summarize the progress being made to find the most resistant steam generator material.

6.2.3.3a. Inconel 600. When used as mill-annealed tubing, Inconel 600 is susceptible to intergranular SCC (Fig. 6-10). Therefore, attempts to improve the SCC resistance have pursued modifying the grain boundary microstructure. Laboratory-scale experiments have been successful[29] and longer-term pot boiler tests are underway.[26] Of the two heat treatment processes used, viz. a high-temperature "purification" anneal and a thermal treatment, the thermally treated tubing had superior caustic SCC resistance to the mill-annealed product in all cases.[29]

Thermal treatment in the carbide precipitation temperature range (above ~870 K) introduced a wide variety of grain-boundary microstructures.[29] Grain-boundary precipitation ranged from fine discrete particles, to a semicontinuous layer, to large discrete particles. Figure 6-12 correlates these grain-boundary microstructures to the caustic SCC resistance of thermally treated tubing. The thermal treatments which imparted the maximum improvement in caustic SCC resistance are shown as solid bars in this diagram. Thermal treatment at 977 K, and for longer times at 866 and 922 K, reduced crack depths by almost an order of magnitude in 10,000-h tests in deaerated 10% NaOH at 588 K (Fig. 6-13). Comparison with the data in the earlier Fig. 6-10 suggests a SCC resistance as good as, if not superior to, that of Alloy 800. The maximum improvement in caustic SCC resistance correlated with a grain-boundary microstructure consisting of a semicontinuous precipitate without an associated chromium-depleted layer but with phosphorus segregated to the boundaries. It should also be noted that in a lead-doped water environment, samples

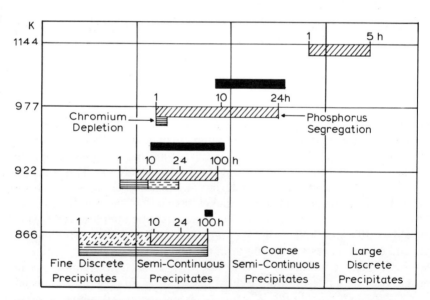

Fig. 6-12. Correlation of grain-boundary microstructure with caustic SCC performance in thermally treated Inconel 600. Solid lines correspond to the maximum improvement in caustic SCC resistance. Annealing times (h) are given on a sliding scale for each temperature (Ref. 29).

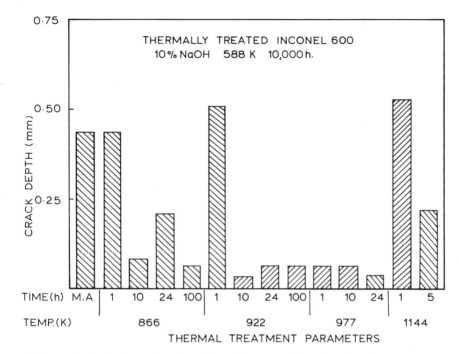

Fig. 6-13. Crack depth as a function of thermal treatment for plastically deformed Inconel 600 C-rings exposed to deaerated 10% NaOH at 588 K for 10,000 h (Ref. 29).

with such a structure had a transgranular crack morphology, indicating that even in this severe environment, the threshold stress for SCC was increased.

Thermal treatment increases the critical stress level above which rapid crack propagation occurs, as well as decreasing the crack propagation rate. Airey[30] therefore, attributes the improved SCC resistance to a reduction in residual stress (retarding crack initiation) and grain boundary structural modification (inhibiting crack growth). The results of this program also indicated that one of the benefits of thermal treatment is to reduce the variability in the caustic SCC resistance. As implied earlier, there is a large scatter in the data on the caustic SCC resistance of mill-annealed Inconel 600 presumably due to the testing of different heats of materials with different thermomechanical processing histories.[20] If the role of thermal treatment is to reduce this scatter, it could also explain some of the conflicts in the data where in some cases thermal treatment improves the caustic SCC resistance and in others it does not.

6.2.3.3b. Alloy 800. Although this alloy is the subject of current evaluation in the U.S., it has been used for several years for steam generator tubes in German designed PWRs, and, as noted earlier, was selected as steam generator tube material for Super Phenix (refer also to Section 6.3). This widely used material was the sole topic of a full-scale international conference, "Alloy 800," in 1978, and much of the discussion in this and the companion LMFBR section will be drawn from the Proceedings.[31]

Alloy 800 is not completely immune from cracking in water containing chloride ions, oxygen, and acid pH or alkaline environments. The laboratory investigations of cracking of Alloy 800 summarized in Fig. 6-10 earlier showed that Alloy 800 has an apparently higher stress threshold for crack initiation in low caustic concentrations than Inconel 600, but that it fails rapidly at low stresses in high caustic concentrations.

It was also noted earlier that the effect of oxygen was also important to the susceptibility of Alloy 800 to SCC. In one study,[23] three metallurgical forms (termed Grades 1, 2, and the "sensitized state") of this alloy were exposed to 1 and 10% NaOH solutions at 623 K, in either the deoxygenated or oxygen-saturated conditions. Grade 1 is the mill-annealed condition, characterized by a fine grain size and carbon precipitation as TiC. Grade 2 is annealed at a higher temperature after working (1423 K versus 1253 K for Grade 1), and then water quenched. This treatment produces a rather coarse grained structure but retains carbon in solution. The "sensitized" condition of Grade 1 is intended to simulate the HAZ that accompanies welding. C rings of these three metallurgical forms were submitted to a strong permanent deformation, the stress being very much greater than the 0.2% elastic limit. Cracks were observed after only 30 days in 1% deoxygenated solutions. This effect, and the strong influence of caustic concentration, is evident from Fig. 6-14. Note also that "Grade 1" is clearly more crack sensitive than "Grade 2," and that the sensitization treatment of "Grade 1" Alloy 800 has almost no effect on crack behavior under caustic conditions.

Although susceptible to cracking in certain chloride environments containing oxygen, Alloy 800 proves to be less suceptible than either the austenitic Types 304 and 316 or Inconel 600. Results of a number of studies have shown: (a) that a chloride concentration threshold exists at 473 K, between 0.5 and 2 ppm, below which stress cracking in Alloy 800 is either not produced or strongly delayed; and (b) that cold work, sensitization, and low Ti/C + N ratios reduce the cracking resistance of Alloy 800.[23]

Moving now to in-service experience, we see that Alloy 800 has proved to be a highly reliable steam generator tube material. In German PWRs the mill-annealed material (Grade 1) is used. This seems a curious choice, in light of the data in Fig. 6-14 which shows Grade 2 to be far superior to Grade 1 in caustic solutions. However, this selection can

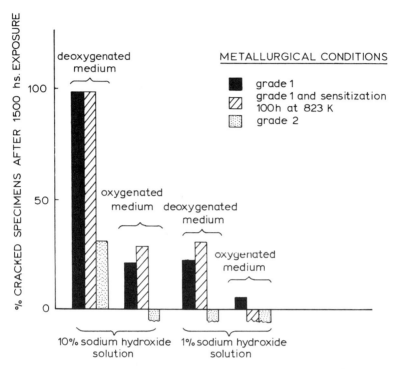

Fig. 6-14. Percent of cracked specimens after 1500-h exposure, all Alloy 800 specimens being taken into account; total number of specimens 216 (adapted from Ref. 23).

perhaps be explained on the basis of defense against chloride attack and mechanical problems (e.g., 0.2 yield stress is higher) rather than caustic cracking.

In any event, a PWR with Alloy 800 steam generator tubes has been operating in Germany for several years now. Careful examination of the tubes failed to reveal any signs of cracking or pitting, but some general corrosion was observed, which resulted in metal loss to a thickness of 60 μm.[32] On the basis of this good experience, KWU has decided to continue to use Alloy 800, and the alloy is being evaluated in model boiler tests in the U.S.

6.2.3.3c. Inconel 690. Inconel 690 is a relatively new high-chromium (30 wt.%), nickel-base alloy, designed to provide a stable austenitic structure with thermal and mechanical properties close to Inconel 600, combined with resistance to corrosion and oxidation in aggressive environments comparable to Alloy 800.

The corrosion resistance of Inconel 690 has been studied in a wide variety of media of interest to PWR steam generators.[33,34] The caustic stress corrosion test data indicated that Inconel 690 is the most resistant of the alloys tested (which included Type 304 stainless steel, Alloy 800, and Inconel 600) in oxygenated 50% NaOH and deoxygenated 10% NaOH, while its resistance is slightly inferior to that of Inconel 600 in deoxygenated 50% NaOH. The cracking resistance of Inconel 690 was also highest in tests in high-temperature water containing (high and low) oxygen, and this property was not impaired by various heat treatments, cold work, or the presence of welds. Besides stress corrosion cracking resistance, this alloy exhibits a high general corrosion resistance (i.e., low weight loss) in oxidizing acids and in flowing high-temperature water environments.

6.2.3.3d. SCR-3 Alloy. A new Ni–Cr–Fe alloy, dubbed SCR-3, has recently been developed by the Japanese for use as PWR steam generator tubing.[35] This alloy contains almost equal amounts of nickel (25 to 26 wt.%) and chromium (24 wt.%), and small amounts of vanadium and titanium; carbon level is ~.02 wt.% and silicon levels range from 1.7 to 2 wt.%. Thus, SCR-3 is lower in chromium than Inconel 690, but is higher than either Inconel 600 (15.6 wt.%) or Alloy 800 (21 wt.%).

The SCC susceptibility of this new alloy has been compared with Type 304 stainless steel, Inconel 600, and Alloy 800 in the "standard" media (i.e., chlorinated water, caustic solutions, phosphate solutions, etc.), and, in all instances it proved to be the most resistant alloy. By way of example, Fig. 6-15 shows the Japanese data on SCC susceptibility in high-temperature, high-pressure water at 573 K.[35]

In closing this discussion of alternative steam generator tubing materials, it is appropriate to comment on the pattern that emerges from the large data base now available from laboratory tests. First, material selection continues to be made strictly on the basis of SCC resistance since all candidate alloys possess adequate strengths and general corrosion properties in the environments of interest (i.e., chloride and hydroxide). Second, there appears to be general agreement that Type 304 stainless steel is the least resistant alloy to stress corrosion, while Inconel 690 and SCR-3 exhibit the most resistance. No intercomparison data have been obtained on the latter two alloys, however, and, since SCR-3 is not commercially available (outside Japan?) it is unlikely to see widespread use. Third, in comparing the currently used Inconel 600 and Alloy 800 materials, it is clear that care must be exercised due to the sensitivity of their SCC susceptibility to metallurgical conditions and test environment. Strictly from the basis of in-service experience one would have to select Alloy 800 (as KWU have). However, the modified heat treatments developed for Inconel 600 hold great promise and are likely to see increased use in the U.S.

6.2.3.3e. Ferritic (Martensitic) Stainless Steels. Whereas improved SCC-resistant tube materials appear to be available, the picture

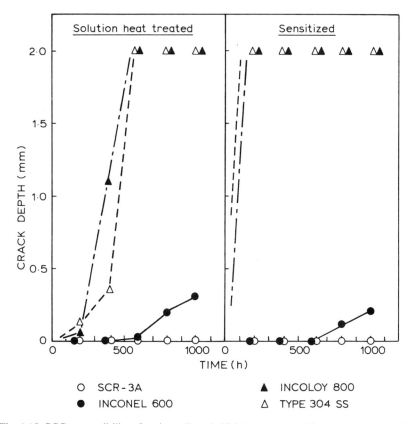

Fig. 6-15. SCC susceptibility of various alloys in high-temperature, high-pressure water; 573 K, 500 ppm Cl$^-$, 8 ppm O$_2$, pH 10 (Ref. 35).

is not nearly as promising for support-plate materials. Ferritic materials are expected to be resistant to SCC but prone to severe general attack and pitting in steam or water environments. Their resistance to general corrosion should, however, improve as the chromium content is increased. Nevertheless, model boiler testing of the 400 series ferritic steels with seawater chemistry has resulted in tube locking within the drilled holes.[26] This could be a result of material deposition in the crevice or corrosion buildup from the support plate itself. However, these preliminary results cast doubt on the adequacy of these steels as a replacement for carbon steel. As will be discussed below in Section 6.3.1, this alloy class is a leading candidate for advanced LMFBR steam generators. Although no adverse properties have been discovered during the first phase of qualification, "prototypic" tests of the type noted above have yet to be conducted.

6.3. LMFBR Steam Generator and IHX Development

In contrast to water–steam heat exchangers, experience with sodium–sodium or sodium–steam heat exchangers is limited. Nevertheless, the operating experience of IHXs and steam generators in test reactors and medium sized prototype LMFBRs is encouraging. Problems that have arisen appear amenable to solution through design or material changes, and with respect to the latter, an intensive worldwise research effort is ensuring that the necessary data base is available.

The reliability of the sodium/water boundary is the major concern in the LMFBR heat transport system. The steam generator design, materials, and fabrication scheme must not only ensure this high reliability but, in the event of failure of this boundary, the system must safely accommodate the resulting sodium–water reactions that could occur. These requirements have necessarily dominated the evolution of steam generator design for LMFBRs, as the following brief review will show.

The EBR-II steam generator utilizes duplex tubes and two separate tube sheets in the superheaters and evaporators to minimize the possibility of a sodium–water reaction.[36] This steam system has operated reliably for the past 15 years; a report in 1975[36] indicated that the $2\frac{1}{4}$Cr–1Mo steel employed in the steam generator had not suffered significant degradation in metallurgical properties. After 9000 h operation in the 575 to 711 K temperature regime, room temperature yield and tensile strengths were reduced 23.3 and 4.5%, respectively, and carbide precipitation resulted in an increase in upper-shelf toughness but little or no increase in transition temperature (ΔTT). Microstructural analysis indicated no evidence of embrittlement as a result of this service. Based, in part, on this good experience, $2\frac{1}{4}$Cr–1Mo steel was selected by the U.S. for the CRBRP demonstration project.[6] The alloy will be prepared by either vacuum arc remelt (VAR) or electroslag remelt (ESR) practices (refer to Chapter 7 for additional details on casting practice) to minimize impurity content and will be used in the most stable metallurgical condition—fully annealed.

The French Phenix reactor steam generator has also operated flawlessly. A 1977 report[37] indicated that, after three years of operation at full load and a number of startups and shutdowns, no leaks were detected in the single wall, $2\frac{1}{4}$Cr–1Mo and $2\frac{1}{4}$Cr–1Mo–1Nb (evaporator) and Type 321 (superheater) steel tubes. However, it was recognized earlier that the modular design adopted for Phenix was not well suited for commercial size plants ($\geq 1,000$ MWe), and so a completely different design/material combination was selected for Super Phenix. Inasmuch as the maximum temperature would be 798 K, it was decided to use Alloy 800 as the tube material throughout the steam generator.[7]

Experience with the Russian BN-350 LMFBR ($2\frac{1}{4}$Cr–1Mo) and the U.K. PFR ($2\frac{1}{4}$Cr–1Mo/316 Stainless Steel) steam generators has not been as good.[3] In the former, massive tubing leaks occurred, apparently from manufacturing defects in two evaporators. During operation of the PFR smaller leaks have occurred in the evaporators and superheaters in two of the three circuits.[38,39] A detailed destructive examination of the PFR tube bundles indicated that stress corrosion cracking of several tubes had occurred. However, the original leak was probably due to a weld defect found in one of the tube-to-tube-plate welds. On pressurization, steam had leaked, creating, through reaction with sodium, caustic and stress corrosion cracking of the austenitic tube plate. The susceptibility of the austenitic stainless steels to caustic cracking (mentioned earlier in this chapter) has been concluded as the primary cause of cracking of tubes and tube-plates in the PFR steam generator.[38] Consequently, in the material selection process for their commercial size LMFBR, the CDFR, the U.K. experts were particularly sensitive to the problem of "caustic cracking after leak." Of the candidate alternative materials, i.e., the higher-nickel alloys such as Alloy 800, and the ferritic steels such as $2\frac{1}{4}$Cr–1Mo and 9Cr–1Mo, they selected the latter alloy in the normalized and tempered condition. The reasons for this choice, provided by Taylor[38] in a recent paper, are of interest in the light of the U.S. and French selections. With respect to $2\frac{1}{4}$Cr–1Mo steel, Taylor states that niobium stabilization is essential to prevent carbon loss in sodium at higher temperatures (CDFR will operate at 760 K) and that this modification creates a somewhat different structure for which creep data are very sparse. Their concern about Alloy 800 was the susceptibility to caustic cracking noted in the earlier discussion of PWR materials.

Thus, it is clear that materials selection for the next generation of LMFBR steam generators has largely depended on the combination of a country's *own* experience base and the design and operating parameters of the system. If generalizations can be made, it is that: the lower temperature and/or recirculating designs can be satisfied by the ferritic steels, with $2\frac{1}{4}$Cr–1Mo steel being the conventional alloy, whereas the higher temperature and/or once-through designs have opted for Alloy 800. Further, the U.S., French, U.K., and German experts (appear to) agree that, while the more advanced ferritic steels, such as the niobium-stabilized version of $2\frac{1}{4}$Cr–1Mo and the ferritic stainless class (9–12 wt.% Cr), look good for LMFBR steam generator service, more data and experience are necessary. Progress in characterizing these materials for service in the sodium-steam/water environment will be described in Section 6.3.1.

Operating experience with intermediate heat exchangers has not been widely reported and so one must conclude that no severe problems have emerged. In a French report[8] it was noted that one of the IHXs of the

Phenix reactor suffered a secondary sodium leak in 1976, and all six units were subsequently removed from service and modified one after the other, with the plant partially operating on two loops during the repair period.

The leak was due to a crack, symmetrical with the IHX axis, which occurred in the vicinity of the weld between the outlet header cut-off plate and the inner shell (refer to Fig. 6-4 earlier, position A). Calculations showed that thermal- and pressure-induced stresses were excessive at the point where the crack occurred.[8] Accordingly, the Super Phenix IHX design was modified to homogenize the temperature of the secondary sodium leaving the tube bundle (i.e., reduce the ΔT and hence the thermal stresses); also, the outlet header design has been improved. The encouraging fact about this experience is the conclusion that the austenitic stainless steel behavior in the IHX is adequate to allow scaleup to the commercial size units. Recent developments in the characterization of these materials for long-term service in sodium will be described in Section 6.3.2.

6.3.1. Steam Generator Materials Properties

Low-alloy $2\frac{1}{4}$Cr–1Mo ferritic steel has been favored internationally for use in the construction of sodium-heated steam generators on the basis of its high thermal conductivity, resistance to stress-corrosion cracking in both chloride- and caustic-contaminated environments, established fabrication techniques, and cost. A major concern for this application, however, is the susceptibility of the steel to decarburization when exposed to high-temperature flowing sodium.[6] The carbon loss from the material leads to a significant reduction in elevated-temperature mechanical strength, which must be accounted for by using initially thicker sections. Furthermore, in sodium systems that contain austenitic and ferritic components, decarburization of the ferritic material leads to carburization of the austenitic stainless steels.

So-called ferritic stainless steels (9–12 wt.% Cr, with 1–2 wt.% Mo) have been proposed as alternative materials for both steam evaporator and superheater units to minimize the extent of carbon transfer. These high-chromium ferritic steels provide a greater resistance to carbon transfer and possess adequate elevated-temperature mechanical properties (refer to Fig. 6-16) and caustic corrosion resistance.[40,41] Unfortunately alloys of this class are not approved for high-temperature nuclear use in the U.S., as per Code Case 1592. Therefore, in the U.S. program they are considered advanced alloys for future generations of LMFBRs and proportionately less effort has been expended on their characterization.[40,42] These alloys are of more near-term interest in the U.K. and Europe.[41,43]

Alloy 800 has a high resistance to stress-corrosion cracking in caustic- or chloride-contaminated aqueous environments (as described in Section 6.2) and has better elevated-temperature strength and resistance to sodium–water wastage damage than ferritic $2\frac{1}{4}$Cr–1Mo steel. Currently, only the coarse-grain, high-carbon version (Alloy 800H) is approved in ASME Code Case 1592 (Fig. 6-16). The French[7] have chosen a fine-grain, intermediate-carbon specification of the alloy for the Super Phenix steam generators, the U.S. established a fine-grained, low-carbon version as the backup to $2\frac{1}{4}$Cr–1Mo for CRBRP,[40,44] while the English[32] have suggested using a low-carbon, coarse-grained variety of Alloy 800 for future generations of LMFBR steam generators. These different viewpoints as to the optimum Alloy 800 specification could reflect both a difference in selection of criteria and/or differences in the available data base for the selection. In any event, the two critical areas that must be addressed are (1) approval of design allowables by the ASME Boiler and

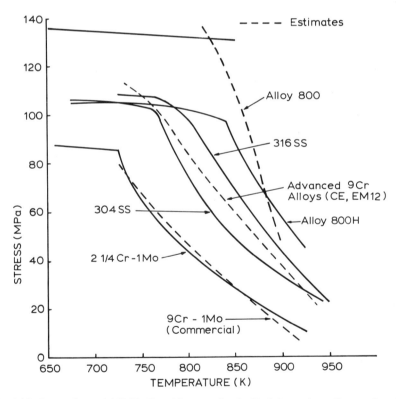

Fig. 6-16. Comparison of ASME allowable stress levels (S_{me}) for various alloys at 3×10^5 h (solid lines), or alloys under consideration for ASME Code Case 1592 (dashed lines).

Pressure Vessel Code (at least for U.S. use), and (2) the limited data on the sodium-corrosion behavior and the effect of sodium exposure on the mechanical properties of Alloy 800. To the first point, Fig. 6-16 compares the current code allowable stress values for Alloy 800 and $2\frac{1}{4}$Cr–1Mo ferritic steel with the *expected* values for the Alloy 800 modifications and the "advanced" 9Cr ferritics. Both show improvements over current materials.

Research on the ferritic steels and Alloy 800 materials in support of LMFBR steam generator application is a worldwide effort that has been reported on regularly. An international conference on "Ferritic Steels for Fast Reactor Steam Generators" was held in the U.K. in 1977,[45] and, as noted earlier, an entire conference was devoted to Alloy 800 in 1978.[31] More recently, Chopra et al.[46] have reviewed the effects of sodium on these two alloy classes, and a report on the status of the now terminated U.S. program on Alloy 800 was released.[47] In the interests of conciseness, therefore, this section will highlight only those data that are associated with the assurance of long-term structural integrity in the sodium–water/steam environment. These will include sodium corrosion effects, water–steam corrosion effects, and tube wastage by sodium–water reactions.

6.3.1.1. Sodium Corrosion

The general corrosion behavior of steels in sodium was introduced in Chapter 3, Section 3.2.4. Flowing high-temperature sodium very slowly dissolves the major metallic components of the steels (iron, chromium, and nickel) from the hot sections of the system, and transports these species to the colder sections where deposition occurs. Flowing sodium also serves as a medium through which the minor elements in the steels can migrate about the system. The element of principal concern is carbon, inasmuch as this species is largely responsible for the high-temperature strength of steels. Besides alloy composition, therefore, the major variables that control corrosion rates and mechanical property changes in sodium include temperature, oxygen level, the presence of other impurities (e.g., C, N), and flow rate. Corrosion is typically measured as a reduction in wall thickness (or load-bearing cross section), or as a penetration depth, in μ/yr, rather than as a weight change (refer to Fig. 3-19). Key properties evaluated include tensile strength and ductility, creep rupture, and fatigue.

The corrosion behavior of $2\frac{1}{4}$Cr–1Mo steel has been studied[45,46] over the temperature range from 720 to 920 K at oxygen levels between ~1 and ~25 ppm. The results show that the corrosion rate increases with an increase in sodium velocity up to ~7 m/s, at which point the corrosion rate becomes velocity independent. At a given temperature the dependence of corrosion rate on oxygen concentration in sodium is of the form $[O]^n$, where n is ~2. After an initial period, which tends to be variable, the metal-loss rates attain a steady-state value and remain constant with time for given sodium-purity and operating conditions. The metallurgical condition of the steel does not have a significant effect on the steady-state corrosion rates.

Chopra et al.[46] point out that the kinetics of carburization/decarburization of 2¼Cr–1Mo in static and dynamic sodium have been studied by numerous investigators, and the general concensus is that the extent of carburization or decarburization (expressed in terms of carbon loss per unit surface area of the steel, M) is a function of temperature and sodium-exposure time t as defined by the parabolic relationship

$$M = kt^{1/2} \qquad (6\text{-}1)$$

where the carburization/decarburization rate constant k has an exponential dependence on reciprocal absolute temperature.

Krankota and Armijo[48] have analyzed available data on the decarburization of 2¼Cr–1Mo steel in sodium and obtained "best-fit" and "upper-limit" curves for the decarburization rate constant. The upper-limit curve as well as some of the more recent data on the decarburization of 2¼Cr–1Mo steel are shown in Fig. 6-17.[46] The scatter in the results arises from differences in the carbon-activity gradients between the steel and the sodium, which are produced by different initial carbon contents in the steel and/or the carbon content of the sodium.

As noted at the outset, the loss of carbon from 2¼Cr–1Mo steel in a sodium environment leads to a reduction in mechanical strength. For components with intermediate section thicknesses (e.g., pipe), the decrease in strength is proportional to the depth of the decarburized layer. Maximum reduction in strength will occur in thin sections, namely, superheater and evaporator tubing. Therefore, for design of components with thin and intermediate section thicknesses, it is essential to establish long-term environmental effects and modify the allowable design stresses to ensure satisfactory performance of components over the expected service life.

Ideally, the effect of sodium environment on the mechanical behavior of structural materials should be evaluated from mechanical-property data on material with specific depths of decarburized or carburized layer, which were established under known conditions, e.g., time, temperature, and sodium purity. However, as Chopra et al.[46] report, the current assessment of the change in mechanical behavior of 2¼Cr–1Mo steel due to decarburization in a sodium environment is primarily based on the mechanical-property data obtained from steels with low initial bulk carbon

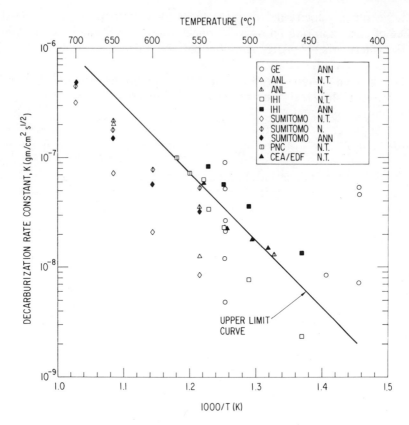

Fig. 6-17. Temperature dependence of decarburization rate constants for $2\frac{1}{4}$Cr–1Mo steel in sodium (adapted from Ref. 46).

contents. These data are in a good agreement with results obtained from specimens which were decarburized in sodium to different bulk carbon contents. The conclusion is that the tensile and yield strengths of $2\frac{1}{4}$Cr–1Mo steel with >0.03 wt.% carbon are adequate while steels with ≤0.01 wt.% carbon exhibit unacceptably poor tensile and yield strengths.

Recent studies of the influence of thermal aging on the mechanical properties of $2\frac{1}{4}$Cr–1Mo steel show that age softening is a dominant characteristic of this steel.[49] Consequently, to establish the effect of a sodium environment on the mechanical properties it is important to evaluate the influence of thermal aging as well as decarburization. Figure 6-18 shows the change in mechanical strength of decarburized $2\frac{1}{4}$Cr–1Mo steel as a function of bulk carbon content. The control data shown in this figure refer to specimens that were thermally aged in an inert atmosphere for identical time and temperature conditions. Decarburization progressively reduces the tensile and yield strengths of the steel. Chopra et al.[46] used the data in Fig. 6-18 and decarburization kinetics of Fig. 6-17 to estimate the strength reduction for typical superheater tubing. For a design life of 10^5 h at 783 K, they computed 0.025 wt.% carbon loss in the ~3mm thick tubes, which translates to a ~10% strength reduction.

The influence of sodium environment on the creep-rupture properties of $2\frac{1}{4}$Cr–1Mo steel is also due mainly to decarburization of the material, and, perhaps not too surprisingly, the reduction in creep-rupture strength for the superheater tubing during a service life of 10^5 h in sodium at 783 K will also be ~10%.[46] The loss in bulk carbon content and the change in creep-rupture strength at different temperatures is plotted in Fig. 6-19 as a function of section or wall thickness of a component. The results show that the 10^5 h creep-rupture strength ratio for material with wall thicknesses >3 mm is >0.9.

In contrast to the tensile and creep properties, where the variation in mechanical strength is primarily due to decarburization rather than the sodium environment *per se,* the fatigue life of $2\frac{1}{4}$Cr–1Mo steel tested in sodium is greater by a factor of 3 to 10 than when tested in air.[50] As for the case of the austenitic steels in pure sodium, discussed in Chapter 4,

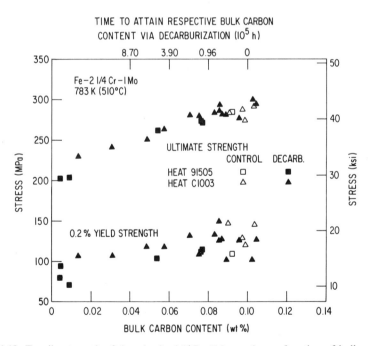

Fig. 6-18. Tensile strength of decarburized $2\frac{1}{4}$Cr–1Mo steel as a function of bulk carbon concentration (Ref. 46).

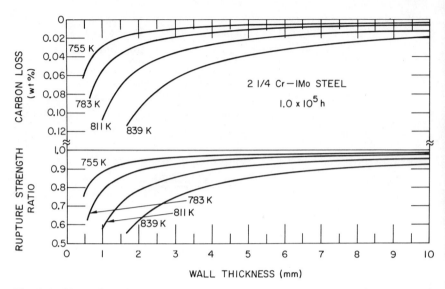

Fig. 6-19. Change in carbon concentration and creep-rupture strength for $2\frac{1}{4}$Cr–1Mo steel as a function of wall thickness (Ref. 46).

the absence of surface oxidation in a sodium environment increases the fatigue life of the material. As can also happen for the austenitic steels, however, the fatigue life of thermally aged or sodium-exposed specimens is 40% lower compared to the as-received material.[50] This is the more important result for long-term use of this material in sodium systems.

The creep-fatigue interaction for $2\frac{1}{4}$Cr–1Mo steel has also been investigated[50] in a sodium environment using a sawtooth waveform in which strain rates during the tensile and compressive half of the fatigue cycle differ by a factor of 100. Representative results from these tests are shown in Fig. 6-20. The fatigue life of isothermally annealed and sodium-exposed specimens tested under fast–slow (F/S) conditions (similar to compressive hold-time tests) is approximately the same as that for specimens tested under continuous cycling (F/F), whereas the fatigue life under slow–fast (S/F) conditions (similar to tensile hold-time tests) is a factor of ~2 lower. This behavior for $2\frac{1}{4}$Cr–1Mo steel in a sodium environment is opposite to that in air, where compressive hold time was found to be more damaging than tensile hold time at low strain ranges (<1%).[46]

In comparison to the data base on $2\frac{1}{4}$Cr–1Mo steels, information on the corrosion behavior of the ferritic "stainless" steels in a sodium environment is limited.[46] In general, the corrosion rate for 9Cr–1Mo steel is lower than that for $2\frac{1}{4}$Cr–1Mo steel; however, other differences in the corrosion behavior of the two steels have been noted. For example, the

9Cr–1Mo material is prone to internal oxidation and to the formation of a surface-oxide film when exposed to high-temperature sodium containing 20–40 ppm oxygen.[51] The surface and internal-oxidation behavior depends on the oxygen level and velocity of sodium. Surface oxides may persist for long periods in sodium at velocities of 3 to 4 m/s. However, at higher velocities, i.e., ≥9 m/s, the thickness of the surface-oxide film and the depth of internal penetration is relatively small. Surface and internal oxidation is not observed in specimens exposed to low-oxygen sodium, or in specimens placed in downstream locations at lower temperatures. Clearly, more work is required to establish the corrosion rate and the operating limits for the ferritic stainless class of steels.

The studies on the carbon-transfer behavior of 9Cr–1Mo steels in sodium show that these steels can either carburize or decarburize depending on the carbon activity in sodium. The carburization/decarburization behavior of higher-chromium ferritic steels is also very sensitive to the microstructure obtained during the initial heat treatment. Studies[52] on the stability of carbide phases in various chromium–molybdenum steels show that the most stable carbides in 9 to 12Cr–1 to 2 Mo steels are $M_{23}C_6$ and M_6C.

Data on the effect of sodium environment on the mechanical properties of the ferritic stainless steels are sparse. The tensile and creep-rupture strengths for various 9Cr–1Mo steels (normalized and tempered) tested in air are shown in Figs. 6-21a and 6-21b, respectively.[53–55] Limited results indicate that the sodium environment has virtually no effect on either the tensile or creep-rupture strengths.[56,57] Carburization of

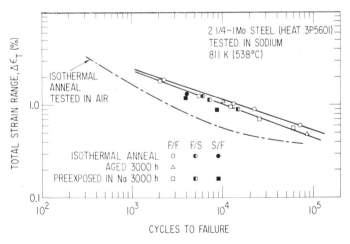

Fig. 6-20. Fatigue-life curves for 2¼Cr–1Mo steel tested in sodium at 811 K. Strain rates F (fast) and S (slow) are 4×10^{-3} and 4×10^{-5} s^{-1}, respectively (adapted from Ref. 46).

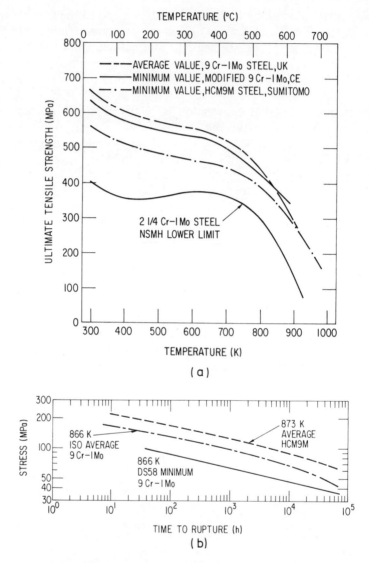

Fig. 6-21. Temperature dependence of (a) ultimate tensile strengths and (b) 866 K creep-rupture strengths of various 9Cr–1Mo steels (Refs. 53–55).

9Cr–1Mo steel in sodium increases creep endurance in short-term tests but causes a significant decrease in rupture ductility.[57] Creep experiments[56] on the 12Cr–1Mo alloy at 833 K in sodium did not show an effect of sodium on creep rate and onset of tertiary creep within a testing period up to 9000 h. From this limited evidence, Wood[56] and Chopra *et al.*[46] concluded that the sodium environment will have no significant effect on the mechanical strength of 9Cr–1Mo steels at temperatures below 820 K.

As noted earlier, the fatigue life of ferritic steels, in general, is better in sodium than in air. Although the fatigue behavior of 9Cr–1Mo steel in sodium has not been investigated, the fatigue life in a helium environment is 10% higher than that in air.[56] Data on the effect of sodium environment on the mechanical behavior of 9Cr–Mo steels are needed to establish the performance limits for components fabricated from this material.

The corrosion behavior of Alloy 800 in flowing sodium is similar to that of the austenitic stainless steels (see Chapter 3).[46] For example, after an initial corrosion period the metal loss is linear with time, i.e., the corrosion rate reaches a steady-state value (Fig. 6-22). For the steam generator temperature range of 720 to 820 K, corrosion occurring in clean sodium is negligible (<500 μm over 30 yrs). Also, there is no rapid, selective grain-boundary attack even under the most severe test conditions. Information on the carburization/decarburization behavior of Alloy 800, however, is sparse,[46] and the data tend to be concentrated outside the expected steam generator temperature regime (<870 K). For instance, the equilibrium carbon concentration for Alloy 800 has been obtained at various carburizing potentials in sodium at temperatures between 865 and 1030 K.[58] These results indicate that an alloy with a nominal carbon

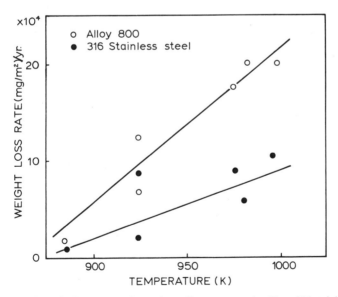

Fig. 6-22. Corrosion of Alloy 800 specimens by sodium compared to Type 316 stainless steel (Ref. 73).

concentration of 0.07 wt.% will decarburize at temperatures above ~890 and 980 K when exposed to sodium containing ≤0.05 and 0.20 ppm carbon, respectively, and it will carburize at lower temperatures. The equilibrium carbon concentrations predicted by this relationship, however, are a factor of 3 lower than the surface carbon concentrations observed in Alloy 800 specimens exposed for 10,400 h at 873 and 973 K in sodium containing 0.2 ppm carbon.[59] In this more recent work,[59] carbon concentrations in Alloy 800 reached maximum values of .1 and 0.4% after exposures at 873 and 973 K respectively. The depth of carburized material is larger at the higher temperature, but both exceeded 60 μm. Inasmuch as surface carburization may cause embrittlement, additional data on carburization/decarburization behavior are clearly required to establish the equilibrium carbon concentration–carbon activity relationship and the temperature dependence of the rate constant for Alloy 800 over the expected steam generator operating range.

Current knowledge of the effect of sodium exposure on the mechanical properties of Alloy 800 is, surprisingly, also quite limited.[46] Based on the air properties, e.g., Fig. 6-16, one would expect properties in sodium to be superior to the ferritics; however, fatigue data have not been reported, and the creep and creep-rupture data have been taken at 798 K and above, again at the upper end or above the steam generator operating regime. Results of the most recent creep-rupture study are reproduced in Fig. 6-23[60] and indicate that, at 923 K, the liquid sodium has a small beneficial effect over air behavior, but at 798 K failures in sodium occurred consistently earlier than in air. Reductions in rupture life of ~50% were observed, with a tendency for the reduction to increase at longer durations. Since the sodium-exposed surfaces exhibited evidence of enhanced crack propagation at both temperatures, it is speculated that crack initiation must be retarded at the higher temperature. In any event, the prospect of increased crack propagation rates in sodium could prove detrimental in a fatigue or creep-fatigue situation. The reader will recall that impure (carbon-containing) sodium accelerated fatigue crack growth in the austenitic steels, described in Chapter 4, Section 4.3.1.

6.3.1.2. Water–Steam Corrosion Performance

As noted earlier, in Section 6.2, the major difference between the (2–12% Cr) ferritic steels and Alloy 800 is that the former are most susceptible to severe general attack and pitting in steam or water environments, while the latter alloys are more likely to fail by SCC.

The general corrosion (oxidation) behavior of $2\frac{1}{4}$Cr–1Mo steel has been studied in aqueous solutions, in pure and impure superheated steam,

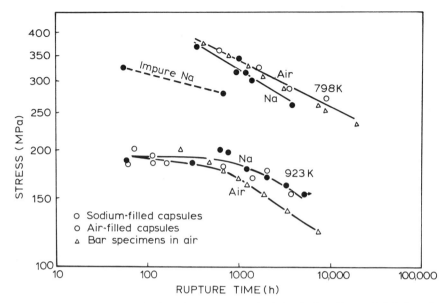

Fig. 6-23. Stress rupture results for Alloy 800, comparing behavior in sodium and air (adapted from Ref. 60).

in saturated steam, and under nucleate boiling conditions, with chloride and oxygen additions.[45,61] Oxidation is typically measured either as a metal weight loss, as an oxide film thickness, or as a wall penetration depth. Oxidation is a function of solution pH, increasing dramatically at low or very high values. However, the isothermal corrosion data for $2\frac{1}{4}$Cr–1Mo steel (as for most ferritic steels—see later) can be adequately represented by a linear rate law and a polynomial expression in the concentration of either hydroxyl or hydrogen ion.

Under laboratory ("normal") conditions a relatively stable protective oxide layer is formed; extrapolation to 30-yr plant lifetimes gives 100 to 130 μm metal consumption in the temperature regime 770–800 K. However, recommended allowances are much higher (510–760 μm) to allow for periodic loss of scale integrity and other causes.[62]

The necessity for the conservatism at this time is clear when one considers the results of recent experiments on (a) the effect of heat flux on corrosion[63] and (b) elevated-temperature environmental fatigue damage[63a] in $2\frac{1}{4}$Cr–1Mo steel.

Experiments conducted under heat transfer conditions (a heat flux of 126 kW/m² compared to the expected 270 kW/m² across a typical LMFBR superheater tube wall) indicate that metal consumption can be almost twice as great as the isothermally exposed specimens after 6000 h in the 780 to 815 K regime[63]; Fig. 6-24 shows the average penetration over the heated length of the specimen calculated from the defilmed weight loss. Three effects, heat flux, growth stresses, and thermal transients, are possible causes of the enhanced corrosion, but a specific mechanism has not emerged to date. This must be resolved because there is a performance, as well as economic, incentive to reduce the current conservatism.

An analysis[63a] of previously reported elevated-temperature (589–866 K) fatigue with hold-time data in air, supplemented by a metallographic examination of the $2\frac{1}{4}$Cr–1Mo steels, attributes the observed reduction in fatigue life to the combination of reduced interaction solid-solution hardening and accelerated fatigue crack initiation due to oxide cracking. Circumferential oxide cracks are found on specimens tested with compressive hold periods, and only random cracking and spalling are observed on both continuously cycled and tension-hold-period specimens. These oxide cracks perpendicular to the applied stress initiate the fatigue cracks in the underlying metal. Thus, they cause a greater reduction in fatigue in the strain range regimes where crack initiation dominates the fatigue life, i.e., low-strain range, high-cycle fatigue life. The effect of hold pe-

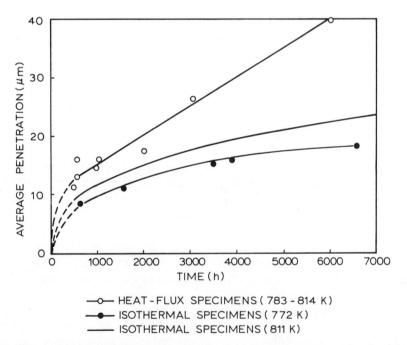

Fig. 6-24. Average penetration of $2\frac{1}{4}$Cr–1Mo steel corrosion specimen exposed to superheated steam (adapted from Ref. 63).

riods is therefore primarily due to an environmental interaction (rather than to a creep interaction) with the fatigue damage process.

Both caustic and salt impurity effects on the corrosion behavior of $2\frac{1}{4}$Cr–1Mo steel have been evaluated.[45,61] Constant temperature and pressure tests in superheated steam saturated with chloride (2 ppm as NaCl) and with a controlled oxygen concentration (<.007 ppm for 4626 h and then 8 ppm for a further 5958 h) revealed no cracking in $2\frac{1}{4}$Cr–1Mo tubes even though they were stressed up to twice the design stress limit. The extent of general oxidation in the absence of oxygen was somewhat less than in relatively pure steam at comparable temperatures. However, microscopic examination showed numerous broad shallow pits ranging from 13 to 25 μm. Operation with 8 ppm O_2 resulted in much heavier oxidation with noticeable oxide spalling.[61]

Cyclic temperature testing has produced mixed results. No tendency for cracking of $2\frac{1}{4}$Cr–1Mo steel was seen under 4 to 10 ppm chloride (added as NaCl) conditions when cycled from 700 or 643 K down to 553 K, even when oxygen concentrations in the steam reached 20 ppm. However, in cyclic tests under chloride conditions (2.7 ppm to 10 ppm) produced by adding $CaCl_2$ and 8 ppm oxygen, all $2\frac{1}{4}$Cr–1Mo tube specimens stressed to 1 to $1\frac{1}{2}$ times design allowable failed after 6475 h of operation due to wall thinning from general corrosion.[61] There was no evidence of cracking or localized attack of the steel.

Tests under boiling conditions, simulating the effects of high concentrations of chloride and caustic impurities (e.g., as in crevices, etc.) confirmed the problem of general corrosion of $2\frac{1}{4}$Cr–1Mo steel, but, significantly, did not reveal any signs of cracking.[61] In a more recent ongoing study with a single-tube model,[64] exposure to 30 ppb Na as NaOH under liquid film dry-out conditions produced no observable pitting nor cracking in approximately 5000 h. A thin, porous corrosion film, ~0.02 to 0.04 mm thick, covered the surface in both the pre- and postcritical heat flux regions.

A detailed program[65] of constant load and constant extension rate (CERT) stress corrosion tests has also been conducted on $2\frac{1}{4}$Cr–1Mo in aqueous solutions (589 K) and superheated steam (755 K). It was found that this alloy is completely resistant to caustic cracking in 3 and 5% NaOH at these temperatures. A single case of "marginal" cracking occurred in 10% NaOH at 589 K in a CERT, but attempts to reproduce this observation were not successful. The rate of general corrosion was found to be significant, but it decreases as the temperature is reduced. Significant differences in general corrosion were also found for different heats of this steel. The ESR alloy in the annealed condition produced the lowest corrosion rate in caustic, while the welded samples gave the highest.

Thus, it appears that cracking of $2\frac{1}{4}$Cr–1Mo steel under "steam generator" conditions is inhibited by the general corrosion that occurs above 589 K; however, cracking of this steel can be achieved in 5–10% NaOH at 505 K,[65] and so it cannot be considered immune to SCC.

The higher chromium ferritic steels are also characterized by general corrosion attack in steam but to a lesser extent than the 2¼Cr–1Mo steel.[66,67] Small additions of silicon have also been found to be beneficial in this respect.[66] In one study[66] there was only a slightly lower extent of attack of 9Cr–1Mo steel over 2¼Cr–1Mo–Nb steel at 823 K when the steels contained 0.35 to 0.4 wt.% Si; however, an increase in the silicon content of the 9Cr–1Mo steel to 0.85 wt.% halved the extent of attack after 8000 h. Hence, silicon was of more significance than chromium in determining the steam corrosion behavior of 9Cr–1Mo. The effect of chromium dominates at higher Cr contents, however. For example, an increased beneficial effect from the chromium was observed when a 12 wt.% Cr steel was tested. Also, the chromium in 9Cr steels can exert a beneficial effect in other environments and, for example, reduced the attack in 15% NaOH at 589 K over that for a 2¼Cr steel by a third.[68]

Antill[69] recently reported on the relative susceptibilities of a 9Cr–1Mo steel (Sandvik HT7) and Alloy 800 (Sandvik Sanicro 30) in 19 different boiling water solutions using simple capsule tests. The data are typical of those obtained in other laboratory studies. In pure water at 623 K, neither material is significantly attacked. In steam at 773 K, the general corrosion of the ferritic steel is significantly greater than for Alloy 800. The addition of neutral chloride (NaCl) at 623 K has no significant influence. Acidic chloride ($MgCl_2$) increases slightly the general attack of 9Cr–1Mo steel, while the addition of large amounts of oxygen with chloride leads to failure of both materials by pitting, with some intergranular cracking of the Alloy 800; the ferritic steel has the shorter failure times in these oxygenated chloride environments. In other studies,[70] acid sulfate and phosphate environments have been reported as far more aggressive to ferritic steels than HCl solutions of equivalent pH.

In contrast to the behavior in the environments discussed above, the 9Cr–1Mo steel was found by Antill[69] to be more resistant than Alloy 800 to caustic environments. As noted earlier in Section 6.2, Alloy 800 fails by intergrangular cracking in high caustic concentrations (refer to Fig. 6-10), but ferritic steel just forms thick surface scales. Indig[65] reported slight cracking of ferritic steel in caustic solutions, but only under conditions of continuous straining or of high imposed potential. In this connection, one must be aware that 9Cr–1Mo steel can be sensitized and, hence, rendered susceptible to cracking in various environments: it must, therefore, be used in the right metallurgical condition and welds must be heat treated by, for example, 1 h at 1063 K.[65]

Much of the data presented to date (i.e., above and in Section 6.2) on Alloy 800 refers to the alloy's susceptibility to SCC. The data in Figs.

6-10 and 6-14 illustrate the relatively high threshold stress for SCC; the only area of concern is the increased susceptibility to cracking that is observed in concentrated caustic solutions, which should not be experienced in an LMFBR steam generator.

The general corrosion of Alloy 800 in high-temperature steam has been studied extensively by Leistikow.[71] Linear corrosion kinetics were observed in deoxygenated steam in the temperature range 820 to 970 K, up to 10,000 h, and a strongly adherent uniform oxide layer develops. Interestingly, the creep resistance of Alloy 800 is not altered by corrosion in this regime.[72] The results of the steam exposure tests are plotted onto the curves derived from tests in argon in Fig. 6-25. An insignificant increase of oxidation was measured, but the oxide scale remained adherent and protective, favoring a slight tendency to material strengthening under long-term oxidizing conditions.

6.3.1.3. Tube Wastage by Sodium–Water Reactions

The relative resistance of materials to wastage or accelerated corrosion by products from the sodium–water reaction, in the event of a leak in the steam generator, is an important area for study in LMFBR steam generator development.[46,69] Ingress of high-pressure steam or water into

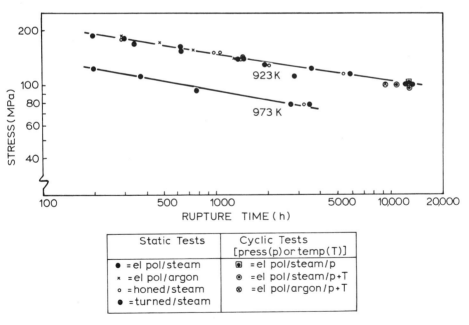

Fig. 6-25. Creep-rupture functions of all constant and cyclic temperature/pressure tests for Alloy 800 in a steam environment (adapted from Ref. 72).

sodium causes a vigorous exothermic reaction and high temperatures (~1400 to 1700 K) in the vicinity of the leak. To ensure safe and reliable operation of sodium-heated steam generators, systems are incorporated for rapid detection of sodium–water reaction (e.g., oxygen and hydrogen meters in the sodium and cover gas), pressure relief (e.g., rupture discs), and cleanup of the reaction products (e.g., cold traps). However, the steam generator tubes must have sufficient resistance to wastage to provide adequate time for detection of a small leak and shutdown of the system to minimize damage to adjacent tubes.

Three distinct damage mechanisms have been postulated during sodium–water reactions, viz. erosion of adjacent tubes due to impingement (i.e., "impingement wastage"), corrosion of the leaking tube (i.e., "self-wastage"), and higher than normal corrosion rates for system materials resulting from circulation of reaction products (i.e., sodium oxide and hydroxide) through the system.[46] Considerable research effort has been expended to develop a quantitative understanding of wastage damage and sodium–water reactions.[73–75] Results show that leak rate, tube spacing, and sodium temperature are important variables in the wastage process. In general, the wastage rate of a material in stagnant sodium increases with an increase in the leak rate over a wide range (e.g., 0.005 to 500 g/s), whereas in flowing sodium, the wastage rate reaches a maximum at a leak rate of ~5 g/s and then decreases at higher leak rates.[76] For instance, the maximum metal penetration rate for $2\frac{1}{4}$Cr–1Mo steel in flowing sodium was determined to be ~80 μm/s at 590 K. The penetration rate decreases with either an increase in the leak-to-target distance (tube spacing) or a decrease in the sodium temperature.[74,76–78]

The relative wastage resistance of Type 304 stainless steel, Alloy 800, $2\frac{1}{4}$Cr–1Mo steel, and 9Cr–1Mo steel studied in water-injection tests indicated that Alloy 800 has a greater resistance to wastage than austenitic stainless steel and that the latter material is more resistant than ferritic steels (Table 6-1).[74,77] The resistance of Alloy 800 and Type 304 stainless

Table 6-1. Penetration Rate and Comparative Wastage Resistance of Steam Generator Materials[a] (Ref. 46)

Materials	Penetration rate, mm sec^{-1}	Mass loss, g	Resistance ratio
Fe–$2\frac{1}{4}$Cr–1Mo	$2.5 \pm 0.3 \times 10^{-3}$	0.356	1.0
Fe–9Cr–1Mo	$2.1 \pm 0.6 \times 10^{-3}$	0.256	1.2
Type 316 SS	$8.0 \pm 3.1 \times 10^{-4}$	0.121	4.0
Alloy 800	$6.3 \pm 3.1 \times 10^{-4}$	0.0144	5.2

[a] All tests performed with controlled sodium–water reaction flames at temperatures between 1173 and 1273 K.

steel increases markedly as the temperature decreases; however, the wastage rate of the $2\frac{1}{4}$Cr–1Mo steel does not exhibit a large temperature dependence.

With a relatively large tube spacing ($\gtrsim 12.5$ mm) and low leak rates ($\gtrsim 0.05$ g/s), impingement wastage may not occur and the primary mode of damage then becomes self-wastage. Under these conditions, small leaks frequently plug from the buildup of reaction products in and immediately around the defect. However, small leaks that self-plug can spontaneously open and grow.[79,80] A plausible mechanism for unplugging involves transport of sodium to the steam side of the tube and the formation of sodium hydroxide, which attacks the base metal around the defect. The corrosion process eventually dislodges the plug and leakage of steam into the sodium occurs at a high rate.[46]

Often, small leaks are present for an extended period of time and then rapidly increase in size.[77,79,80] The results of U.S. and French studies indicate that the self-wastage rate for austenitic stainless steel is higher than that of the $2\frac{1}{4}$–1Mo alloy at high leak rates, but is lower at very low leak rates. The difference in the self-wastage behavior of the two materials under high-leak-rate conditions has been attributed to the poor thermal conductivity of the stainless steel.[77] A relatively low thermal conductivity would result in a higher temperature, and consequently a higher wastage rate. When the leak rate is very low, the heat-generation rate is small for both materials and the wastage rate is determined primarily by the kinetics of the corrosion process. The sudden enlargement of small leaks has been explained on the basis of caustic attack of the steel.[77,78] As the leak continues, a conical pit forms in the tube wall. When the pit reaches the steam side, the cross-sectional area of the leak changes rapidly from that of the original defect and gives rise to the observation of a sudden increase in the leak rate.

6.3.2. IHX Materials Properties

The austenitic stainless steels, Types 304 and 316, are universally accepted for use in the intermediate heat exchanger units in LMFBRs based on their sodium corrosion resistance (refer to Chapters 3 and 4) and high allowable design stresses (see Fig. 6-16, for example).

The general corrosion behavior of austenitic stainless steels in a liquid sodium environment was described in Chapter 3, in relation to LMFBR core materials. Based on the combined data set presented there (Fig. 3-19), the corrosion allowance anticipated for long-term IHX operations can be computed with a fair degree of reliability.[46]

Of more concern to IHX performance, however, is the transfer of interstitial elements such as carbon and nitrogen in the liquid sodium/

stainless steel system and the resulting effect on long-term mechanical behavior (e.g., creep-rupture and fatigue properties). As noted earlier, this is of particular importance if a ferritic steel is used for steam generator tube material.

The kinetics of the carburization/decarburization process in austenitic stainless steels are well understood.[46] The conditions of temperature and carbon concentration in sodium that result in either carburization or decarburization of Type 304 and 316 stainless steel (with nominal initial carbon contents) are shown in Fig. 6-26. The results indicate that in a sodium environment containing 0.2 ppm carbon Types 304 and 316 stainless steel would carburize at temperatures below 940 and 900 K, respectively.

The mechanical properties of austenitic stainless steels in sodium

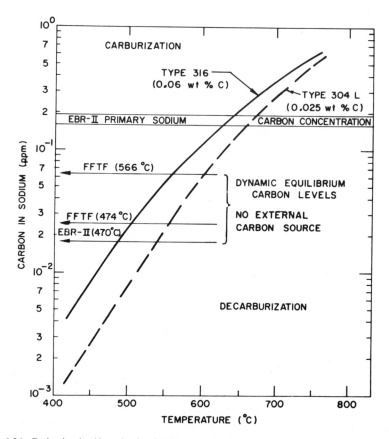

Fig. 6-26. Carburization/decarburization regimes for Types 316 and 304 stainless steel relative to temperature and carbon concentration in sodium (adapted from Ref. 46).

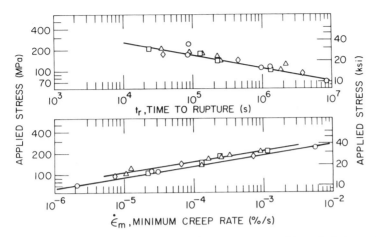

Fig. 6-27. Variation of time-to-rupture and minimum creep rate with applied stress at 923 K for various carburization depths in Type 316 stainless steel. △, 100 μm; □, 200 μm; ◇, 375 μm; ○, solution annealed (adapted from Ref. 84).

have been reviewed by Chopra et al.[46] and by Lloyd.[81] Fatigue and fatigue crack growth data were presented in Chapter 4, in relation to the LMFBR pressure boundary, and the same data are applied to the design of IHXs. Additional important property data are provided here to complete the overall picture of stainless steel behavior in sodium.

As might be expected, the influence of carburization/decarburization on the mechanical properties of austenitic stainless steels has been the subject of numerous investigations.[46,81] Tensile and creep-rupture data have been obtained from materials which were carburized in sodium to produce specific carbon-penetration depths. At temperatures between 670 and 1070 K the ultimate tensile elongation decreases with an increase in carbon concentration in the material.[82-84] For these tests, the depth of carbon penetration in the specimens varied from 0.1 to 0.3 mm and the increase in average bulk carbon concentration of these specimens was up to 0.2 wt.% for Type 316 and 0.15 wt.% for Type 304 stainless steel. The effect of sodium exposure on the tensile properties is greater at lower temperatures because of the greater pickup of carbon at these temperatures.

Creep-rupture data for austenitic stainless steels carburized in a sodium environment show an increase in rupture life and a decrease in minimum creep rate and rupture ductility when compared with annealed material. Type 316 stainless steel exposed to sodium at ~820 K is an exception to this behavior, i.e., a small decrease in rupture life occurs due to carburization.[85] The influence of carburization on the creep-rupture properties of Type 316 stainless steel is illustrated in Fig. 6-27 for the temperature of 923 K. Similar data exist for both Types 316 and 304 stainless steel over the temperature range 820 to 790 K, and, overall, they show an increase of 15 to 30% in the 10^5-h creep-rupture stress for carburized material.[46] Microstructural examination of the creep-tested specimens indicated that carburization of the material inhibits intergranular failure that is generally observed in austenitic steels at low strain rates.

The changes in microstructure and mechanical properties that result when stainless steels undergo decarburization in sodium are quite similar to those observed when low-carbon heats of these steels are exposed to the same thermal treatment. Decrease in carbon and nitrogen concentrations in Types 304 and 316 stainless steel during exposure to high-temperature sodium result in a 25 to 35% reduction in the 10^5-h creep-rupture strength.[46]

Finally, investigations of the effect of thermal aging on the microstructural changes in stainless steels have shown that the stability of intermetallic phases, such as sigma phase, increases with a decrease in either carbon or nitrogen concentrations in austenitic stainless steels.[86,87] These results and data on the influence of interstitial elements on the mechanical properties have provided a basis for comparison of the tensile and creep-rupture properties of stainless steels in air, inert gas, and sodium environments.[46,81] The overall conclusion is that sodium *per se* does not alter the tensile and creep-rupture properties; any effects of sodium exposure reported are due either to carbon transfer, oxygen contamination, or temperature.

Investigations of the effect of sodium on the low-cycle fatigue properties of stainless steels were described in Chapter 4. In summary, Fig. 4-23 showed that the effect of the sodium environment has a beneficial effect on fatigue life in relation to results in air. Thermal aging or sodium exposure under mildly carburizing conditions has a minimal effect on the fatigue life of Type 316 stainless steel in sodium above ~820 K. In Type 304 stainless steel, whereas the thermal aging increased fatigue life in sodium, prior sodium exposure reduced fatigue life above 0.8% strain range and increased it below this value. It might also be added that no influences of sodium upon the high-cycle fatigue resistance of austenitic steels have been observed with the very limited results to date.[81]

The creep-fatigue interaction behavior of stainless steels in a sodium environment has not been evaluated, but behavior should be similar to that in air. For instance, the fatigue life for Type 304 stainless steel at a strain rate of 4×10^{-5} s^{-1} in sodium at 873 K is shown in Fig. 6-28; increasing the strain rate to 4×10^{-3} s^{-1} increases the fatigue life in sodium, which is similar to the air results.

The effect of sodium environment on fatigue-crack growth of Type 304 stainless steel was also described in Chapter 4. To reiterate, the

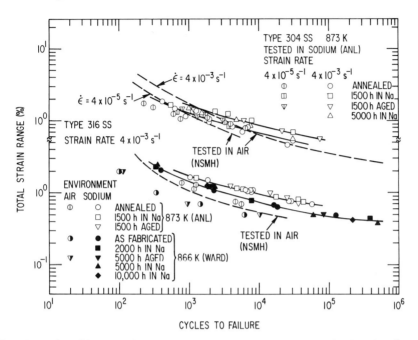

Fig. 6-28. Fatigue-life curves for Types 304 and 316 stainless steel tested in air and sodium at 873 K (Ref. 46).

results show that the fatigue-crack growth rates in pure sodium at 700 and 811 K are considerably lower than in an air environment at the same test temperatures (Fig. 4-29). However, crack growth in carburizing or decarburizing sodium is accelerated. Whether or not this latter observation is a reflection of carbon addition to or removal from the highly stressed crack tip material is not known, and represents a major area of uncertainty for future attention.

6.3.3. Closing Remarks

Information on the behavior of materials in high-temperature sodium and steam/water environments has increased significantly during the past ten years. With respect to sodium, an adequate "laboratory" data base for metallic and nonmetallic element transfer has been developed for sodium circuits constructed of austenitic stainless steels (Types 304, 316, and 321) and the $2\frac{1}{4}$Cr–1Mo ferritic steel. Quantitative data on the compositional and microstructural changes in these materials as well as correlations between the corrosion rate and temperature, sodium purity, and sodium velocity have been obtained. However, additional compatibility data are required to augment the data base for Alloy 800 and 9 to 12Cr–1Mo ferritic steels relative to their performance in a high-temperature sodium environment.

Additional mechanical property data in sodium are required for all materials, but again Alloy 800 and the ferritic stainless steels are the weakest in this area. The primary needs are (1) longer-term creep-rupture data on sodium-exposed materials (2) longer-term fatigue, creep-fatigue, and crack-propagation data from tests in flowing sodium, (3) fatigue and creep fatigue properties of weldments in a sodium environment, and (4) high-cycle (~1 Hz) fatigue properties related to possible flow-induced vibration effects.

On the basis of relevant sodium effects, clearly Types 304 and 316 stainless steel and $2\frac{1}{4}$Cr–1Mo steel are the most characterized heat exchanger materials. The choice of Alloy 800 by the French, therefore, must be based on either the good waterside corrosion resistance and/or the higher initial strength, which would allow a certain amount of in-service degradation without loss in performance.

The situation is somewhat reversed when considering waterside corrosion effects. This area should be of particular concern to LMFBR designers in light of recent PWR steam generator experience. The data base discussed above indicates that the significant corrosion of $2\frac{1}{4}$Cr–1Mo steels will limit their temperature of operation. Improvements are possible through the addition of niobium (i.e., as in the stabilized version) or additional chromium and silicon (i.e., as in the ferritic stainless class) but, nevertheless, the 9–12%Cr alloys appear to offer distinct advantages here. The general low corrosion of Alloy 800 and the relatively high resistance of the alloy to SCC justify its selection as a candidate for LMFBR steam generators. However, the susceptibility to caustic and chloride SCC necessitates extra care to ensure that locally high concentrations of sodium or chloride ions will not accumulate on the secondary side of the steam generator.

The possible effects of heat flux, oxygen content of the water, and hydrodynamic factors in the boiling/dry-out zones need to be evaluated for Alloy 800 and the ferritic stainless steels. The heat transfer effect on $2\frac{1}{4}$Cr–1Mo corrosion is significant; although no data exist on the other alloys, it could clearly affect the cracking and spalling of scales in steam. As seen earlier for the PWR steam generator, hydrodynamic factors, together with the physical structure of scales/cruds and heat transfer, influence the concentration of salts in boiler water. Recent work with a miniature once-through boiler rig[88] has shown that when boiler water is dosed with 20–80 ppb NaCl the salt is hydrolyzed. Both the sodium and chloride ions can move independently into the superheater and become bound to Alloy 800 superheater surfaces at a level approximating a monolayer. The take-up of sodium is faster than that of chloride ions but the chloride is more strongly bound. A test section of 9Cr–1Mo steel takes up sodium similarly but not chloride ion. Sodium from NaOH behaves similarly to that in NaCl.

The significance of these findings on boiler behavior has yet to be assessed. More of these boiler tests need to be conducted. Unfortunately they are time consuming and expensive, but they will probably be the only way of arriving at an optimal steam generator material since, at this stage, no single material can either be discarded or, on the other hand, selected as the primary material. The waterside aspects of the problem clearly must take precedence over the sodium side inasmuch as the data to date for the latter indicate a more understandable pattern of property changes with time.

6.4. Condenser Experience

The experience base on condenser performance is drawn as much from fossil power plants as from LWR nuclear power plants. Very little experience has been gained with condensers in LMFBR systems. While the turbine cycle for nuclear power plants is different from that for fossil-fueled power plants, the essential function of the condenser is the same, particularly with respect to material selection and performance, and so there is some value in including fossil plant experience in a general review of condenser experience and associated materials problems.

Condenser tube leakage is the primary concern with condenser performance, and it has an impact on the operating cost, availability, and useful service life of power generating facilities. Failure of condenser tubes, which has been significant in the past, introduces circulating cooling water into the pure condensate and may thereby contaminate the entire steam–water cycle. The degree of contamination is a function of the rate of inleakage and the composition of the cooling water.

Two trends in power plant design and operation have further intensified the impact of condenser tube leakage[13]:

1. Most new (and some existing) power stations have been forced to use cooling towers or ponds for main condenser cooling in place of once-through cooling. The resultant higher dissolved solids in the cooling water and the generally higher pressure on the condenser waterside magnify the contamination introduced by condenser tube leaks.
2. The use of only volatile chemicals (AVT) is now recommended for use in PWR secondary or turbine cycle water treatment (refer

to Section 6.2 earlier). Consequently, the introduction of circulating cooling water into the condensate by condenser inleakage will result in scale formation and/or increased corrosion in the steam generators. Only minute quantities of inleakage can be tolerated before the stringent limitations on feedwater or steam generator water quality are exceeded and the plant must be shut down. Condensate demineralization can extend the rate of inleakage that can be tolerated before plant shutdown is required, but adds significantly to capital and operating costs. There is also concern that, if condensate demineralizers malfunction, they themselves will adversely affect water quality.

Nuclear units have experienced capacity factor losses totaling 0.74% from problems with condensers, and a minimum loss of 0.53% from condenser tube problems alone over a "commercial age" period of eight years or less.[14d] Obviously, tube failures and condenser retubing have been one of the most frequent, widespread, and consequential problems affecting condensers. Tube failures usually result only in a reduction of the unit output while the tubes are repaired or plugged. The more lengthy outages have been associated with retubing of the entire condenser.[14d] This experience, together with the future trends, indicates how important it is to take all reasonable measures to minimize condenser tube leaks. Qualification of improved tubing materials is a critical part of this effort.

Condenser tubes fail as the result of many different mechanisms,[13] involving both the tube material and the service environment (e.g., service sector within the condenser, steam-side environment, waterside environment, etc.). One of the better guides as to how condenser tube materials will perform in the future is the way they have performed in the past. Michels et al.[89] have combined several survey or experience reports to produce the chart in Table 6-2 for tube failures in once-through salt water cooling units. Tube failure rates are reported as

$$\text{Failure rate } F\ (\%) = \frac{\left(\frac{\text{number of tubes plugged}}{\text{total number of tubes in unit }(T)}\right)}{\text{total actual operating time (years)}}$$

Tube failure rates are shown together with total operating period (denoted by the extent of each box projected on the time scale) for the number of units or tubes in the sample studied.

Some useful generalizations with respect to individual materials performance in salt and fresh water units were made in a recently published Bechtel report on this subject.[13] A selection of these are reproduced below; subsections will then treat interesting aspects of the performance

Table 6-2. Performance of Tube Materials in Units with Once-through Saltwater Cooling[a] (Ref. 89)

Material	Service	Data
70/30 Cu–Ni, 0.6 Fe	Ship	40 units, $T = 2216$; $F = 0.007$
	Ship	40 units, $T = 25{,}616$; $F = 0.045$
	Power	13 units, $T = 97{,}500$; $F = 0.053$
70/30 Cu–Ni, 2Fe2Mn	Desalination	$T = 79{,}864$; $F = 0.007$
90/10 Cu–Ni	Power	13 units, $T = 97{,}500$; $F = 0.044$
	Power	13 units, $T = 97{,}000$; $F = 0.096$
	Desalination	$T = 46{,}927$; $F = 0.43$
	Power	$F = 0.4$–0.5 (estimated)
Al brass	Power	55 units, $T = 412{,}500$; $F = 0.17$
	Power	79 units, $T = 592{,}500$; $F = 0.29$
	Desalination	$T = 46{,}927$; $F = 0.3$
	Power	$F = 0.45$ (estimated)
	Power (Japan)	Estimated $F = 0.4$ in 1965 period, $F = 0.02$ in 1968 period
Admiralty brass	Power	37 units, $T = 227{,}500$; $F = 0.83$
Newer materials — Titanium	Desalination	3 units; $F = 0.02$
	Power	13 units; Vibration failures reported
AL-6X	Power	5 units; No failures yet reported

Actual operating time, years → 0, 5, 10, 15, 20

[a] Failure rate in percentage per year of actual operating time; T, number of tubes or units in sample studied.

of the materials observed during the detailed survey conducted by the Bechtel team.

1. In the main condensing section [(16) in Fig. 6-6] of condensers cooled by fresh water in a once-through mode, all of the commonly used condenser tube materials give acceptable service. For admiralty brass, 90–10 Cu–Ni (Alloy 706), and Type 304 stainless steel there is a better than 90% probability that less than 10% of the tubes will need to be plugged after 40 years of service, or 5% plugged after 20 years.
2. In the air removal section [(8) in Fig. 6-6] of condensers cooled by once-through flow of fresh water, some severely high failure rates have been encountered when admiralty brass tubes have been used. In this service, the probability that admiralty brass will have a total failure rate less than 0.333 (10% plugs in 40 years) is only 26.5%. The apparent reason for this poor performance is the presence of ammonia.

3. In the air removal section of freshwater-cooled condensers, Type 304 stainless steel has a 96.2% probability of serving 40 years with less than 10% failures.
4. The very limited data bank for closed cooling-water cycle condensers does not allow formulation of valid conclusions. However, the available data indicate that in the condensing section Admiralty brass may have a lessened service life as compared to open-cycle freshwater applications, but Type 304 stainless steel performs almost as well in closed-cooling water cycles as in freshwater.
5. Until recently, the traditional condenser tube alloys used in the condensing section of saltwater-cooled condensers were aluminum brass and aluminum bronze. Neither alloy can be expected to last the plant lifetime without retubing. The probability of 40 years service with less than 10% tube failures is 33.3% for aluminum brass and 50% for aluminum bronze.
6. Alloy 706 (90–10 Cu–Ni) has been used in the condensing section of saltwater-cooled condensers in recent years. Its anticipated service life is about the same as for aluminum brass/bronze alloys. The probability that 90–10 Cu–Ni will serve 40 years with less than 10% tube failures is 40%.
7. Titanium, the newest condenser tube material, is the only material included in the survey that can be expected to last the plant lifetime in the condensing section of saltwater-cooled condensers. Titanium was found to have a 90% probability of lasting 40 years with less than 10% tube failures. This probability would be nearly 100% if it were not for guard plugs placed in two tubesets that were experiencing vibration problems. All of the actual failures of titanium tubes were attributable to mechanical causes, predominantly vibration. Not one instance of corrosion or erosion failure was found during the field survey.
8. In air-removal service in saltwater-cooled condensers, a significant data base was generated for only 70–30 Cu–Ni. In such service, 70–30 Cu–Ni has a 69% probability of serving 40 years with less than 10% tube failures.

6.4.1. Copper-Base Alloys

Copper-base alloys have historically been used for condenser tubing materials because of their good heat transfer coefficients, antibiofouling characteristics, and relatively good corrosion resistance.[13,89,90] Table 6-2 lists the copper-base alloys used in condensers.

Historically, admiralty brass condenser tubing has been used for freshwater service while the copper–nickels, aluminum brass, and aluminum bronze are used predominantly in seawater cooled plants.[90] In the past decade, the copper–nickels, and especially 90–10 Cu–Ni, have been specified more often as tube material for seawater cooled plants. The copper–nickels, i.e., 70–30 Cu–Ni and 90–10 Cu–Ni, have lower corrosion rates in seawater and are more resistant to ammonia attack than the other copper-base alloys.[89,90] All of the copper alloys are highly susceptible to erosion-corrosion. Erosion-corrosion includes degradation caused by air release and locally turbulent, high-velocity water.

A chromium-bearing copper–nickel alloy, IN838 (CA722), has been developed, which reportedly has significantly increased resistance to erosion-corrosion in seawater.[91] However, no condensers have yet been tubed with CA722 and thus no operating data are available.

6.4.1.1. Admiralty Brass

Admiralty brass is the most common tube alloy for freshwater service in condensers, and an excellent performance record has been posted.[13] Waterside corrosion accounts for most of the failures recorded in this material, and the mechanisms range from general corrosion to stress corrosion cracking. Admiralty brass is highly susceptible to ammonia attack, and also to ammonia-induced stress corrosion cracking. Since high concentrations of ammonia can exist in the air-removal sections of surface condensers,[92] it is not surprising that tube performance is poor in this region.

6.4.1.2. 90–10 Copper–Nickel (CL and CA706)

The 90–10 Cu–Ni alloys are primarily considered for seawater-cooled plants, but performance has been relatively poor.[13] Problems arise from erosion-corrosion and localized (pitting) corrosion on the waterside.

With respect to the first phenomenon, the data of Efird and Anderson,[90] reproduced in Fig. 6-29, illustrate the deleterious effect of

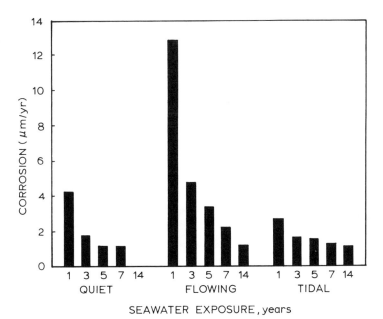

Fig. 6-29. Change in corrosion rate with time for 90–10 Cu–Ni alloy in quiet, flowing, and tidal-zone seawater (adapted from Ref. 90).

flowing seawater on the 90–10 Cu–Ni alloy. The initial weight loss in flowing seawater is much higher than the other two exposures, reflecting a lower rate of protective film formation under these highly aerated and somewhat turbulent conditions. A second point of interest is the time required for the corrosion rate to become linear. Examination of Fig. 6-29 indicates that this occurs after ~4 yrs exposure in quiet seawater and ~8 years exposure in the tidal zone, but it is still decreasing and has not stabilized after 14 years exposure in flowing seawater.

Obviously, the observed diminishing corrosion rate with time is advantageous but, in a real system, it is evident that velocity-related problems and, in particular partial tube blockage, superpose on the normal cooling water velocity to aggravate the erosion-corrosion problem.

Other erosion-corrosion problems that occur with 90–10 Cu–Ni and other copper alloys are related to the release of air bubbles from the cooling water or to intermittent cavitation.[93] The impingement of rapidly moving water, particularly where air bubbles are present or where intermittent cavitation occurs, probably results in breakdown of protective films and subsequent severe localized or generalized attack of the tube surface. This action commonly occurs near tube inlet ends where turbulence is highest, but may extend along the entire length. Severe impingement attack on or near the tube outlet can also occur when there is excessive downstream vacuum.

A primary weakness of 90–10 Cu–Ni and most other copper alloys is their low resistance to sulfide containing waters. Sulfides generally result from pollution by organic wastes. Decomposition of these wastes lowers the dissolved oxygen content of the water and allows anaerobic sulfate-reducing bacteria to convert dissolved sulfate to hydrogen sulfide. Sulfide films form on copper alloys in polluted waters. These films are considerably more noble than the bare metal. Therefore, breaks in the film can cause severe localized corrosion. Nosetani et al.[94] have shown that 0.05 ppm sulfide can markedly increase corrosion on aluminum brass tubes. Similar effects will occur in 90–10 Cu–Ni, although the pattern of attack may be more uniform than the other copper-base alloys such as admiralty brass, aluminum brass, or aluminum bronze.[13] Experience[89] has demonstrated that copper–nickel alloys are superior to other copper alloys in tolerating polluted water. INCO† has shown, in short-term tests, that 2 to 4 ppm H_2S in seawater will increase the corrosion rate of 90–10 Cu–Ni from 76 μm per year (in clean seawater) to 0.8 mm per year.[13] These corrosion rates were apparently based on weight loss; therefore, the pitting rate may have been significantly higher.

The above discussion indicates that very low amounts of dissolved

† International Nickel Company.

hydrogen sulfide (i.e., less than 1 ppm) can cause significant corrosion. It is therefore possible that many of the pitting failures are due to sulfide corrosion. Pitting corrosion was cited 50% of the time as a contributing failure mechanism of 90–10 Cu–Ni in seawater.[13]

The presence of ferrous ions in the cooling water reportedly increases the corrosion resistance of copper-base alloys.[95] The ferrous ions, which result from corrosion of upstream components, e.g., inlet water boxes, or from the deliberate addition of ferrous compounds, e.g., ferrous sulfate, cause the formation of an adherent hydrated ferric oxide (hematite) scale. Tests have shown that tubes which are initially covered with a film of ferric hydroxide are even resistant to polluted seawater (clean seawater injected with 0.1 ppm sulfide ion) for at least four to five months.[13,89] However, the ferric hydroxide deposition is dependent on the initial film already present; e.g., it will not deposit on tubes already covered with a sulfide film.

6.4.1.3. 70–30 Copper–Nickel (CN and CA715)

70–30 Cu–Ni alloys are used primarily for the air-removal section of seawater-cooled condensers. However, the aforementioned survey[13] indicates that this alloy is not well suited for long service in seawater. Using the 10% tubes plugged criterion, 70–30 Cu–Ni is shown to have less than 70% probability of providing a 4-yr service life. This is, nevertheless, a better performance rating than the 40% probability for 90–10 Cu–Ni under the same conditions.

The 70–30 Cu–Ni alloys suffer the same problems as the 90–10 Cu–Ni alloys. Erosion-corrosion is cited as a contributing cause of many tube failures. The data of Efird and Anderson,[90] comparing the 14-year corrosion behavior of the two alloy classes in seawater (Fig. 6-30), indicate very little difference either between weight loss or corrosion rate. In light of the actual performance experience cited earlier, one must therefore be cautious in applying coupon exposure data to the prediction of condenser tube lifetime.

70–30 Cu–Ni is highly susceptible to corrosion in water containing sulfide ions; a tubeset in one plant that became contaminated with an unknown concentration of hydrogen sulfide lasted less than one year.[13] Also, as for the 90–10 Cu–Ni alloys, ferrous sulfate additions to the water increase 70–30 Cu–Ni tube performance.

6.4.1.4. Aluminum Brass and Bronze

Excessive impingement-attack failures of admiralty brass tubes in turbine condensers used in the British navy prompted the research programs that led to the development of aluminum brass, as well as to the

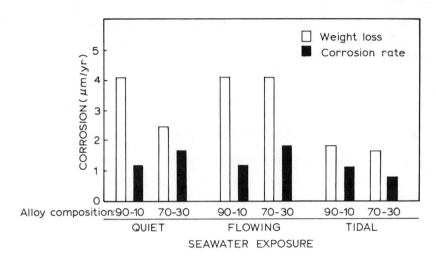

Fig. 6-30. Fourteen years' corrosion rate data for 90–10 and 70–30 Cu–Ni alloys in seawater (adapted from Ref. 90).

copper–nickel alloys. The goal was to provide greater reliability in turbine condenser applications on ships, seashore power plants, and other heat exchangers using seawater for cooling. However, experience with the fossil and nuclear power plants[13] indicates that this alloy is unsatisfactory for service that requires high reliability and a long service life. Although aluminum bronzes are commonly used in seawater, very few applications involve condenser tubes, and experience with these has been poor.

The aluminum brass and bronze alloys suffer the same problems as the Cu–Ni alloys, namely erosion-corrosion and sulfide attack. INCO test data indicate that aluminum brass has a susceptibility similar to that of the Cu–Ni alloys to polluted seawater. Ferrous sulfate additions should improve corrosion resistance, but no data are yet available to support this.[13]

6.4.1.5. Chromium-Modified Copper–Nickel Alloys (IN-838 or CA-722)

Two chromium-modified Cu–Ni alloys have been developed which provide high levels of impingement corrosion resistance in flowing seawater systems.[91] Although no condenser experience is available yet, it is nevertheless of interest to review the evolution of these alloys as an example of the type of alloy optimization work being conducted for condenser materials.

During an extensive systematic survey[91] of the effects of alloying additions on seawater impingement-corrosion resistance of copper–nickel alloys, the addition of 0.3% Cr to standard 70–30 Cu–Ni was found to be particularly beneficial. This observation prompted a detailed study of the effects of Cr additions of up to 1% for a range of nickel levels. A jet test apparatus was used to assess the effectiveness of alloying additions to reduce impingement attack. This test device permits the exposure of test samples in quiescent seawater while at the same time each specimen is subjected to a 2-mm-diameter jet of aerated seawater impinging against a small area of the specimen surface, creating a locally turbulent condition. These test conditions assure a highly turbulent and reproducible test condition which has been found useful for evaluating alloys with respect to impingement corrosion resistance associated with flowing seawater systems.

Data for a number of laboratory-produced Cr-modified copper–nickel alloys, ranging from 5 to 38% Ni, along with commercially produced Cu–Ni alloys, are shown in Fig. 6-31. The comparatively high impingement corrosion rates realized with the commercial Cu–Ni alloys provide an indication of the stringent test conditions involved and reflect a temperature sensitivity that was not as apparent with the chromium-containing alloys. It is of interest to note that under these conditions of severe turbulence, 90–10 Cu–Ni generally provides a slightly higher level of impingement corrosion resistance than 70–30 Cu–Ni.

The jet tests indicated optimum impingement corrosion resistance for a series of Cr-modified alloys containing 15–38% Ni (Fig. 6-31a).[91] Studies were than conducted to determine the optimum Cr levels, concentrating on the upper and lower limits of the optimum nickel content range. The results of this study are shown in Fig. 6-31b. A lower limit of 0.3% Cr is indicated in both instances. Although impingement corrosion resistance is maintained for up to 1% Cr, the higher Cr levels tend to reduce corrosion resistance under less turbulent exposure conditions.

As a result of these studies, two basic alloys containing 16 and 30% Ni were selected for more extensive seawater corrosion studies. These alloys are identified as IN-838 (16% Ni, 0.4% Cr) and IN-848 (30% Ni, 0.4% Cr). The C, Si, and Ti levels are limited to $<.05$ wt.% because their presence in excess will adversely affect corrosion resistance.

Figure 6-32[91] provides a comparison of results of the various seawater corrosion tests, which indicate the relative velocity effects for the alloys of interest with a curve for unalloyed copper included for reference. The stress corrosion cracking resistance of these alloys in *clean* seawater was also high; exposure of U-bend specimens for over two years revealed no signs of cracking.

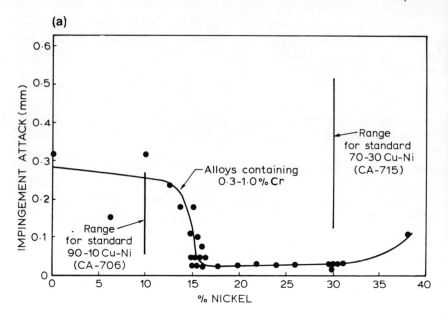
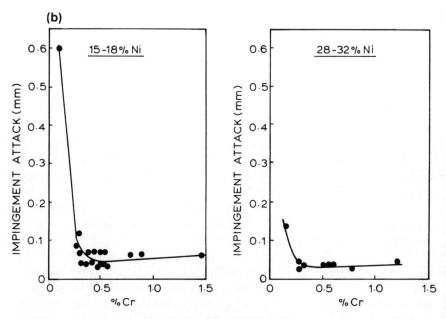

Fig. 6-31. Effect of (a) nickel and (b) chromium content on seawater impingement corrosion resistance of Cr-modified Cu–Ni alloys (adapted from Ref. 91).

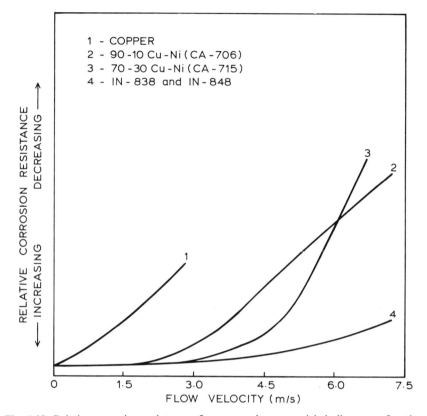

Fig. 6-32. Relative corrosion resistance of copper and copper–nickel alloys as a function of seawater velocity (Ref. 91).

In addition, this same study[91] evaluated the effect of seawater exposure on welded specimens. After sixth month exposures in "quiet" seawater no adverse corrosion effects were seen on specimens with autogenous, TIG, and manual-covered electrode welds using Monel filler metal (70–30 Cu–Ni) and an experimental 70–30 Cu–Ni filler metal containing 0.6% Cr.

IN-838 (or CA-722 as it is now generally known in the U.S.) is now undergoing extensive field testing in power plant condensers, with encouraging results to date.[91]

6.4.2. Iron–Chromium Base Alloys

Since the early 1960s, increasing interest has been shown in stainless steel tubing for large-surface condensers. In the aforementioned survey,[13] 29% of the total tubesets were stainless steel as indicated and 20%

of the tubes and 20% of the total condensing surface area in the surveyed plants were fabricated from either Type 304 or Type 316 stainless steel. Type 304 accounts for 99.6% of the stainless steel condenser tubing surface area.

Stainless steels show much greater resistance to erosion-corrosion than copper-base alloys. There is no recommended maximum water velocity design limit for stainless steel tubing in condenser service. Erosion by cooling waters has not been a problem. In fact, erosive waters such as those containing large amounts of suspended solids actually improve the performance of stainless tubing by maintaining a cleaner surface. In many cases, stainless steel tubing has been used for the steam impingement areas in order to take advantage of its erosion resistance where a distinct impingement zone could be determined.

Stainless steels are also immune to ammonia attack and to ammonia-induced stress corrosion cracking. For this reason, they are also used in air-removal sections where the ammonia concentration may be high.

It is well known that chloride ions increase the susceptibility of stainless steel to pitting and crevice corrosion, but interestingly, chloride-induced stress corrosion cracking has not been observed in stainless steel tubing used in power plant condensers.[13] This is probably due to the fact that the cooling–water temperature in most condensers is below the so-called critical temperature for stress-corrosion cracking, i.e., 320 to 330 K in practical applications. Type 304 stainless steel appears to be resistant to cooling-water chloride concentrations of ~100 ppm when condenser operating temperatures are ~311 to 317 K. Stress corrosion cracking perhaps could occur when cooling towers are used which result in higher temperatures (to 320 K). However, it is more likely that stainless steel will fail first by other mechanisms, e.g., pitting corrosion.

One weakness of Type 304 and 316 stainless steels is their susceptibility to pitting corrosion.[96] This susceptibility is increased when the cooling water has a high chloride concentration, e.g., ocean water or some cooling tower waters. Pitting attack is aggravated by stagnant or low flow conditions in which solids can settle on the tube surface. Small concentrations of manganese have been identified as the cause of pitting corrosion in at least one case.[13]

The addition of molybdenum to stainless steels increases their resistance to pitting corrosion. Type 316 stainless steel has a molybdenum content of 2 to 3%. This amount of molybdenum apparently is not required in most condensers cooled by freshwater with a low chloride concentration, but it is too low for ocean and brackish cooling waters which have a high chloride concentration. Type 316 does not appear to be well suited for seawater or brackish water service.[13] In the survey,[13] pitting corrosion accounted for 59% of the failures and erosion-corrosion for 30%. The "super austenitic," Allegheny Ludlum alloy 6X (25N–20Cr–6.25Mo) shows promise in overcoming the pitting problem in high-chloride waters. Test tubes have given up to six years of service in seawater-cooled condensers without a failure. Recently, two seawater-cooled condensers have been completely retubed with AL6X.[13] The Allegheny Ludlum alloy 29(Cr)–4(Mo)–C Alloy is also being touted as a possible improved tubing alloy. This ferritic alloy is cheaper than AL6X and exhibits somewhat better corrosion resistance.

Overall, iron–chromium base alloy experience in condenser tube application is good. Only 2% of the failure mechanisms cited in the survey were related to waterside corrosion; 89% of these failures were related to steam-side mechanical failures and 9% to faulty tube rolling. Clearly, condenser design and fabrication procedures require attention in order to gain benefit from the improved material performance.

6.4.3. Titanium-Base Alloys

Titanium as a tubing material for domestic steam power plant surface condensers first was investigated in 1959 when seven test tubes were installed. The first condenser with all-titanium tubes was put into service in May 1972.[97]

Commercially pure titanium is unique in that it is the only condenser tube material that is used in an unalloyed form. Seamless titanium has been tested by some utilities in the U.S., but it is not cost competitive with seam-welded tubing. The Japanese use seam-welded tubing in their condenser applications and, in the U.K., seamless titanium tubing has been used in power station condensers.[13] Normally the tubing is installed by retubing an existing condenser. Installation in new condensers designed for titanium tubes has occurred only recently.

Tubing is typically manufactured to ASTM specification B-338, Grade 2, and is 22.2 mm or 25.4 mm O.D., and seam welded. Tube-to-tube-sheet joints are rolled, often with grooved tube-sheet holes. Muntz metal (or aluminum bronze) tubesheet material (most tubesheets were existing in condensers retubed with titanium) is normal.

The tubing is exposed to circulating water containing a high concentration of dissolved solids, often low in dissolved oxygen and with high pollutant concentrations. Under these general conditions the performance of titanium has been very good. The field survey data[13] show that titanium condenser tubes have a high reliability. Titanium tubes have an 87% probability of operating for 40 years with not more than 1% plugged tubes. The median failure rate value of the tubesets surveyed was zero. However, these probabilities and statistics should be used with caution since the experience base is limited; for example, the age of the oldest titanium installation surveyed was $5\frac{1}{2}$ yrs.

A number of technical factors relate to the use and performance of titanium tubing in surface condensers. These include vibration, tube-to-tubesheet joints, heat transfer/fouling, hydriding/galvanic corrosion, and other corrosion and erosion factors. Approximately one-half of the relatively few failures of titanium tubing are directly attributable to vibration damage (compare the OTSG situation in Section 6.2). The damage itself is manifested by (1) fatigue failure of the tube where it passes through a support plate or (2) flattened regions on the outside surface of the tube along its length due to collision with adjacent tubes, which can cause eventual rupture.

This failure mechanism is by no means peculiar to titanium. Instances of vibration damage have been observed in admiralty brass, aluminum brass, and stainless steel tubing.[13] Usually, tube vibration is caused by high shell-side cross-flow steam velocities at the top or upper sides of the tubing bundles. The major excitation force causing condenser tube vibration results from the cross-flow steam velocity. To avoid impact with adjacent tubes at their midspan, the tube vibration amplitude must be restricted. For a given steam flow condition, amplitude reduction can be accomplished by: (1) decreasing tube span; (2) increasing the modulus of elasticity; or (3) increasing the moment of inertia. The natural frequency will simultaneously increase with any of these changes as well, thus indicating an increase in tube stiffness.

Comparing titanium to the copper-base and stainless steel tubing alloys under identical conditions shows that it is more susceptible to this type of damage because: (1) titanium has a relatively low modulus of elasticity; and (2) thinner-walled titanium tubing has a lower moment of inertia. Both of these result in a tube with higher vibration amplitudes which can be reduced by decreasing the tube span.[98]

In addition to the failure of the tube itself, the other potential major source of circulating water inleakage into the condenser is the tube-to-tube-sheet joint. In power industry condensers, this joint is typically made by expanding the tube into the tube sheet—forming an interference fit. Some condensers have welded or packed joints. The incidence of detectable leakage in titanium condensers is very low; therefore, either joint integrity is sufficiently good to prevent leakage or the amount of leakage is too small to be revealed by the leak detection and location methods used. Welded joints have seen little use in surface condensers with titanium tubing. Of principal concern when fusion-welding titanium is the maintenance of a complete inert gas atmosphere to prevent severe embrittlement from atmospheric contaminants (see also Chapter 7, Section

7.3.2). Present-day condensers require field erection because of their large size, and this will present a problem in maintaining cleanliness. Modular construction of condensers using shop-welded joints of lighter-weight titanium tubing should provide a solution to this dilemma.[99]

When comparing the performance of titanium to the copper-base alloys in particular, heat transfer and fouling characteristics are frequently mentioned.[13] The copper-base alloys have a higher thermal conductivity than titanium and a natural, surface poisoning effect on biofouling organisms. Various design features are being used or proposed to enhance the thermal performance of titanium.[99] These include thinner-walled tubing and increased tube water velocity from the normal 2.2 m/sec to 3 m/sec or higher. Nevertheless, the thermal performance of titanium appears to be good. In the U.K., for example, it was found that, although 70–30 copper–nickel tubes had an initially higher overall heat transfer coefficient than titanium, after 14 months of operation the titanium performance was better than that of copper-nickel due to lower fouling; and the Japanese found that after 8 months of operation without tube cleaning, titanium tubing in a small, industrial, steam condenser was still performing at the original 90% design cleanliness.[13]

A fourth factor relating to the use and performance of titanium is the electrical potential differences developed between titanium and other metals that can result in either: (1) the adsorption of atomic hydrogen onto the titanium, which may lead to hydrogen embrittlement[100]; or (2) galvanic wastages of the dissimilar metal materials used for tube-sheets or water boxes.

The Japanese have reported adsorption of hydrogen and hydriding of the titanium below the surface of the tubes after 1 to 2 years of operation in surface condensers.[13,99] The reported hydriding occurred on the inside surface in the first 60 mm of length from each end of the tubes to a maximum depth of about ~0.1 mm. The amount increased toward the ends, and the outlet end experienced a higher hydrogen adsorption than the inlet end. No failures of the titanium were reported, however. A case of subsurface hydriding was observed on the outside of a tube at (or very near) a support plate location in a U.S. condenser.[13] This tube had been removed for inspection following failure by vibration-induced damage, and laboratory analysis revealed hydriding. However, it is not known whether the hydrogen contamination occurred before or after installation in the condenser.

The hydrogen embrittlement phenomenon has been investigated directly in the laboratory by cathodically charging a variety of titanium samples in a synthetic sea-salt solution at ambient temperatures.[100] In addition, samples of test titanium tubing from in-service condensers were removed for examination. This study showed that it is very difficult to charge hydrogen into titanium or even to form a surface hydride layer at ambient temperature in a neutral pH solution. Therefore, Jacobs and McMaster[98] were led to conclude that, since power plant surface condensers operate at a low temperature of about 310 K, it seems highly unlikely that hydrogen embrittlement of the titanium tubes would occur even if cathodic protection did form a hydride film on the surface. The examination of titanium tubes after years of service in polluted seawater under cathodic protection seems to support this conclusion inasmuch as no evidence of hydrides or deterioration of mechanical properties could be detected.[13,99]

Titanium is cathodic to all alloys typically used for condenser tubesheets and water boxes. Its use in a condenser increases the normal uncoupled corrosion rate of any dissimilar metal used in such services. Some likelihood for galvanic corrosion exists for any dissimilar metal couple in condensers, but the likelihood is higher with titanium because reducing reactions with oxygen and hydrogen occur more readily on titanium. Impressed current cathodic protection systems must therefore be well controlled when used with titanium. The use of coatings and more galvanically compatible materials (i.e., aluminum bronze tube sheets) can also provide improved protection of materials coupled with titanium.[13]

Titanium is cathodic to all alloys typically used for condenser tubesheets and water boxes. Its use in a condenser increases the normal uncoupled corrosion rate of any dissimilar metal used in such service. Some likelihood for galvanic corrosion exists for any dissimilar metal couple in condensers, but the likelihood is higher with titanium because reducing reactions with oxygen and hydrogen occur more readily on titanium. Impressed current cathodic protection systems must therefore be well controlled when used with titanium. The use of coatings and more galvanically compatible materials (i.e., aluminum bronze tube-sheets) can also provide improved protection of materials coupled with titanium.[13]

Titanium is also apparently immune to stress corrosion, corrosion fatigue, and fretting corrosion in polluted cooling waters and in high-purity chloride-contaminated or ammoniated steam.[99] Again, the protective surface oxide film prevents initiation of critical cracks. However, under laboratory conditions at least, if the protective film is ruptured and prevented from reforming, or if a fatigue crack is present, stress corrosion cracks can form in aqueous media. Wanhill[101] has reviewed the fracture modes, metallurgical and environmental influences, and possible mechanisms of aqueous SCC in titanium alloys, and finds strong evidence to support the view that the damaging species is diffusible hydrogen. Clearly, therefore, hydrogen does *not* play an insignificant role in the performance of titanium in aqueous media, as would be implied by the data discussed earlier.[98-100] The problem of elucidating the exact mechanism is, however, complex and not open to easy resolution.

Finally, the steam-side corrosion resistance of titanium results in minimal metal ion carryover to downstream steam-generating equipment. However, Sato[99] has indicated that some difficulties may exist with hydrogen attack from the steam side of BWR units and a probability for crevice corrosion under heavy water-side deposits. The second point is also brought out by Jacobs and McMaster,[98] that if a solid chloride salt deposit forms on the titanium surface, the concentrating and insulating effect of the deposit may increase the surface temperature ("hot wall effect") and salt concentration sufficiently to cause pitting. This is clearly an area where more studies are warranted.

6.4.4. Closing Remarks

Table 6-3 is an attempt to summarize the various advantages and disadvantages of the condenser tube materials, which have been raised in surveys[13,89] and meetings[99] on this topic. Of the copper-base alloys, the new chromium-modified alloys appear to offer the best prospects for

Table 6-3. Summary of Advantages and Disadvantages of Condenser Tube Materials

Condenser tube materials	Advantages	Problems
Copper-base alloys	Noble material Prevent biofouling	General attack: $Cu + NH_3$ SCC: (1) $NH_3 + O_2$ (2) S species Erosion of softer alloys Attack under deposits Galvanic sensitivities Narrow acid—base oxide stability
Iron–chromium-base alloys	Broad stability in acid–base Intermediate nobility	Pitting Attack under deposits Crevice attack SCC: only in very acid chloride at room temperature
Titanium–base alloys	Broad range of stability of protective oxide films	Susceptible to vibration-fatigue Hydriding attack in: reducing conditions, or crevice Inherently active material May not prevent biofouling?

trouble-free operation in the future, but the stainless steel and titanium tubing experience justify their replacement of copper-base alloys where feasible.

Overall, the titanium alloys offer a significant advantage over traditional condenser tube alloys where severely aggressive cooling water conditions are present (e.g., saltwater cooling). The extent of their future use will, however, depend upon economic factors, and these have to be judged upon individual circumstances.

References

1. Ph. Berge and J. R. Donati, An Evaluation of PWR Steam Generator Tubing Alloys, *Nucl. Energ.*, **17**, 291–299 (1978).
2. R. Garnsey, Corrosion of PWR Steam Generators, Central Electricity Generating Board, U.K. Report RD/L/N/4/79, Job No. VF163 (March 1979).
3. C. R. Brinkman and M. Katcher, Materials Technology for LMFBR Steam Generators, *Met. Prog.*, 54–61 (July 1979).
4. J. M. Kendall, LMFBR Steam Cycles—Is Efficiency the Ultimate Goal?, paper presented at 1979 Annual Meeting of American Nuclear Society, Georgia, June 3–7, 1979; summary in *Trans. Am. Nucl. Soc.*, **32**, 564–565 (1979).
5a. Pool Type LMFBR Plant 1000 MW(e) Phase A Extension 2 Designs, Electric Power Research Institute Reports NP-1014-SY (GE); NP-1015-SY (AI) and NP-1016-SY (W).
5b. B. E. Dawson, Preliminary Design: Duplex Tube Low-Pressure Saturated Steam Generator for Large LMFBR Plant, NP1219, EPRI Research Project 620-29 Final October 1979 report.
6. J. C. Whipple and C. N. Spalaris, Design of the Clinch River Breeder Reactor Plant Steam Generators, *Nucl. Technol.*, **28**, 305–314 (1976).
7. M. G. Robin, Careful Attention to Detail Was Necessary in Developing the Super Phenix Steam Generators, *Nucl. Eng. Int.*, 46–48 (May 1977).
8. M. Pierrey, M. Antonakas, M. D'Onghia, M. Peyrelongue, and M. Pini Prato, French Fast Breeder Reactor Main Components: The Intermediate Heat Exchangers, in: European Nuclear Conference, Hamburg, Germany, May 6–11, 1979; Abstract in *Trans. Am. Nucl. Soc.*, **31**, 619–621 (1979).
9. UWMAK-I, A Wisconsin Toroidal Fusion Reactor Design, Fusion Technology Progress Report UWFDM-68 (Vol. II), University of Wisconsin (May 1975).
10. R. G. Hickman, Tritium-Related Materials Problems in Fusion Reactors, in: *Critical Materials Problems in Energy Production*, Charles Stein, ed., Academic Press, New York (1976), pp. 189–220.
11. UWMAK-II, A Conceptual D-T Fueled, Helium-Cooled Tokamak Fusion Power Reactor Design: Fusion Technology Progress Report UWFDM-112, University of Wisconsin (May 1975).
12. SOLASE—A Laser Fusion Reactor Study, Fusion Engineering Program Report, UWFDM-220, University of Wisconsin (December 1977).
13. Steam Plant Surface Condenser Leakage Study, EPRI NP-481, Project 624-1, Vol. 1, final report, March 1977, prepared by Bechtel Corp.
14a. O. S. Tatone and R. S. Pathania, Steam Generator Tube Performance: Experience with Water-Cooled Nuclear Power Reactors during 1978, Atomic Energy of Canada Report AECl-6852 (February 1980).

14b. D. G. Eisenhut, B. D. Liau, J. Strosnider, Summary of Operating Experience with Recirculating Steam Generators, USNRC Report NUREG-0523 (January 1979).
14c. B. D. Liau and J. Strosnider, Summary of Tube Integrity Operating Experience with Once-through Steam Generators, USNRC Report NUREG-0571 (March 1980).
14d. R. H. Koppe and E. A. Olson, Nuclear and Large Fossil Unit Operation Experience, Report for Electric Power Research Institute, EPRI 1191 (1979).
15. W. H. Layman, L. J. Martel, S. J. Green, G. Hetsroni, C. Shoemaker, and J. A. Mundis, Status of Steam Generators, paper presented to American Power Conference, Chicago, Illinois (April 1979).
16. D. Van Rooyen and J. R. Weeks, Denting of Inconel Steam Generator Tubes in Pressurized Water Reactors, final report, BNL-NUREG-50778 (January 1978) Brookhaven National Laboratory, New York.
17. K. E. Stahlkopf, R. E. Smith and T. U. Marston, Nuclear Pressure Boundary Materials Problems and Proposed Solutions, *Nucl. Eng. Des.* **46**, 65–79 (1978).
18. E. C. Potter and G. M. W. Mann, The Fast Linear Growth of Magnetite on Mild Steel in High-Temperature Aqueous Conditions, *Br. Corros. J.*, **1**, 26–35 (1965).
19. Stress Analysis of Pressurized Water Reactor Steam Generator Tube Denting Phenomena, EPRI NP-828, Project 700 interim report (July 1978), Failure Analysis Associates.
20. G. S. Was, A Review of the Mechanical, Metallurgical, and Environmental Factors Affecting Cracking of Inconel 600, interim report, EPRI Project RP 1166-3 (October 1978).
21. T. S. Bulischeck, Y. S. Park, and D. Van Rooyen, Stress Corrosion Cracking of Inconel 600 Tubing in Deaerated High-Temperature Water, BNL-NUREG-51027 (June 1979), Brookhaven National Laboratory, New York.
22. I. L. W. Wilson, F. W. Pement, R. G. Aspden, and R. T. Begley, Caustic Stress Corrosion Behavior of Fe–Ni–Cr Nuclear Steam Generator Tubing Alloys, *Nucl. Technol.*, **31**, 70–84 (1976).
23. J. Blanchet and H. Coriou, Review of the Corrosion Resistance Properties of Alloy 800 in High-Temperature Steam, in: *Proceedings of the Petten International Conference on Alloy 800*, Petten, Netherlands, March 14–16, 1978, W. Betteridge, R. Krefeld, H. Krockel, S. J. Llyod, M. Van de Voorde, and C. Vivante, eds., pp. 241–262.
24. R. S. Pathania, Caustic Cracking of Steam Generator Tube Materials, *Corrosion*, **34**, 149–156 (1978).
25. Ph. Berge, J. R. Donati, B. Prieux, and D. Villard, Caustic Stress Corrosion of Fe–Cr–Ni Austenitic Alloys, *Corrosion*, **33**, 425–435 (1977).
26. S. J. Green, Pressurized Water Steam Generators, paper presented at Nuclear Heat Exchanger Session, American Society of Mechanical Engineers, Nuclear Division Conference, August 18–21, 1980, San Francisco, California.
27. C. M. Chen and G. J. Theus, Chemistry of Corrosion-Producing Salts in Light Water Reactors, EPRI Research Project 967-1, final report, October 1978.
28. Ph. Berge, H. D. Bui, J. R. Donati, and D. Villard, Residual Stresses in Bent Tubes for Nuclear Steam Generators, *Corrosion*, **32**, 357–364 (1976).
29. G. P. Airey, Optimization of Metallurgical Variables to Improve the Stress Corrosion Cracking Resistance of Inconel 600, EPRI NP final report, Project 621-1 (1980), Westinghouse Corp.
30. G. P. Airey, The Effect of Carbon Content and Thermal Treatment on the SCC Behavior of Inconel Alloy 600 Steam Generator Tubing, *Corrosion*, **35**, 129 (1979).
31. *Alloy 800: Proceedings of the Petten International Conference*, March 1978, W. Betteridge, R. Krefeld, H. Krockel, S. J. Lloyd, M. Van de Voorde, and C. Vivante, eds. North Holland Publishing Company (1978).

32. S. F. Pugh, A. Status Review of Alloy 800 in Nuclear Application, *J. Br. Nucl. Energ. Soc.*, **14**, 221–226 (1975).
33. A. J. Sedriks, J. W. Schultz, and M. A. Cordovi, Inconel Alloy 690—A New Corrosion-Resistant Material, paper presented at Meeting of Stress Corrosion Committee, Japan Society of Corrosion Engineers (April 28, 1978), *Boshoku Gijutsu*, **28**, 82–95 (1979).
34. F. W. Pement, I. L. Wilson, and R. G. Aspden, Stress Corrosion Cracking Studies of High-Nickel Austenitic Alloys in Several High-Temperature Aqueous Solutions, Paper No. 50, Corrosion 79, March 12–16, 1979, Atlanta, Georgia.
35. M. Kowaka, H. Fujikawa, and T. Kobayashi, Development of New Alloy SCR-3 Resistant to Stress-Corrosion Cracking in High-Temperature, High-Pressure Water, paper presented at Golden Gate Metals and Welding Conference: International Advances in Materials, San Francisco, January 31–February 2, 1979.
36. J. A. Shields, Jr., and K. L. Longua, The Effect of Ten-Years Experimental Breeder Reactor II Service on $2\frac{1}{4}$Cr-1Mo Steel, *Nucl. Technol.*, **28**, 471–481 (1976).
37. M. G. Robin, French Steam Generator Experience—Phenix and Beyond, *Nucl. Technol.*, **28**, 482–489 (1976).
38. D. Taylor, Operation of Prototype Fast Reactor Steam Generators Led Directly to Commercial-Size Design, *Nucl. Eng. Int.*, 49–53 (May 1977).
39. A. D. Evans, A. M. Broomfield, J. A. Smedley, and J. A. Bray, Operating Experience with the PFR Evaporators, in: *Ferritic Steels for Fast Reactor Steam Generators, Proceedings of the International Conference*, S. F. Pugh and E. A. Little, eds., Vol. 1, pp. 3–15, British Nuclear Energy Society, London (1978).
40. P. Patriarca, S. D. Harkness, J. M. Duke, and L. R. Cooper, U.S. Advanced Materials Development Program, *Nucl. Technol.*, **28**, 516–536 (1976).
41. C. Willby and J. Walters, Material Choices for the Commercial Fast Reactor Steam Generators, in: *Ferritic Steels for Fast Reactor Steam Generators, Proceedings of the International Conference*, S. F. Pugh and E. A. Little eds., Vol. 1, pp. 40–49, British Nuclear Energy Society, London (1978).
42. G. C. Bodine, Jr., B. Chakravarti, S. D. Harkness, C. M. Owens, B. Roberts, D. Vandergriff, and C. T. Ward, The Development of a 9Cr Steel with Improved Strength and Toughness, in: *Ferritic Steels for Fast Reactor Steam Generators, Proceedings of the International Conference*, S. F. Pugh and E. A. Little, eds., Vol. 1, pp. 160–163, British Nuclear Energy Society, London (1978).
43. M. G. Robin and J. Birault, Design Philosophy and Functional Requirements of a Sodium-Heated Steam Generator Made of Ferritic Steel, in: *Ferritic Steels for Fast Reactor Steam Generators, Proceedings of International Conference*, S. F. Pugh and E. A. Little, eds., Vol. 1, pp. 50–54, British Nuclear Energy Society, London (1978).
44. J. M. Duke, C. E. Sessions, and W. E. Roy, Qualification of Alloy 800 for Sodium-Heated Steam Generators, in: Proceedings of the International Conference on Liquid Metal Technology in Energy Production, Seven Springs, Pennsylvania, T1D-CONF-760503-P1P2.
45. *Ferritic Steels for Fast Reactor Steam Generators, Proceedings of the International Conference*, Vols. 1 and 2, (S. F. Pugh and E. A. Little, eds., British Nuclear Energy Society, London (1978).
46. O. K. Chopra, J. Y. N. Wang, and K. Natesan, Review of Sodium Effects on Candidate Materials for Central Receiver Solar-Thermal Power Systems, Argonne National Laboratory Report, ANL-79-36 (1979).
47. Status of Incoloy Alloy 800 Development for Breeder Reactor Steam Generators, compiled by J. R. DiStefano, ORNL/Sub-4308/3 (December 1978).
48. J. L. Krankota and J. S. Armijo, The Kinetics of Decarburization of $2\frac{1}{4}$Cr–1% Mo Steel in Sodium, *Nucl. Technol.*, **24**, 225–233 (1974).

49. R. L. Klueh, Thermal Aging Effects on the Mechanical Properties of Annealed 2¼Cr–1Mo Steel, ORNL-5324 (1977), Oak Ridge National Laboratory.
50. O. K. Chopra and K. Natesan, Mechanical Properties Test Data for Structural Materials; quarterly progress report for period ending April 30, 1978, ORNL-5416, p. 78.
51. A. G. Crouch and P. R. Bussey, Corrosion of Ferritic Steels in Flowing Sodium, in: *Ferritic Steels for Fast Reactor Steam Generators, Proceedings of the International Conference*, S. F. Pugh and E. A. Little, eds., pp. 258–263, British Nuclear Energy Society, London (1978).
52. J. Orr, F. R. Beckitt, and G. D. Fawkes, The Physical Metallurgy of Chromium–Molybdenum Steels for Fast Reactor Boilers, in: *Ferritic Steels for Fast Reactor Steam Generators, Proceedings of the International Conference*, S. F. Pugh and E. A. Little, eds., pp. 91–109, British Nuclear Energy Society, London (1978).
53. D. S. Wood, A. B. Baldwin, F. W. Grounds, J. Wynn, E. G. Wilson, and J. Wareing, Mechanical Properties Data on 9% Cr Steel, in: *Ferritic Steels for Fast Reactor Steam Generators, Proceedings of the International Conference*, S. F. Pugh and E. A. Little, eds., pp. 189–192, British Nuclear Energy Society, London (1978).
54. T. Yukitoshi, K. Yoshikawa, K. Tokimasa, T. Kudo, Y. Shida, and Y. Inaba, Development of 9Cr–2Mo Steel for Fast Breeder Reactor Steam Generators, in: *Ferritic Steels for Fast Reactor Steam Generators, Proceedings of the International Conference*, S. F. Pugh and E. A. Little, eds., pp. 87–90, British Nuclear Energy Society, London (1978).
55. G. C. Bodine, B. Chakravarti, C. M. Owens, B. W. Roberts, D. M. Vandergriff, and C. T. Ward, A Program for the Development of Advanced Ferritic Alloys for LMFBR Structural Application, US-ERDA Report ORNL-Sub 4291/1 (1977).
56. D. S. Wood, Effects of a Sodium Environment on the Mechanical Properties of Ferritic Steels, in: *Ferritic Steels for Fast Reactor Steam Generators, Proceedings of the International Conference*, S. F. Pugh and E. A. Little, eds., pp. 293–299, British Nuclear Energy Society, London (1978).
57. W. Charnock, J. E. Cordwell, and P. Marshall, The Influence of High-Temperature Sodium on the Structure and Mechanical Properties of 9Cr Steel, in: *Ferritic Steels for Fast Reactor Steam Generators, Proceedings of the International Conference*, S. F. Pugh and E. A. Little, eds., pp. 310–314, British Nuclear Energy Society, London (1978).
58. Sodium Technology Program—Progress Report, WARD-3045Ti-3, Westinghouse Electric Corporation Advanced Reactors Division, Madison, Pennsylvania (1972).
59. H. U. Borgstedt, G. Frees, and A. Marin, Corrosion and Carburization of Incoloy 800 in Liquid Sodium up to 973 K, in: *Alloy 800; Proceedings of the Petten International Conference*, W. Betteridge, R. Krefeld, H. Krockel, S. J. Lloyd, M. Van de Voorde, and C. Vivante, eds., pp. 291–295, North Holland Publishing Company (1978).
60. R. S. Fidler, The Effect of Liquid Sodium on the Creep and Rupture Properties of Alloy 800, in: *Alloy 800; Proceedings of the Petten International Conference*, W. Betteridge, R. Krefeld, H. Krockel, S. J. Lloyd, M. Van de Voorde, and C. Vivante, eds. pp. 297–302, North Holland Publishing Company (1978).
61. J. H. DeVan and J. C. Griess, Clinch River Breeder Reactor Environmental Effects—General Waterside Corrosion, *Nucl. Technol.* **28**, 398–405 (1976).
62. J. S. Armijo, J. L. Krankota, C. N. Spalaris, K. M. Horst, and F. E. Tippits, Materials Selection and Expected Performance in Near-Term LMFBR Steam Generators, in: *Proceedings of the International Conference on Fast Reactor Power Stations*, pp. 189–203, Nuclear Energy Society, London (1974).
63. J. C. Griess, L. V. Hampton, J. H. DeVan, and K. D. Challenger, Corrosion of 2¼Cr–1Mo Steel under Superheat Heat Transfer Conditions, in: *Ferritic Steels for Fast*

Reactor Steam Generators, Proceedings of the International Conference, Vol. 2, S. F. Pugh and E. A. Little, eds., pp. 367–370, British Nuclear Energy Society, London (1978).
63a. K. B. Challenger, A. K. Miller, and C. R. Brinkman, An Explanation for the Effects of Hold Periods on the Elevated Temperature Fatigue Behavior of 2¼Cr–1Mo Steel, Submitted to the *J. Eng. Mater. Technol.* (1980).
64. P. Cohen, T. Padden, D. Schmidt, and L. E. Efferding, Accelerated Corrosion Testing of a Sodium-Heated 2¼Cr–1Mo Steam Generator Tube, in: *Ferritic Steels for Fast Reactor Steam Generators, Proceedings of the International Conference*, Vol. 2, S. F. Pugh and E. A. Little, eds., pp. 399–402, British Nuclear Energy Society, London (1978).
65. M. E. Indig, Stress Corrosion Studies for 2¼Cr–1Mo Steel, in: *Ferritic Steels for Fast Reactor Steam Generators, Proceedings of the International Conference*, Vol. 2, S. F. Pugh and E. A. Little, eds., pp. 408–412, British Nuclear energy Society, London (1978).
66. P. Hurst and H. C. Cowan, The Oxidation of 9CrMo and Other Steels in 6.9 MPa Steam at 748 and 823 K, in: *Ferritic Steels for Fast Reactor Steam Generators, Proceedings of the International Conference*, S. F. Pugh and E. A. Little, eds., pp. 371–377, British Nuclear Energy Society, London (1978).
67. M. I. Manning and E. Metcalfe, Oxidation of Ferritic Steels in Steam, in: *Ferritic Steels for Fast Reactor Steam Generators, Proceedings of the International Conference*, S. F. Pugh and E. A. Little, eds., pp. 378–382, British Nuclear Energy Society, London (1978).
68. G. J. Bignold, Review of Waterside Performance of 9Cr–1Mo boiler tube Material, in: *Ferritic Steels for Fast Reactor Steam Generators, Proceedings of the International Conference*, S. F. Pugh and E. A. Little, eds. pp. 342–345, British Nuclear Energy Society, London (1978).
69. J. E. Antill, Corrosion of Materials for Sodium-Cooled Fast Reactors, *Nucl. Energ.*, **17**, 313–319 (1978).
70. R. Garnsey, The Generation of Corrosive Conditions in Sodium-Heated Steam Generators, in: *Ferritic Steels for Fast Reactor Steam Generators, Proceedings of the International Conference*, S. F. Pugh and E. A. Little, eds., pp. 383–398, British Nuclear Energy Society, London (1978).
71. S. Leistikow, A Study of the Corrosion of Alloy 800 in High-Temperature Steam, in: *Proceedings of the International Conference on Fast Reactor Power Stations*, British Nuclear Energy Society, London (1974).
72. S. Leistikow and R. Kraft, Creep-Rupture Testing of Incoloy 800, in: *Alloy 800, Proceedings of the Petten International Conference*, W. Betteridge, R. Krefeld, H. Krockel, S. J. Lloyd, M. Van de Voorde, and C. Vivante, pp. 263–270, eds., North Holland Publishing Company (1978).
73. L. Champeix, High-Temperature Corrosion and Mechanical Properties in Sodium Environment, in: *Alloy 800, Proceedings of the Petten International Conference*, W. Betteridge R. Krefeld, H. Krockel, S. J. Lloyd, M. Van de Voorde, and C. Vivante eds., pp. 283–289, North Holland Publishing Company (1978).
74. R. Anderson, Analysis of Experimental Data on Material Wastage by Sodium–Water Reaction Jets, *Nucl. Energ.*, **18**, 333–342 (1979).
75. J. F. B. Payne, Production of Metal Wastage Produced by Sodium–Water Reaction Jets, *Nucl. Energ.*, **18**, 327–331 (1979).
76. H. V. Chamberlain, J. H. Coleman, E. C. Kovacic, and A. A. Shoudy, Wastage of Steam-Generator Materials by Sodium–Water Reactions, in: *Proceedings of the International Conference on Sodium Technology and Large Fast Reactor Design*, ANL-7520, (1968), pp. 384–409.

77. D. A. Greene, Small Leak Damage and Protection Systems in Steam Generators, in: *Proceedings of the International Conference on Liquid Metal Technology in Energy Production*, CONF-760503-P1 (1976), pp. 233–241.
78. K. Tregonning, Calculation of Wastage by Small Water Leaks in Sodium-Heated Steam Generators, in: *Proceedings of the International Conference on Liquid Metal Technology in Energy Production*, CONF-760503-P1 (1976), pp, 218–225.
79. F. A. Kozlov, G. P. Sergeev, A. R. Sednev, and V. M. Makarov, Studies on Some Problems of Leaks in Sodium–Water Steam Generators, in: *Proceedings of the International Conference on Liquid Metal Technology in Energy Production*, CONF-760503-P1 (1976), pp. 202–210.
80. D. W. Sandusky, Small Leak Shutdown, Location and Behavior of LMFBR Steam Generators, in: *Proceedings of the International Conference on Liquid Metal Technology in Energy Production*, CONF-760503-P1 (1976), pp. 226–232.
81. G. J. Lloyd, Mechanical Properties of Austenitic Stainless Steels in Sodium, *At. Energ. Rev.* **16**, 155–208 (1978).
82. K. Natesan, D. L. Smith, T. F. Kassner, and O. K. Chopra, Influence of Sodium Environment on the Tensile Behavior of Austenitic Stainless Steels, in: *ASME Symposium on Structural Material for Service at Elevated Temperatures in Nuclear Power Generation*, MPC-1 (1975), pp. 302–315.
83. O. K. Chopra and K. Natesan, Representation of Elevated-Temperature Tensile Behavior of Type 304 Stainless Steel in a sodium Environment, *J. Eng. Mater. Technol.*, **99**, 366–371 (1977).
84. K. Natesan, O. K. Chopra, and T. F. Kassner, Influence of Sodium Environment on the Uniaxial Tensile Behavior of Titanium-Modified Type 316 Stainless Steel, *J. Nucl. Mater.*, **73**, 137–150 (1978).
85. K. Natesan, O. K. Chopra, and T. F. Kassner, Effect of Sodium on the Creep-Rupture Behavior of Type 304 Stainless Steel, in: *Proceedings of the International Conference on Liquid Metal Technology in Energy Production*, CONF-760503-P1 (1976), pp. 338–345.
86. B. Weiss and R. Stickler, Phase Instabilities During High Temperature Exposure of 316 Austenitic Stainless Steel, *Metall. Trans.*, **3**, 851–866 (1972).
87. G. F. Tisinai, J. K. Stanley, and C. H. Samans, Effect of Nitrogen on Sigma Formation in Cr–Ni Steels at 1200 F (650 C), *Trans. AIME*, **200**, 1259–1267 (1954).
88. A. M. Pritchard, C. F. Knights, G. P. Marsh, K. A. Peakall, R. Perkins, B. L. Myatt and J. E. Antill, Corrosion Behavior of Iron–Chromium Alloys used as Boiler Tube Materials in Fast Reactor Steam Generators, in: *Ferritic Steels for Fast Reactor Steam Generators, Proceedings of the International Conference*, S. F. Pugh and E. A. Little, eds., pp. 360–366, British Nuclear Energy Society, London (1978).
89. H. T. Michels, W. W. Kirk, and A. H. Tuthill, The Role of Corrosion and Fouling in Steam Condenser Performance, *Nucl. Energ.*, **17**, 335–342 (1978).
90. K. D. Efird and D. B. Anderson, Sea Water Corrosion of 90-10 and 70-30 Cu–Ni: 14 Year Exposures, *Mater. Perform.*, **14**, 37–40 (1975).
91. D. B. Anderson and F. A. Badia, Chromium-Modified Copper–Nickel Alloys for Improved Seawater Impingement Resistance, *J. Eng. Power*, Ser. A, **95**, 132–135 (1973).
92. F. W. Fink and W. K. Boyd, Corrosion Problems in Power Plant Steam Condensers, Topical Report to Copper Development Associations, May 1971, Battelle Columbus Laboratories.
93. C. A. Gleason, Condenser Corrosion, in: *The Corrosion Handbook*, H. H. Uhlig, ed., pp. 545–559, John Wiley and Sons, Inc., New York (1948).
94. T. Nosetani, S. Sato, K. Kazama, Y. Yamaguchi, and T. Yasui, Effects of Various Factors on the Performance of Copper Alloy Condenser Tubes, Sumitomo Light Metal Technical Reports, Vol. 12, No. 2, April 1971.

95. T. W. Bostwick, Reducing Corrosion of Power Plant Condenser Tubing (with Ferrous Sulfate), *Corrosion,* **17,** 12–19 (1961).
96. R. J. Brigham and E. W. Tozer, Temperature as a Pitting Criterion, *Corrosion,* **29,** 33–36 (1973).
97. Installation List, TIMET Codeweld Condenser Tube, Titanium Metals Corporation of America (April 1976).
98. R. L. Jacobs and J. A. McMaster, Titanium Tubing: Economical Solution to Heat Exchanger Corrosion, *Mater. Protect. Perform.* **11,** 33–38 (1972).
99. Performance of Condensers in Nuclear and Fossil Power Plants, Vols. 1 and 2, R. Stahle, ed., Ohio State University (June 1975).
100. L. C. Covington, W. M. Parris, and D. M. McCue, The Resistance of Titanium Tubes to Hydrogen Embrittlement in Surface Condensers, in: *Corrosion/76*, NACE, Houston Texas (March 1970), Paper No. 79.
101. R. J. H. Wanhill, Aqueous Stress Corrosion in Titanium Alloys, *Br. Corros. J.,* **10,** 69–78 (1975).

7

Steam Turbine Materials

The modern steam turbines rotate at 1800 or 3600 rpm and produce a "shaft output" of 800–1300 MW. Less than 50 years ago, a machine with the same rpm was rated at less than 200 MW. The underlying reason for the trend to large size is the economy of scale, i.e., the lower capital cost in dollars per kilowatt as size increases.

This large increase in size has, understandably, increased the demands on the materials of construction. For example, if the rotor of one of the modern turbines is of integral, one-piece design, the required forging is very large, approaching 200,000 kg. Also, the increasing stresses and temperatures have called for higher-strength, high-toughness alloys. For the near future generation of 1200 MWe nuclear plants, rotor forgings will be above 400,000 kg, which exceeds the capacity of most steel companies. For instance, only one steel company (Japan Steel) was able to supply generator shafts for the 1200 MWe nuclear plant at Biblis, West Germany.[1]

High reliability is essential for a steam turbine system because, since there is neither backup nor system redundancy, when the turbine suffers a failure event or is shut down for maintenance, the plant is off-line. Data gathered and analyses conducted in recent years[2,3] continue to show that the turbine system is one of three leading causes of nuclear (and fossil) plant outages, both forced and scheduled (see, for example, the nuclear plant data in Table 1-2, Chapter 1).

Lost availability attributable to turbine generators results from a variety of problems. In one category are relatively frequent events of moderate outage duration (i.e., a few hundred hours); the single most prominent contributor, 19%, is turbine blade failures.[2] Of the U.S. units,

Westinghouse turbines appear to be more prone to blade failures than, say, the General Electric turbines. A recent study[2] showed that ~4% of the nearly 5% capacity factor loss attributed to Westinghouse turbine problems was due to blade failures; in contrast, blade problems in General Electric turbines constituted only a small fraction (0.23%) of the 1.2% capacity factor loss attributed to turbines, and generally appear to be minor, being associated with either vibration or balancing.

A second prevalent turbine problem is disc cracking which appears to be on the increase, at least in the U.S., and is therefore cause for some concern. Within the past two years, several U.S. PWRs have suffered outages due to the need to replace discs having cracked rims.[2,3] Disc cracking is also a recognized (and heavily researched) problem in the U.K., following a catastrophic burst failure of a disc in a nuclear plant in 1969.[4]

This brings up the second category, the infrequent, but catastrophic, failures which can remove a unit from service for periods of months to years. Such cases include the bursting of a turbine or generator rotor or shrunk-on disc and the fatigue failure of a shaft.[5] Bush[5] computed a low probability value of 3×10^{-5} to 3×10^{-4} per turbine year for this latter category, from an analysis of 30 fossil turbine failures over the period 1950 to 1977.

Apart from the consequences of actual failures, however, concern over the risk and consequences of failures of rotors and discs also impacts outage time due to the stringent requirements for in-service inspection and recommendations for retirement based upon extremely conservative interpretations of inspection results. In one evaluation[3] of rotor forging integrity, for example, 50% of all turbine rotors that had been inspected were found to contain defects, and 12% of inspected rotors had been recommended for retirement. Replacement and/or refurbishment costs are, clearly, of high importance to the affected utilities inasmuch as a typical low-pressure turbine rotor costs about $6 million.

Confronted with such impressive failure probabilities and their economic impact, the industry (and nuclear in particular) has embarked on a significant program to improve the reliability of steam turbines. The approach is similar to that discussed in earlier chapters, viz. improved understanding of failure mechanisms, improved structural integrity analyses (i.e., through both analysis and property data base improvements), and improved materials. The following sections will present the more recent developments that have evolved, primarily from the detailed "failure analyses" conducted by U.S.[3,6-9] and U.K.[4] groups. These discussions are prefaced, as usual, by a description of the basic (turbine) design and typical materials of construction.

7.1. Design and Materials of Construction

The main design features of nuclear reactor turbines are based upon the vast experience obtained with conventional fossil turbine designs. The addition of moisture separators and steam reheaters, however, is unique to nuclear wet steam turbines.

The simplifed diagram in Fig. 7-1 shows that a typical modern 1800 rpm steam turbine unit might be comprised of three subunits, viz. a high-pressure (HP) turbine and two low-pressure (LP) turbines. The first of the LP turbines is sometimes dubbed the intermediate pressure (IP) turbine, or the IP/LP (i.e., integral) turbine. Also, there can be three LP units on some systems. Steam enters the HP stages of the turbine first, following which it passes through a moisture separator and, if used, a reheater, and enters the LP turbine.

7.1.1. Major Components

From a materials viewpoint the components of interest to turbine reliability are the rotor, the discs, and the blades. As indicated in Table 1-1, Chapter 1, the first two are manufactured from low-alloy (Ni, Cr, Mo, V) bainitic steels, while the latter is typically 12Cr martensitic stainless steel.

7.1.1.1. Rotors

The large rotors (sometimes called spindles) required for modern nuclear plants are machined from forging ingots that can weigh from 200,000 to 400,000 kg. Typical rotor configurations are shown in Fig. 7-2.[8] Not all manufacturers agree to the need (or desirability) of a centerline bore hole (Fig. 7-1 compared with Fig. 7-2a), and one supplier produces the rotor by welding together large discs (Fig. 7-2c); welds are shown on the figure as dark lines at various peripheral locations.

Vacuum-arc casting technology (electric furnace melting and vacuum degassing) is in general use to produce these large rotor forgings, replacing the air melting practice of the 1950's. The problem with this process is that as the ingot size increases, the yield drops (from 80% with 10,000- to 20,000-kg ingots to less than 60% with 300,000 kg ingots) due to gross segregation and inhomogeneous distribution of nonmetallic inclusions upon solidification.[1] The ingot is given a long high-temperature thermal soak to minimize this chemical segregation. Subsequently, the ingot is forged, not only to obtain the desired size and shape, but also to gain the

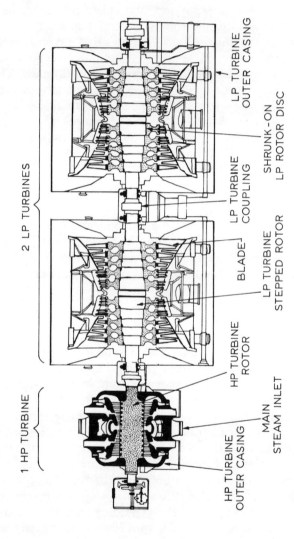

Fig. 7-1. Schematic longitudinal section of modern 1800 rpm four-flow nuclear turbine.

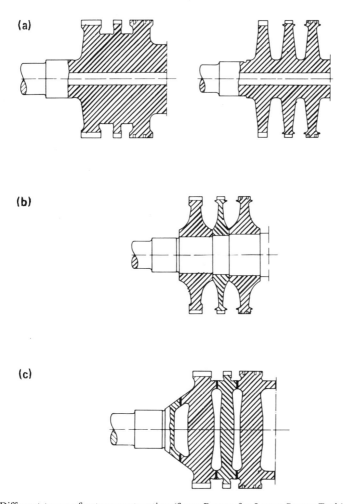

Fig. 7-2. Different types of rotor construction (from *Rotors for Large Steam Turbines*, H. Hohn, Brown, Boveri and Company). (a) Integral forging construction; (b) shrunk-on discs construction; (c) welded disc construction.

advantage of the mechanical work which is beneficial in the breaking up of the cast microstructure and in consolidating the normal ingot porosity. A double normalizing and temper treatment is then applied to large forgings to refine the coarse as-forged microstructure, and finally a further heat treatment is used to attain the desired mechanical properties. For further details on the manufacture of large rotor forgings, the reader is directed to the paper by Smith and Hartman presented at the 1970 International Forgemasters Meeting.[10]

7.1.1.2. (Shrunk-on) Discs

Inasmuch as very large forgings are required to produce the size of LP turbine rotors necessary for today's high-output units, some manufacturers fabricate smaller, stepped-diameter shafts, and a series of discs are thermally expanded to fit these diameters (Fig. 7-3). Upon cooling, the discs and underlying shafts form a series of shrink-fit surfaces. To prevent rotation of the disc on the shaft at operating speeds and temperatures, keyways and locking pins are provided (typically three or four) at this interface. Keyways may be round, square, or of other geometries such as arch-shape. Care must be taken in the shape of these keyways in order to minimize their stress enhancement effect. Also, the choice of a blind or a through-thickness keyway can influence the possibility that undesirable chemical species will be entrapped within this volume (see Section 7.2.1).

Figure 7-1 is a typical example of shrunk-on disc construction in the LP turbines, whereas the smaller-diameter HP turbine is of the integral-rotor type. (Note that HP and some LP rotors are made with solid shafts.) The number of blading stages on a disc varies with the disc's position. The inlet disc of the LP turbine in Fig. 7-3a contains three blade rows, the next disc carries two, and the remaining three discs have one each. The blades are attached at the disc periphery; each manufacturer has his preferred design for the attachment configuration (see, for example, Fig. 7-3b).

7.1.1.3. Blades

The blade (or vane) is the transducer for extracting energy from the steam and producing a force on the turbine shaft to cause rotation (Fig. 7-3). The aerodynamic forces of the high-velocity steam produce a general bending of the blade configuration and excite a series of vibrational modes. Fatigue-associated failures are, therefore, not uncommon, and severe problems are encountered when the blade (or a blade group) is not sufficiently design detuned from the fundamental harmonic resonances encountered during each startup and from the higher-mode running frequencies.

Being in the main steam flow path, the blades are subjected to whatever environment prevails within the turbine. The superheated and saturated steam may carry undesirable chemical contaminants from, for example, condenser tube in-leakage and/or demineralizer resin breakdown, which can precipitate on the blades as corrosive salts. Chemical pitting of the blade surfaces is frequently observed, and the growth of fatigue cracks may also be enhanced by these chemicals. The steam can

Steam Turbine Materials

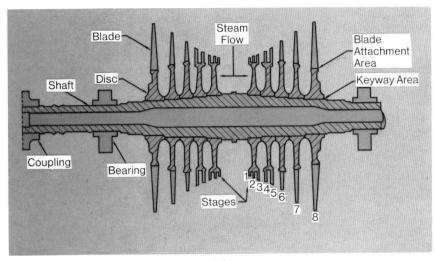

(a)

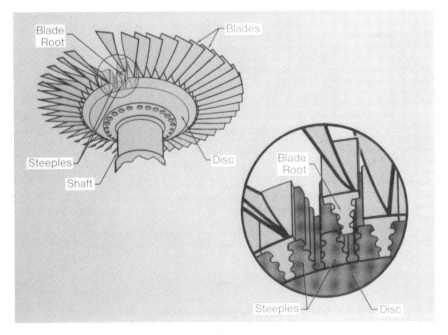

(b)

Fig. 7-3. Typical LP turbine configuration showing (a) discs and (b) blades. (courtesy M. Kolar, EPRI).

also carry particulate matter (e.g., iron oxide, silica, exfoliation products from boiler tubes) which cause solid particle erosion of the vane. Hard-faced surfaces are applied, or special heat treatments are given to the leading edge of the last-stage blades to stem the erosion process, but these approaches themselves have often been the source of blading problems, possibly due to high residual stress.

7.1.2. Material Characteristics

As noted earlier, low-alloy (Ni, Cr, Mo, V) ferritic (bainitic) steels are used for most of the major turbine components. Three basic compositions are used for practically all of the large turbine rotors and disc forgings in the U.S. ASTM specifications A469, A470, and A471 define their chemistry (Appendix C), and some of their mechanical properties are given in Table 7-1.[11] For casing casting and forged valve chests where weldability is essential, a $\frac{1}{2}$Cr–Mo–V alloy containing 0.15% C is used (air-cooled to a predominantly ferritic microstructure). For the blades, a forged 12% Cr stainless steel (AISI Type 403) is used to provide creep strength along with oxidation and corrosion resistance.

Desired properties of the low-alloy steels are obtained by controlling heat treatment parameters and alloy chemical composition.[11] The bainitic microstructure, namely a fine dispersion of carbides combined with a high density of dislocations, imparts high strength and toughness. Molybdenum has the major effect in the development of the bainitic structure; when present with sufficient boron (0.002 to 0.0045%), molybdenum retards ferrite nucleation at austenite grain boundaries. Continuous cooling from the austenitizing temperature either by normalizing (i.e., air cooling) or oil or water spray quenching is used commercially to obtain the bainitic structure. The TTT diagram (Fig. 7-4) indicates that the slower cooling rates of the rotor core can result in some proeutectoid ferrite in addition to bainite. Also, segregation of impurities such as P, As, Sn, and Sb at the prior austenite grain boundaries can result in the phenomenon known

Table 7-1. Mechanical Properties of Rotor Steels (Ref. 11)

Property	A469 (1971) Class 8	A470 (1974) Class 8	A471 (1970) Class 8
Tensile strength, min. (MPa)	829	725–860	1105
Yield strength, min. (MPa)	691	585	1000
Elongation in 50 mm, min. (%)	16	14–17	13
Reduction of area, min. (%)	45	38–43	37
Ductile–brittle transition temperature, max. (K)	277	394	283
Room temperature impact energy, min. (J)	55	8.2	41

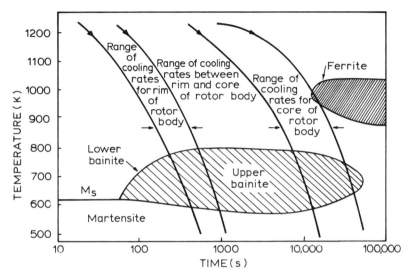

Fig. 7-4. Idealized continuous-cooling transformation diagram for 1% Cr–Mo–V rotor forgings (Ref. 11).

as temper embrittlement (see Section 7.3.1.2). Both ferrite formation and temper embrittlement are to be avoided if possible, because evidence to be described later indicates that fracture toughness is reduced and susceptibility to stress corrosion cracking is increased in those structures.

The strength and toughness of the bainitic structure is controlled mainly by the fineness of the microstructure. Alloying elements such as nickel, manganese, chromium, molybdenum, and vanadium help achieve this. Their amounts must be controlled, however, to avoid increasing the susceptibility to temper embrittlement; chromium, nickel, and manganese additions enhance susceptibility, for example, while additions of molybdenum up to ~0.7 wt.% inhibit embrittlement.

The subsequent temper at about 980 K serves both to eliminate any temper-embrittled regions (i.e., "deembrittlement") and to develop the best combination of creep rupture strengths. The creep-rupture strengths are determined by the size and distribution of precipitated carbides and their stabilities, and among these, the equilibrium carbide, V_4C_3, plays an important role. In order to precipitate this carbide in the most suitable size and distribution in the bainitic structure, the contents of the carbide-forming elements C, Cr, Mo, and V have to be well balanced.

Changing from air-melting to the current vacuum degassing process has reduced the size and number of MnS inclusions and produced a microstructural change from polygonal bainitic packets to smaller acicular bainitic packets. These changes improve the tensile ductility, elevate the upper shelf energy in standard Charpy tests, and greatly improve the fatigue endurance limit of smooth rotating beam specimens. In each case, the intermetallic inclusions play a major role. On the other hand, variation in inclusion content and microstructure of 1Cr–Mo–V rotor steels are shown to have little effect on both the yield and ultimate tensile strengths over the temperature range of 77 to 811 K. This insensitivity results because, in lath type microstructures, the flow properties are controlled by the dislocation substructure, not the bainite packet size or inclusions. Similarly, the fatigue crack growth rates of 1Cr–Mo–V rotor steels are insensitive to inclusion content, microstructure, and test temperature. Also, prior and intermittent strain aging does not affect the fatigue crack growth rates.

A recent experimental and analytical program, utilizing tensile, Charpy, and compact tension fracture testing specimens,[11] showed that in this class of steels, the finer bainitic microstructure resulting from the current vacuum degassed heat process improves fracture toughness at temperatures below the nil-ductility transition temperature and reduces the transition temperature (Fig. 7-5). These improvements result from a higher microscopic cleavage fracture strength, associated with the smaller bainitic packet size, which controls the effective microcrack size. A series of experiments with notched specimens showed that the fracture toughness is not strongly dependent on crack sharpness and that the critical stage in fracture is cleavage crack propagation through the boundaries between bainite packets.

7.2. Turbine Damage Mechanisms

This section will be focused on a consideration of the potentially catastrophic nature of turbine failures, in other words, damage to turbine discs and rotors. Blade failures from corrosion fatigue, high-cycle fatigue, or SCC appear to be limited to a particular vendor design,[2] and are therefore more amenable to solution through design and/or material changes (see, for example, Section 7.3.2). On the other hand, the disc and rotor failure mechanisms are apparently generic, and a complete understanding of the complex interplay between material, design, and operation is not yet available.

Two turbine failure events, more than any others, have strongly influenced the directions taken to improve turbine reliability. In 1969 a LP steam turbine at the Hinkley Point A nuclear plant in the U.K. failed catastrophically due to stress-corrosion cracking at the keyway of one of the shrunk-on discs.[4] The second event of significance is the bursting

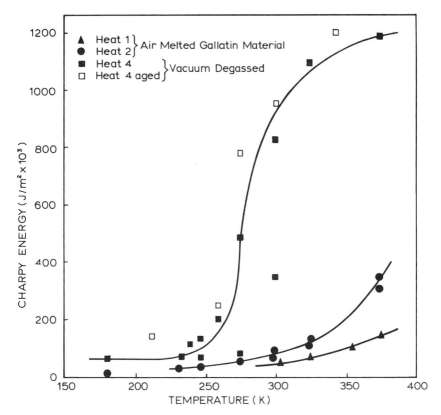

Fig. 7-5. Effect of test temperature on the normalized impact energy for precracked Charpy specimens of the air-melted and vacuum-degassed Cr–Mo–V steels (Ref. 11).

fracture in June 1974 of the TVA† Gallatin (fossil) plant, Unit No. 2, IP rotor, reportedly from subsurface inclusions.[3,6] Possible crack growth mechanisms include low-cycle and thermal-fatigue cracking, creep crack growth, and their interactions.

The evidence from the U.K. experience has revealed that SCC at disc keyways in nonreheat turbines can be a major hazard, and as a result there is an increasing worldwide awareness of the possibility of environmentally induced failures of Ni–Cr–Mo–V steels in steam turbines. More recent U.S. experience points to potentially serious SCC problems in LP turbine discs.[2,3,8,12,13] Of particular note are the disc burst incidents at the Shippingport (1974) and Yankee Rowe (1980) PWRs, and in the Oak

† Tennessee Valley Authority.

Creek fossil plant (1977), and incidences of disc cracks found in the Rancho Seco (1977), Arkansas Nuclear One-Unit 1 (1977), and H.B. Robinson (1978) PWRs. No turbine missiles developed as a result of the PWR incidents, but missiles external to the turbine were produced in the fossil plant. Interestingly, stress-corrosion cracking has not yet been reported in turbine discs in BWR units, nor in any LWRs operating in Germany or Japan.[12,13]

The Gallatin rotor incident prompted one turbine vendor to recommend that rotor forgings manufactured prior to the mid-1950's be ultrasonically examined for flaws that might result in similar catastrophic failures. Needless to say, other flawed rotors were found, some of which were retired, others being the subject of detailed fracture mechanics analyses, to justify continued operation.

The failure analysis work and associated materials characterization and fracture mechanics support that resulted from some of these incidents will be described in the following two subsections. The final subsection will comment on a recurring problem with turbine blades, that of moisture erosion.

7.2.1. Stress-Corrosion Cracking

Cracks in turbine discs[4,12,13] have been attributed to stress-corrosion cracking. The detailed U.K. study[4] concluded that intergranular stress corrosion cracking of the turbine steels could occur even in "pure" wet steam; in other words, an aggressive, concentrated environment, such as caustic, is not necessary for crack initiation, although its presence would accelerate cracking. A similar view is shared by those who conducted the failure analysis of the disc cracking in the H. B. Robinson LP turbine,[13] because no salts or contaminents of any type were found, despite a fracture appearance that was typical of intergranular SCC. On the other hand, in the case of two other U.S. PWR disc cracking incidents, it was concluded that caustic stress-corrosion cracking did occur.[12,13] For the Oak Creek Unit 3 fossil turbine disc failure it was concluded that the "mainly transgranular" fracture was the result of SCC, in this instance caused by a mixture of caustic and sulfide, and with propagation probably occurring by a hydrogen-induced mechanism.[13] Other earlier incidences of cracking in rotor-related equipment have also been shown conclusively to have been caused by caustic SCC.[14] It is therefore clear that this class of steel, like the austenitic stainless and nickel-base alloys, is susceptible to stress-corrosion cracking in steam/water environments. The details that follow, primarily from the papers by Hodge and Mogford[4] and Lyle et al.,[13] describe progress in the general understanding of this phenomenon in the turbine environment.

The U.K. program[4] involved a comprehensive examination of actual cracked discs, characterization of the fracture patterns and the expected service environment, and supporting laboratory tests. It was established that disc cracking occurred only in a wet steam environment (i.e., at turbine stages where saturated steam conditions are achieved). However, a unique cause of the cracking was not established. Exposure of a highly stressed crevice is essential and crack initiation appeared to be promoted by local small keyway surface irregularities such as poor surface finish and the presence of nonmetallic inclusions. The severity of cracking was also highest at the highest temperature regions where wet steam was encountered, and a relationship was established between average crack growth rate and steam temperature (Fig. 7-6). However, no correlation with chemistry was obtained. In general the levels of soluble ions like Na^+, Cl^-, and SO_4^{-2} were low, which probably reflects leachout during shutdown. Other deposits of significance included silicon, sulfur, and molybdenum. It therefore appears that the only common ingredient was a wet environment.

From detailed metallographic examinations of cracked discs it was concluded[4] that metallurgical structure appeared to have little effect on the actual cracking, i.e., large-grained lower bainite cracked in a similar way to small-grained fine bainite. Although cracking was predominantly intergranular, immediately adjacent to the keyway it was usually trans-

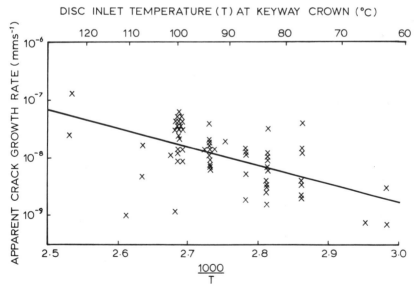

Fig. 7-6. Apparent crack growth rate in Ni–Cr–Mo–V alloys from maximum crack depth and rotor life, plotted against disc operating temperature (Ref. 4).

granular and changed to intergranular within the space of one or two grains. When cracks were traced back to their origin, the initiation site was found to be associated with either nonmetallic inclusions, corrosion pits, machining defects, or heavily worked surface layers.

Stress-corrosion cracking experiments were conducted in both site environmental cracking rigs and in laboratory apparatus.[4] In the former, stressed turbine steel specimens were exposed to LP steam (superheated dry steam, wet steam and a dry–wet steam cycle) in stainless steel chambers. Cracking always occurred in 3% Cr–Mo and 3% Ni–Cr–Mo–V specimens in wet steam, and sometimes in the wet/dry chambers; the cracking morphology was similar to the cracking found in service at disc keyways and bores. Crack growth rates of 2 to 40×10^{-8} mm/s, obtained at 383 to 403 K in the steam rig tests, were consistent with the value of 5×10^{-8} mm/s, obtained at 393 K from Fig. 7-6. These results are in general agreement with other studies that show the high cracking susceptibility of turbine steels in low-oxygen-content solutions. A data base obtained from both laboratory and in-service conditions over the temperature range 330 to 530 K[15] indicates that maximum crack growth rates occur in the 420 to 470 K regime; thus, the U.K. experience is in a less susceptible temperature range.

The U.K. laboratory tests[4] confirmed the power station rig tests in that SCC initiated in plain specimens loaded to 50–70% of their yield stress when exposed to wet steam. However, the ease with which cracks initiated depended on the surface condition of a particular sample, and the research substantiated the initiation sites observed on cracked discs (e.g., nonmetallic inclusions, pits, heavily worked surface layers, etc.).

The susceptibility of the turbine steels to temper embrittlement was considered as a possible contributing mechanism to SCC of the discs. Temper embrittlement develops in Ni–Cr–Mo–V and Cr–Mo–V alloys when cooled slowly or when isothermally heated in the temperature range of 670 to 840 K, and results in a reduction in fracture toughness and increased susceptibility to SCC.[16] Intergranular cracking along prior austenitic grain boundaries is favored. As noted earlier, temper embrittlement can occur in large forgings during the fabrication stage, but generally if it is detected it can be reversed by reheating the steel above ~870 K. The more worrisome problem is that temper embrittlement can occur in-service. Many of the discs examined by the U.K. were temper embrittled to some degree as determined from the wide range of fracture toughness values.[4]

Several studies of temper embrittlement in Ni–Cr–Mo–V steels have shown that for a given impurity level (P, As, Sn, and Sb) the susceptibility to embrittlement depends upon the major alloying elements (Cr, Ni, Mn, and Si). More will be said about this in Section 7.3.1.2. The discs in the

U.K. turbines were mainly made from 3% Cr–Mo (≤0.1 Ni), 3% Ni–Cr–Mo–V (close to A471), or 3½ Ni steels (no Mo or V), and both acid open hearth (AOH) and basic electric (BE) steel making processes were used; in general the latter produces cleaner steel with lower residual elements, such as sulfur and phosphorus, and better toughness. Temper-embrittled steels were found to be more susceptible to initiation of stress corrosion cracks in the U.K. laboratory tests. However, temper embrittlement was also shown not to influence significantly stress-corrosion crack growth. No correlation could be found between disc cracking and alloy chemistry, except that the 3½% Ni steel appears to be immune.[4]

The U.S. disc failure analysis work[13] that followed this U.K. study appears to substantiate most, if not all, of their conclusions. Thus, the consensus on disc cracking is that:

- CrMo and NiCrMoV steels are susceptible to SCC, but 3½% Ni steel apparently is not.
- Steels produced by both acid open hearth and basic electric processes are susceptible.
- Temper embrittlement is not a prerequisite for SCC. Both normalized-and-tempered and quenched-and-tempered NiCrMoV steels with yield strengths of about 800 MPa or higher are susceptible.
- Cracking occurs only in wet steam at crevices and areas of limited steam access (i.e., at the location of the so-called Wilson line).
- Cracking is usually intergranular, but may be transgranular, depending apparently upon contaminants present.
- Hydroxides and/or sulfides appear to have been responsible for most cracking incidents, but SCC may also occur in high-purity wet steam.
- The period of operation before cracking occurs appears related to the level of contamination present in the system and possibly to the type of water chemistry control employed.
- There appears to be no significant correlation between alloy grain size and susceptibility.

From the foregoing it should be evident that any highly stressed crevice region in CrMo and Ni–Cr–Mo–V turbine components exposed to wet steam conditions is susceptible to SCC. This is not just restricted to the keyed disc design in built-up rotors (refer to Fig. 7-3), because highly stressed crevices can also be associated with the peripheral blade attachment regions. If cracks form, they may grow at a significant rate. However, it would be possible to tolerate such a situation if the critical crack depth for fast fracture (i.e., $K > K_{Ic}$) were larger than the maximum depth to which a crack would grow during service. The completely safe situation, of course, demands a guarantee that cracks could never initiate, but in practice it may be necessary to accept the possibility of cracking at a known rate and ensure that the steel is sufficiently tough (or the stresses sufficiently low) to tolerate the cracks. Conservatively neglecting initiation lifetime, the U.K. investigators[4] estimated the time for cracks to grow to critical size to vary from one to ten years (refer to Fig. 7-6), depending upon the fracture toughness. However, at least two important factors add to the uncertainty of such an SI analysis being used as a basis for reliability assessment (sometimes called "retirement for cause").

First, the large rotor forgings are subject to segregation and mechanical property variability, which demands control of the minimum fracture toughness. In other words, although the maximum toughness that can be achieved with modern rotor steels is quite adequate, the variability in toughness values would necessitate use of a lower bounding curve (e.g., Fig. 1-13), which could lead to low predicted lifetimes and associated high retirement rates.

Also an issue is steam purity and the effect of impurities on crack growth rates. Such a wide variation of turbine chemistries is possible that the selection of a conservative "severe" environment might result in unacceptably high predicted crack growth rates, or conversely, lead to a false sense of security if the worst environment has not yet been encountered.

More than any other effect, it is this uncertainty of the influence of chemistry on the cracking process that has, in the U.S., precluded a quantitative failure risk assessment of the type developed by the U.K.

The magnitude of the problem can be appreciated when one considers the origins of the steam entering the turbine. Steam purity is entirely dependent upon the quality of the feedwater entering the steam generation system. The factors most important in determining the impurity concentration of the feedwater, and therefore of the steam, are the type of steam supply system used (e.g., direct cycle like the BWR, and recirculating steam generator or once-through steam generator of the PWRs), condenser leakage, and transient operational conditions.[13]

A survey[18] of over 100 steam supplies for modern large steam turbines revealed several instances of stress-corrosion cracking damage that, in most cases, could be related to upset conditions of steam purity and water chemistry. Caustic, chlorides, and sulfite (which degenerates into hydrogen sulfide) were identified as the most serious contaminants. The presence of sodium hydroxide and sodium chloride in the vicinity of the cracked regions was confirmed in the examination of the Rancho Seco turbine discs.[12,13] This unit had operated with water chemistry outside accepted limits during early operation. Although the cracking in the Arkansas turbine occurred in the same region, and caustic, chloride, and copper contamination was present, this unit experienced no seriously abnormal chemistry.[12,13] Traces of chlorides, sulfate, and caustic were also found near the cracks in the Shippingport disc but were not positively identified as their cause.[12] In contrast, no salts or contaminants of any type were found at the failure sites in the H. B. Robinson unit.[13] Nevertheless, the importance of chemistry control to turbine integrity is demonstrated by the fact that: (a) the Germans and Japanese impose stricter standards on steam/water quality and this appears to be manifested in lower instances of turbine failures[19]; and (b) caustic stress corrosion of BWR turbine components is rarely observed, presumably due to the overall lower sodium ion concentration and small potential for caustic formation.[12]

Thus, one must conclude that the influence of chemistry on turbine cracking, while real, is subtle, and that control to very low impurity levels in the steam, while essential, may not provide the entire answer to the problem. Numbers as low as 1 ppb sodium in the steam have been cited as highly desirable goals to reach, but levels this low may not always be practical in large systems.[12] For comparison purposes, the "stringent" German standards are 3 ppb Na during normal operation, with a 6 ppb operating limit, and 10 ppb shutdown limit. Clearly then, the water purity requirements in the steam generator for controlling stress corrosion of the turbine materials may prove to be even more stringent than those required to minimize steam generator corrosion (Chapter 6 earlier), especially with the OTSGs.

Under such a wide range of chemical conditions, stress-corrosion crack growth rates could vary by several orders of magnitude. A qualitative indication of this can be obtained from the trend in failure times observed in the U.S. PWR turbine disc cracking incidents, i.e., Rancho Seco (6 mo.) < Arkansas (24 mo.) < H. B. Robinson (51 mo.).[13] In the case of SCC in caustic this variation has been quantified in the laboratory.[4] With Cr–Mo steel the growth rate in 35% aqueous NaOH at 360 K is up to 100 times faster than in steam, and decreases with decreasing NaOH concentration until at 6% the rate is indistinguishable from that in steam/water.

Other laboratory data[19] also show that, in caustic solutions over the temperature range 355 to 400 K, stress corrosion cracks can develop at applied stresses of the order of 25 to 30% of the yield stress in 25,000 h in these materials, which values are substantially below the design stress levels. Thus, if caustic cracking were a primary cause of turbine disc failures, crack growth rates of 10^{-6} to 10^{-5} mm/s might be expected, which would necessitate the use of alloys with much higher toughness in order to predict (and achieve) acceptable component lifetimes.

The physical picture of SCC in turbine alloys is also clouded by the subtlety of the chemistry effects. Two mechanisms have been proposed[13]: film rupture, and hydrogen-induced cracking or hydrogen embrittlement. The essential feature of the film rupture model is that a protective surface film is ruptured by localized plastic deformation, permitting crack propagation by anodic dissolution of the exposed metal. Crack propagation is only possible, according to this model, if the rate of repassivation is neither too fast nor too slow. Between these two extremes, i.e., perfect passivation and general corrosion, respectively, each rupture event causes an increment of crack extension without excessive crack blunting. Hydrogen-induced cracking is caused by the accumulation of a critical concentration of hydrogen at highly stressed regions of the alloy. For cracking to occur, dissociated hydrogen must be available at the specimen surface and must be able to diffuse rapidly enough to the regions of high stress to maintain a critical hydrogen concentration at these regions.

Each of these mechanisms will operate and be favored in a particular operating regime. One can establish a "corrosion potential" map for a particular alloy, alloy condition, and environment, and mark the SCC regimes that are to be avoided. Lyle et al.[13] suggest that, for turbine steels, there should typically be three potential regions (refer to Fig. 7-7)—two associated with loss of oxide film passivity, and the third being where atomic hydrogen is available—within which SCC is likely, but outside of which there should be immunity. Unfortunately, zone boundaries are apt to be shifted, in a manner not too well understood, by small changes in chemical composition (e.g., NaOH), by introduction of "new" impurities (e.g., S), or by local alloy microstructural changes (e.g., temper embrittlement). Therefore, it is extremely difficult to define a safe operating regime with the current state of knowledge; clearly much more information is required on turbine chemistry and alloy microstructure effects. Additionally, the onus is placed on the designer to develop modifications to exclude crevices and reduce overall stresses, and on the materials specialist to develop new or modified materials with much improved resistance to stress corrosion. Progress in this latter area will be discussed in Section 7.3.

7.2.2. Thermal-Mechanical Crack Growth

There is much concern in the industry about the risk of catastrophic rotor bursts from the subcritical growth of subsurface flaws.[3,5] Although the probability of such failures is very low,[5] the consequences are much greater (i.e., low-frequency, high-impact failure—refer to Chapter 1), and therefore increasing attention has been focused on this problem. Inasmuch as environmental effects are excluded, attention is focused on "mechanical"

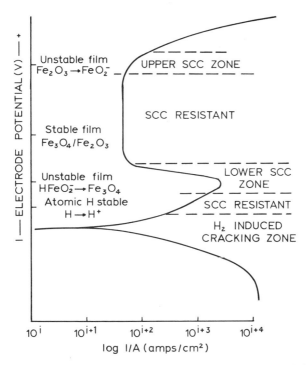

Fig. 7-7. Schematic polarization curve for steel corroding in highly alkaline solution (adapted from Ref. 13).

mechanisms of crack growth, including low-cycle or thermal fatigue, creep crack growth, or creep-fatigue interactions.

The analysis of the Gallatin fossil plant rotor failure[3,6–9] has provided a significant amount of information on the thermal-mechanical behavior of large rotor forgings. The quantitative explanation of this rotor burst has strongly influenced the failure modes included in the (SI) lifetime prediction analyses now being conducted for other turbine rotors.

The Gallatin IP/LP rotor was forged from a 1954 state-of-the art Cr–Mo–V grade alloy steel most closely conforming to the modern ASTM A470 Grade 8 material (refer to Appendix C) but without the stress rupture and impact data, and bore nondestructive examination demanded by modern practice.[6] A basic electric steel making process was used for alloy preparation, but vacuum degassing was not available at the time of manufacture. Forging practice, heat treatment, thermal stability, and machining practices were, however, generally similar to current techniques. It is important to note in passing that the ASTM A470 steel continues to be the standard for HP and IP fossil turbine rotor forgings, but the HP and LP rotors of nuclear turbines and LP rotors of fossil turbines, including shrunk-on discs, are fabricated from ASTM A471 ($3\frac{1}{2}$% Ni–Cr–Mo–V) forgings. The latter are not subject to creep crack growth or embrittling effects, owing to their lower service temperature (generally less than 570 K), but bending and torsional fatigue loadings are still present.[9]

The Gallatin rotor analysis program was conducted on similar lines to the U.K. disc cracking study in that the failure analysis was supported by both mechanical property measurements,[6,11] attempts to simulate the fracture in the laboratory,[9] and SI analyses.[3,7,9]

From detailed metallographic examination of the "event-originating" flaws, it was concluded that subcritical crack growth took place over a 20-year period in a 140 × 10 mm region of banded segregation near the bore.[6] This area, in contrast to the rest of the rotor, contained extremely fine recrystallized ferrite and massive grain boundary carbides. In addition, the initiation zone contained numerous clusters of manganese sulfide (MnS) inclusions. This was also the axial location where the sum of the centrifugal and transient thermal stresses reached its maximum value.

Crack initiation was attributed to a high density of platelike MnS inclusions. Initial crack growth was intergranular and a region of intergranular cavitation and inclusion-matrix decohesion was observed adjacent to the fracture surface, indicative of a significant stress rupture component of damage.[6] A typical example of inclusion decohesion and linkup is shown in the cross section in Fig. 7-8.

Ultimately, intergranular subsurface flaws developed that were of sufficient size to propagate by brittle cleavage when superimposed centrifugal and thermal stresses developed during a start-up of the turbine. The cleavage crack extended radially to about 350 mm where it arrested in the much warmer rim area of the rotor. A subsequent cold start resulted in the catastrophic bursting of the rotor under primarily centrifugal stresses while the rim was colder.[6]

Extensive mechanical property testing was carried out on the Gallatin (and a sybling rotor) material to determine fracture toughness as a function of service exposure and test temperature, fatigue crack growth rates, and the severity of defect distributions in low-cycle fatigue.[6,20] Supplementary creep-rupture and low-cycle fatigue tests were conducted to compare the base properties with results on 1Cr–Mo–V forging steel produced by modern vacuum melting methods, and creep-fatigue tests were run in an attempt to duplicate the intergranular subcritical crack growth mechanism identified for the Gallatin rotor. The complete details of these tests have been reported; the salient results and conclusions are given below.[11,20,21,22]

Figures 7-9 and 7-10 summarize the low-cycle fatigue and stress rupture properties, respectively, of the Gallatin material and compare them with results obtained for a vacuum-degassed 1Cr–Mo–V rotor forging.[21,22]

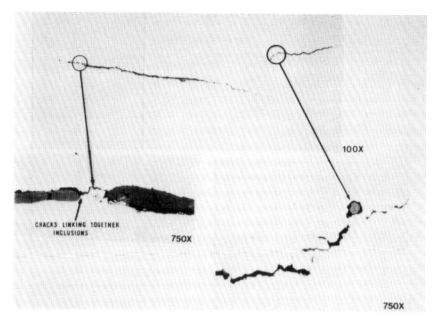

Fig. 7-8. Photomicrograph of fracture surface from Gallatin turbine showing cracks linking together MnS inclusions (Ref. 6). (Figure reduced by 25% for reproduction processes.)

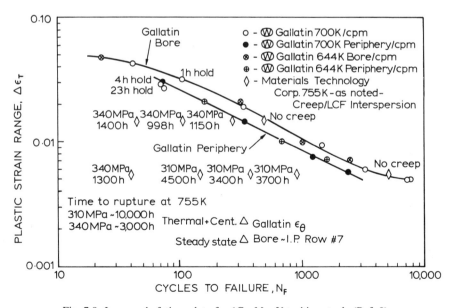

Fig. 7-9. Low-cycle fatigue data for 1Cr–Mo–V turbine steels (Ref. 9).

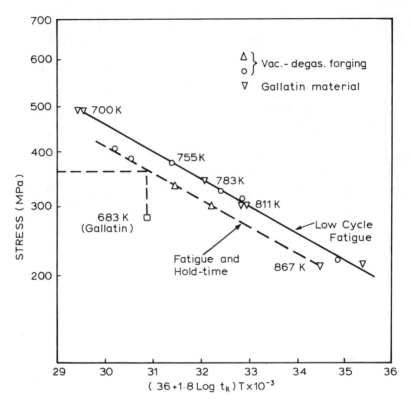

Fig. 7-10. Larson–Miller representation of 1Cr–Mo–V steel stress-rupture data (adapted from Ref. 9).

Note in Fig. 7-9 that a small, but significant, increase in fatigue life is exhibited by the near-bore Gallatin material. No reason for this difference has been verified, although it has been suggested that the presence of ferrite grains in that region provides a mechanism for strain accommodation and crack retardation. Some of the Gallatin tests, as indicated in Fig. 7-9, were held at the tensile strain limit for several hours in each cycle to introduce creep damage. While small reductions in cyclic life were obtained, the results do not differ significantly with hold times up to 24 h. Only the 24-h hold time fatigue specimen exhibited the intergranular fracture that was characteristic of the actual rotor failure, although some intergranular features were reported in a low-temperature fracture in the Auger spectrometer in the course of measuring the concentration of trace elements.[20]

Also plotted in Fig. 7-9 are the results of the creep-fatigue tests on the more modern forging alloy.[22] These tests combine axial strain cycles with periods of constant load, thus producing easily separable combinations of nominal fatigue and creep damage. The significance of these data is that a large component of creep in each cycle is required to explain the initiation of fracture in the Gallatin rotor in the absence of macroscopic defects.

The creep stress-rupture properties of near-bore Gallatin material are compared with the data on the vacuum-degassed alloy in Fig. 7-10 on the basis of the Larson–Miller parameter. The degree of scatter is small and the values are in agreement with other published data from similar forgings.[23] The creep-fatigue test results show about a threefold reduction in rupture lifetime relative to conventional test data; however, the base material data still do not predict the Gallatin fracture under the steady-state stress, time, and temperature conditions at the origin.

Some of the discrepancy between the stress-rupture data and the Gallatin failure point might be attributed to the difference in crack growth mechanism. In contrast to the Gallatin fracture, metallographic examination of the vacumm-degassed alloy specimen failed to reveal any evidence of intergranular cracking or intergranular cavitation. Most probably, the relatively high area fraction of MnS inclusions distributed in clusters along the subcritical crack dominated the failure mode in the Gallatin material. If the presence of inclusions were assumed to increase the local stress in the ligaments, the true stress experienced by the rotor materials might be higher. In the study[9] a value of 8% was used, which raised the local stress to 330 MPa, still ~40MPa lower than would be required to predict rupture from the creep-fatigue data in Fig. 7-10.

The fracture toughness data indicated only a small amount of temper embrittlement in the Gallatin rotor, which did not differ appreciably between hot and cold sections of the rotor. Fatigue crack growth data obeyed the general relationship [Eq. (1-8), Chapter 1]

$$da/dN = C(v, T) \Delta K^{2.7} \qquad (7\text{-}1)$$

over the frequency range $0.0017 \leq v \leq 1.0$ Hz. However, creep-fatigue crack growth results (i.e., hold-time tests) produced a ΔK exponent of 4.54, and lower values of the coefficient C. Thus, since the bulk of structural life is spent growing small cracks, the hold-time data would predict longer lives than the continuous-cycling data.

The laboratory data have guided the mechanical analyses that have been undertaken to explain the Gallatin failure.[3,7,9] Two approaches have been reported. In the first analysis,[7] independent fatigue, creep, and creep-fatigue life calculations were made using the linear damage rule and stress-temperature histories derived from a detailed steady-state and transient elastic analysis and a creep stress analysis. In the more recent analysis,[3,9] a fracture mechanics approach (i.e., SI analysis) was adopted. Qualitatively, the results are similar, since both conclude that, for the starting flaw size assumed, a creep-fatigue interaction is necessary to grow the crack to the critical size for brittle failure (i.e., $K > K_{Ic}$). The details of the calculations differ considerably, however, as should be evident from the following details.

Weisz[7] approached the problem from the simpler point of view of calculating life fractions consumed by creep and fatigue damage. Using the laboratory data on the Gallatin material, he first considered that the two mechanisms operated independently and computes $D = 0.133$ for low-cycle fatigue and $D = 0.17$ for creep damage [refer to Chapter 1, Eq. (1-11), for explanation]. In other words, only 13% of the fatigue life and 17% of the creep life was consumed at the time of failure. Clearly, neither of the two mechanisms alone can explain the failure.

Assuming creep-fatigue interaction, however, he was able to arrive at a $D = 1.2$ by computing cumulative creep and fatigue damage during 288 hot starts (480 K to operating temperature) and fatigue damage alone during 105 cold starts (room temperature to operating temperature). According to Weisz this means that by this time small cracks will have initiated around the inclusions (refer to Fig. 7-8). As they link together they produce the primary flaw that was observed on the fracture surface. It is not only the presence of inclusions, but the presence of inclusions sufficiently close that they could link together, that resulted in the initiation and growth of the flaw by the creep-fatigue mode to the critical size for fast brittle fracture.

Weisz showed that the flaw was of critical size by comparing the calculated K_I to the K_{Ic} value (versus rotor radius). Figure 7-11 shows that both parameters vary with radius due, respectively, to thermal stress and temperature variations through the rotor during the cold start-up. More importantly, Fig. 7-11 also shows that when the cracks propagating from the inclusions link up to reach ~100 mm, a critical size is attained and the crack will "pop in" to a radius of ~280 mm and then arrest. Inasmuch as a value of 350 mm was obtained from the metallographic examination of the fracture, Weisz regards the reasonable agreement sufficient justification for the analytical approach selected. He concludes that, on the next cold start the crack is now so large that it will result in complete fracture of the rotor.

In the fracture mechanics approach,[3,9] both fatigue and creep crack growth mechanisms are recognized as contributing to failure but, unfortunately, insufficient data on the latter precluded a calculation. With respect to fatigue crack growth, however, the results were similar to the earlier fatigue life calculations. Calculation of crack growth from small

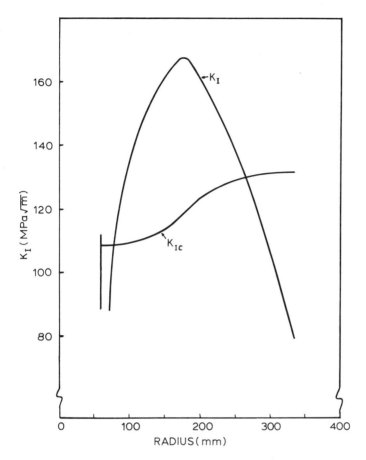

Fig. 7-11. Plot of K_I versus radius; crack arrest calculation at peak stress during cold start of Gallatin turbine rotor (adapted from Ref. 7).

inclusions via LEFM [Eq. (1-8), Chapter 1] showed that the amount of growth predicted for 1000 cold starts (a conservatively large number compared to that used in the first calculation) is negligible. For example, if the maximum stress and temperature are assumed to coincide at the initiation site of the Gallatin rotor, the crack growth from an ~5 mm inclusion in 1000 cycles using the maximum growth rate observed in the laboratory is only 0.23 mm. It should be noted in passing that this SI analysis is part of a computerized rotor lifetime prediction system to assist utilities in performing run/retire (or "retirement for cause") analyses. This program, called STRAP (steam turbine rotor analyses programs), is still under development[9] and, as noted, the fracture mechanics code

(FRAC), in particular, needs to be improved by adding a model for creep crack growth accompanied by intermittent stress-temperature cycling.

To summarize, an explanation for the rotor fracture can be arrived at through a combination of fatigue and creep crack growth damage. This mechanism, of course, necessitates operation of the rotor in the creep regime, which is probably valid for the fossil turbines such as Gallatin that operate at 820 K, but is not the case for nuclear turbines that operate at much lower temperatures, 640 K. If fatigue crack growth alone were considered, the Gallatin data would predict an extension of only 0.66 mm to a 5-mm starting flaw after 10,000 cycles or a 30 yr lifetime. At first sight, this result is very encouraging for the nuclear turbine rotor, but it rests on two fundamental assumptions: first, the detection/selection of the largest starting flaw, and second, the choice of the most probable crack growth mechanism. An entirely different picture would emerge if a significantly larger starting flaw were assumed and/or if a faster crack growth mechanism were adopted. This latter aspect of the problem has already been covered—as shown by Weisz[4] a creep-fatigue damage function can achieve the desired conclusion. It is the first item that has caused concern.[24] A remaining unanswered question for the Gallatin rotor failure analysis is whether a crack in the 200 to 300 mm range, introduced during fabrication or installation, could have escaped detection, and subsequently grown to critical size. The answer has obvious important ramifications with respect to nuclear turbine integrity. That large flaws can exist in a semi-stable manner in turbine rotors is evidenced by the report from Takhar et al.[25] of Detroit Edison Company. Following the Gallatin incident they concluded an NDE examination of the bore of the Connors Creek No. 16 LP turbine rotor and located a 127 mm deep by 406 mm long crack. The operability of the rotor with such an extensive bore crack was explained by the absence of embrittlement and the high fracture toughness ($K_{Ic} = 156$ MPa$\sqrt{m}$), both of which are unusual for a rotor material over 20 years old (compare to Fig. 7-11, for example).

This particular study is of additional interest because it produced a set of recommendations for continued operation, based on a fracture mechanics SI analysis. The large flaw was first removed by machining. The allowed number of thermal (stress) cycles before remaining flaw indications would attain critical size (i.e., $K_I \gtrsim 156$ MPa$\sqrt{m}$) was then computed using LEFM and fatigue crack growth data on NiCrMoV materials of similar vintage; the worst assumed flaw would not come to the surface, theoretically, before 125 cycles of shutdown and start-ups. Accordingly, it was decided to place the turbine back into service on the basis of revised start-up procedures that would minimize thermal stresses and maximize K_{Ic} (e.g., prewarming treatment), limited overspeed tests,

7.2.3. Moisture Erosion

Water cutting and moisture erosion of turbine blades have been a problem in turbine design and plant operation for many years.† This problem has been aggravated by the advent of higher tip velocities of longer last-stage turbine blading and the poorer steam condition of nuclear plants.[26]

An investigation to classify and minimize the effects of water erosion in turbine last-stage or "end" blading was undertaken by Allis Chalmers Power System, Inc. (ACPSI).[26] The study provides further insight into the formation of moisture within turbine staging and offers a measure (the erosion coefficient) to help evaluate the erosion characteristics for varying design and operating conditions. It has been shown that water droplets of a definite size range are responsible for last-stage erosion. Water films on the concave surface of the stationary blade are developed when water droplets centrifuge out of the steam flow because of a change in direction of the steam path. Thus, erosion of last-stage blades depends on the moisture content of the steam entering the preceding stationary blades. The tip speed of turbine blades is a major factor in erosion; and while steam velocity increases with tip speed, the relative velocity of water droplets leaving the stationary blade's trailing edge is essentially zero. If the acceleration of the water droplets spraying off the stationary blade is the same for all blade lengths, the erosion increases directly as a function of tip velocity. Steam density is also a factor affecting the acceleration and atomization of the sprayed water droplets from the trailing edge of the stationary blades. High steam density leads to high acceleration and smaller water droplets. This influence is directly related to the steam pressure between the stationary and rotating blades.

An empirical equation to evaluate the risk of or exposure to erosion for a given set of design and operating parameters was developed by ACPSI.[26] Thus,

$$E = (1 - X_1)^2 (Dn/3600)^3 (1/P_1) \qquad (7\text{-}2)$$

where X_1 is the steam quality ahead of the last stage, D is the tip diameter of the last-stage blades in inches, n is the shaft speed in rpm, and P_1 is the pressure ahead of the last stage (inches of Hg, absolute).

---

† Evidence of water cutting in discs of nonreheat BWR turbines was first observed in early 1980. However, no details had been reported up to June 1980.

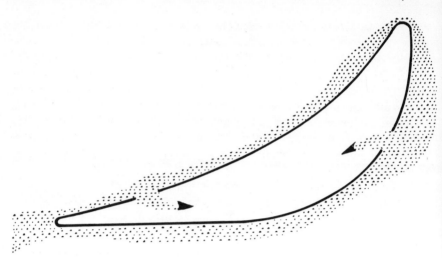

Fig. 7-12. Hollow stationary blade design of Allis Chalmers power systems (adapted from Ref. 26).

The higher the value of the "erosion coefficient" E, the greater is the risk of or exposure to erosion. ACPSI suggest that, at guaranteed steam-flow rates, values of E should not normally be greater than 3600 for nuclear turbines (compared to 2400 for fossil units).

Other devices that can improve the erosion characteristics[26] are: (1) providing larger clearance between the stationary and the rotating blades, so that the steam has time to accelerate the moisture particles, (2) hardening the leading edges of rotation blades to improve erosion resistance, and (3) removing water entrained in the steam path. Steam reheating is now common, as are various methods to centrifuge moisture to the outer casing areas, where it can be drained from the turbine.

Hollow stationary blades are a novel approach proposed by ASPSI to remove moisture before it reaches the endangered rotating blades. The inside of the hollow blade is connected to the condenser, and the resulting vacuum removes the water film from both concave and convex blade surfaces (Fig. 7-12).

7.3. Improvements in Turbine Materials

Both improvements in the quality of the standard Ni–Cr–Mo–V ferritic steels and the qualification of new materials are being pursued in the attempt to, respectively, eliminate possible catastrophic failures of rotors and discs and to improve the overall reliability of turbine blades.

7.3.1. Low-Alloy Ferritic Steels

Two approaches to general improvement in quality and properties of the low-alloy ferritic class of turbine steels are being pursued. The first deals with reduction of impurity segregation by improved ingot casting practice; the second addresses the problem of temper embrittlement. Both approaches lead to significant improvements in fracture toughness, fatigue and creep-rupture strengths, and SCC resistance. Also of importance is the improved inspectability that is a result of increased homogeneity.

7.3.1.1. Impurity Reductions

Only recently has it come to be realized that great improvements are possible in the high-termperature performance of alloy steels, providing that impurity effects can be brought under control. This was first shown clearly by Tipler who compared the properties of pre-1972 vintage commercial and laboratory (vacuum cast) $1Cr-1Mo-\frac{1}{4}V$ and $\frac{1}{2}Cr-\frac{1}{2}Mo-\frac{1}{4}V$ steels.[27a] The performance of the high-purity Cr–Mo–V steels depicted in Fig. 7-13, in terms of rupture strength and ductility, is equivalent to

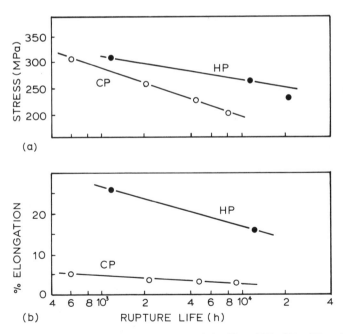

Fig. 7-13. Comparison of creep rupture strength and ductility of $\frac{1}{2}Cr-\frac{1}{2}Mo-\frac{1}{4}V$ steel of commercial purity (CP) and high purity (HP) both austenized at 1323 K and air cooled (Ref. 27a).

that of a new ferrous "super" alloy developed for that temperature regime. The impurity effect becomes even more striking when it is noted that the high-purity specimens failed by the transgranular cup-and-cone mode associated with lower temperatures. The extent of grain-boundary cavitation during creep testing is also significantly reduced for high-purity Cr–Mo–V steels (Fig. 7-14). Later work by Tipler and co-workers[27b] verified this effect using commercial steels of more recent vintage, but also showed the importance of microstructure as well as purity in improving key mechanical properties.

Thus, the main emphasis in improving rotor forgings has to do with the reduction of center porosity, nonmetallic inclusion content, and alloy segregation at the ingot stage. Figure 7-15,[28] which is a schematic of the typical defects found in a conventional ingot, illustrates the ample "opportunity" for improvement. Several advanced methods for producing

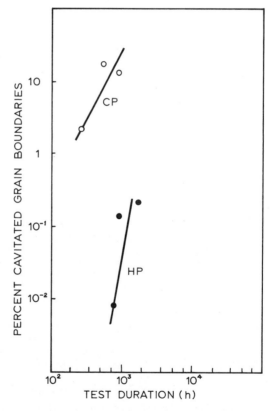

Fig. 7-14. Comparison of extent of grain-boundary cavitation during creep testing of $\frac{1}{2}$Cr–$\frac{1}{2}$Mo–$\frac{1}{4}$V steel of commercial purity (CP) and high purity (HP) (Ref. 27a.)

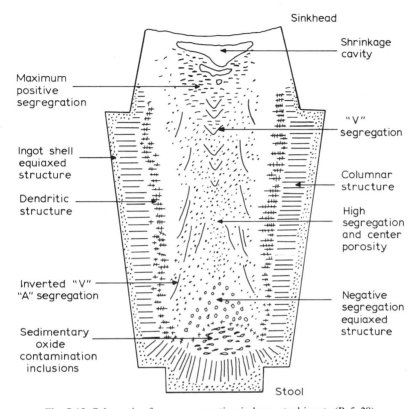

Fig. 7-15. Schematic of macrosegregation in large steel ingots (Ref. 28).

high-temperature Cr–Mo–V steam turbine rotor forgings up to about 109,000 kg are being evaluated.[29] Most attention is being focused on two technologies: improved conventional ingot making practices, and electroslag processing via either electroslag remelting (ESR), electroslag hot topping (ESHT), or central-zone remelting (CZR). The following details are taken from a report prepared recently for the Electric Power Research Institute[29] and a review of fossil turbine materials by Jaffee.[30]

Improved Conventional Processes. A ladle ("external") desulfurization treatment using calcium and magnesium-based reagents can be practiced, which can effectively reduce the sulfur content from present levels of 0.008 to 0.015% S to 0.001 to 0.0005% S. Thus, even if "A" segregation occurs (refer to Fig. 7-15), the formation of MnS inclusions can be minimized if the sulfur content is low enough.[30] Also, the Japanese have pioneered the so-called vacuum carbon deoxidation (VCD) practice, in which the molten steel is deoxidized in the vacuum ladle after the degassing treatment, utilizing the carbon content in the molten steel. It is not necessary to use silicon–aluminum deoxidation with the VCD practice, and therefore the silicon content (and the susceptibility to temper embrittlement)[31] in the steel is reduced. Apparently the smaller dendrite size in the VCD process reduces the "V" and "A" segregation as compared to the conventional Si–Al deoxidation treatment. Excellent uniformity in mechanical properties and freedom from property degradation in the regions associated with segregates have been found in huge 224,000-kg VCD forgings.

Large ingots (400,000 to 500,000 kg) have been prepared by Japan Steel using modified conventional practices.[29] A multipour process is adopted wherein heats from 100,000 kg electric furnaces and holding furnaces are used to sequentially pour a large ingot. One of the problems encountered as the ingot size is increased is that proportionally more of the ingot must be discarded, due to large shrinkage cavities formed at the top and nonmetallic inclusions that have segregated to the bottom (refer to Fig. 7-15). This is overcome in the Japanese practice by maintaining a "hot top"; that is, to slow down cooling, and hence reduce segregation, etc., by keeping the top surface at its melting point. Ingot yields are believed to average about 50% through the use of large-diameter hot tops, enhanced insulation methods, and, in some cases, exothermic materials.[29]

Electroslag Remelting. Cold-mold melting with consumable electrodes has been adopted for the production of large steel ingots in an attempt to minimize convection currents and resulting "A" and "V" segregation (Fig. 7-15). In the consumable electrode method, the electrode prepared by a previous melting operation is melted off in vacuum (vacuum arc remelting or VAR) or through a refining slag (electroslag remelting or ESR). The ESR method seems to be preferred for steel making, because it is adaptable to larger ingots and can be practiced in air, and the refining stage reduces sulfur content.[1] ESR ingots are fully columnar, and the radial and axial segregation, seen in Fig. 7-15, are negligible. Electroslag remelting produces the best quality of any steel making practice. However, it is limited in that the maximum-size ESR ingots are about 90,000 kg at present, although larger size facilities have been designed. If ESR making is done in air, moisture can enter the slag and result in excessive levels of hydrogen. Therefore, it is necessary to practice ESR melting in dry air or in a dry, inert atmosphere.

Electroslag Hot Topping. ESHT combines ESR with conventional ingot making to achieve the "ideal" hot top. This method was pioneered in the U.S. by the Kellogg process of the 1950's and is now being developed commercially in Europe. Following the conventional practice of electroslag furnace melting and vacuum casting of the ingot, molten slag is added to the melt surface, and a consumable electrode of size roughly equal to the 4% solidification shrinkage is melted off through the refining electroslag while the ingot is solidifying. The melt-off rate is adjusted to approximate the shrinkage of the ingot. The coninuous heat input maintains the surface molten and provides, in effect, "ideal" hot-top action during solidification. Because there is little electroslag metal added, sulfur refinement is confined to a relatively small cone near the top of the ingot. Thus, if the ingot is to have low sulfur content throughout its center region, it is necessary to accomplish the desulfurization during the electric furnace melting stage.

The major benefit of ESHT is derived from its effect on the solidification characteristics of large ingots, and the forging yields of ESHT ingots are substantially larger than for conventional ingots because of the reduced top and bottom cropping. ESHT has been developed up to 54,000 kg size, but there is no apparent size limit on ESHT ingots, and melting facilities for producing ESHT ingots up to 315,000 kg are contemplated.

Central-Zone Remelting. Reference to Fig. 7-15 shows that both porosity and segregation is concentrated in the central zones of conventionally poured ingots. The principle of the central-zone remelt or CZR process, therefore, is to remove the entire center region of conventional ingots and replace it with "cleaner" electroslag metal. This approach is being developed in Russia and western Europe.[29] Starting with a conventionally poured ingot, the CZR process removes the center core by hot trepanning. Using the cored ingot as a mold, the center is filled with high-quality electroslag metal by consumable arc melting an electrode, weighing perhaps 10 to 15% of the original ingot, under a refining slag. Compared to ESHT the refining action extends over the entire height of the ingot. A significant portion of the ingot shell also is melted, so that perhaps one-third to two-thirds of the ingot is modified by the CZR melting operation, and improved in both purity and structure. Thus, the CZR process is intermediate between the ESHT and ESR processes in its electroslag metal content.

Like the ESHT, the CZR process appears to be capable of producing the largest size ingots for rotor forgings. Ingots as large as 180,000 kg have been produced and processed into rings for nuclear reactor vessels. However, there has been relatively little evaluation of CZR ingots processed in rotor form.

In summary, based on the available information,[29,30] it appears that either the improved conventional or the electroslag-based ingot processes have the potential to produce better quality rotor forgings than present commercial processes. It appears as though Japanese steel makers prefer improved conventional techniques for large forging production, while alternative methods are being developed in Europe and Russia. Both approaches are of interest in the U.S. For example, in a U.S. industry–government program 50,000 kg VCD conventional and ESR ingots have been produced for rotor forgings that will subsequently be installed as IP/LP rotors in new Westinghouse turbine designs.[8,30] The service performance of these rotors, due for installation in 1980, will be followed in great detail. Mechanical property evaluation of the forging material will also be undertaken.

Finally, before selecting one or another of the advanced forging options, it will be necessary to balance the benefits against costs. The following cost increment for producing 120,000-kg forgings compared to conventional was calculated in one study[29]: ESHT 7%, CZR 11%, ESR 20%. The cost for the advanced low-sulfur conventional practice developed in Japan was not included in this comparison. Jaffee[30] feels that this practice is probably somewhat more costly than the conventional because of the additional desulfurization step and because the use of vacuum carbon deoxidation (VCD) requires holding the melt in a ladle lined with a costly basic refractory. In any event, these metallurgical developments should certainly achieve the goal of better-quality, cost-effective rotor forgings for the utility industry.

7.3.1.2. Elimination of Temper Embrittlement

The earlier discussion noted the susceptibility of rotor forgings to temper embrittlement, due to cooling problems associated with large castings, coupled with the temperature range experienced in service.

Inasmuch as temper embrittlement occurs only in steels of commercial purity, and not in comparable alloys of high purity,[16,17] the techniques discussed above will go a long way toward minimizing this problem. However, "ultra high" purity is unlikely to be achieved in the very large ingots due to equipment and economic limitations. Consequently, ways to eliminate temper embrittlement through alloy modifications, heat treatments, or other postcasting practices have received a lot of attention.

The susceptibility to temper embrittlement of different rotor steels has been studied in terms of both steel-making practice and subsequent processing,[16,17,32-35] and the basic understanding of the phenomenon is reasonably well advanced.

It is now well established that temper embrittlement represents a reduction in the cohesion of prior austenite boundaries due to impurity segregation, although the precise nature of the segregation process is not fully understood yet. It has been suggested that the process of temper embrittlement is one of the equilibrium segregation of impurity elements to the prior austenite grain boundaries during prolonged heating at temperature below about 870 K.[16] However, the results of Auger electron microscopy have shown that impurities and alloying additions segregate together.[35]

The major compositional factors involved in the phenomenon are summarized in Table 7-2, developed by McMahon et al.[33] Some major points are as follows: Phosphorus is generally the most important impurity as far as temper embrittlement is concerned and levels below 0.005% must be attained. Tin concentrations should be at or below 0.01%, particularly in steels containing nickel. The use of silicon should be discouraged and replaced where possible by techniques such as VCD, discussed earlier. Since manganese also promotes phosphorus segregation, special precautions are called for when manganese is used at levels > 0.1%; these include stringent control of phosphorus, use of the scavenging properties of molybdenum and/or niobium, use of special scavengers (perhaps rare earths), or limitations on the chromium content. Because of its attributes, the potential hazards from the use of nickel, namely its capacity to cosegregate with tin, silicon, and antimony, must be mitigated through control of these elements rather than by reduction in nickel. Molybdenum is a very effective scavenger of phosphorus, tin, and antimony, as long as it is not tied up in carbides. If molybdenum is added beyond about 0.7%, the tendency seems to be to form Mo-rich carbides at the expense of Cr-rich carbides during the tempering treatment. Therefore, the view is that the molybdenum concentration should

Table 7-2. Compositional Factors Involved in Temper Embrittlement of Turbine Steels (Ref. 33)

Important impurities:
- Sn, P: Ni–Cr-based steels (e.g., 3.5 Ni Cr MoV)
- P: Cr–Mo-based steels (e.g., 2¼Cr, 1Mo; Cr MoV)

Main effects of alloy elements:
- Ni: raises inherent resistance of steel to brittle fracture
  promotes segregation of Sn, Si (and Sb, if present)
- Cr: imparts hardenability
  imparts some resistance to softening at elevated temperatures
  promotes segregation of P
- Mn: imparts hardenability
  scavenges sulfur
  promotes segregation of P
- Si: deoxidizes
  promotes segregation of P
- Mo: imparts (bainitic) hardenability
  imparts resistance to softening
  scavenges P and Sn
- V: imparts resistance to softening
  aids in grain refinement
- Nb: imparts resistance to softening
  scavenges P
  aids in grain refinement

be limited to 0.7%, unless the chromium content is raised above ~2.3%. Finally, with respect to carbon, the general approach should be to use as little carbon as possible while maintaining sufficient hardenability, and to compensate for reductions in carbon content by reducing the tempering or stress relief temperature.

In addition to control of chemical composition, it is important to control the microstructure of a steel to achieve a minimum susceptibility to temper embrittlement. For a given level of impurity segregation, the fracture toughness transition temperature increases with hardness and with grain size.[33] For the former, reduction of both the carbon content and tempering temperature would improve toughness while maintaining the necessary strength; for the latter, refinement of grain size can be achieved, not only by use of grain refining elements such as V, Nb, and Al, but also by thermal processing, e.g., a low-temperature austenitization treatment.[33]

With the foregoing as a basis, rules or correlations are being developed which can be used as a guide to minimizing temper embrittlement. For example, Murakami et al.[31] have correlated susceptibility to temperature embrittlement, as measured from the fracture toughness parameter FATT,† to a "J-factor," defined as $(Si + Mn)(P + Sn) \times 10^4$. The other known contributors to temper embrittlement, As and Sb, are less effective in $2\frac{1}{4}$Cr–1Mo steels when they are controlled under 0.02 and 0.004% respectively. As shown in Fig. 7.16, good correlations are observed despite such manufacturing variables as thickness, heat treatment, and other chemical compositions. In this particular study, the Si content in the $2\frac{1}{4}$Cr–1Mo steel was considered most significant, and controlling levels to less than 0.1% (through use of the VCD process discussed above) virtually eliminated the shift in FATT.

McMahon and co-workers[33] are developing an embrittlement equation that involves three variables; hardness, grain size, and impurity segregation. Preliminary results are encouraging; successful correlations were obtained between the ductile–brittle transition (from Charpy tests) and these three variables for three laboratory-made 3.5% Ni–1.7% Cr steels, each doped with either 0.06% P, 0.06% Sn, or 0.6% Si.

While the first approach is limited to influencing the specifications for new turbine components, the latter has the potential for providing a nondestructive method of estimating the damage to the component caused by in-service embrittling mechanisms; for example, small samples extracted from a steam turbine rotor still in service can be periodically

---

† Fracture appearance transition temperature, FATT, is used in this class of materials as a ductile–brittle transition indication. The FATT is the temperature at which the fracture exhibits 50% cleavage and 50% ductile fracture. It is generally higher than DBTT.

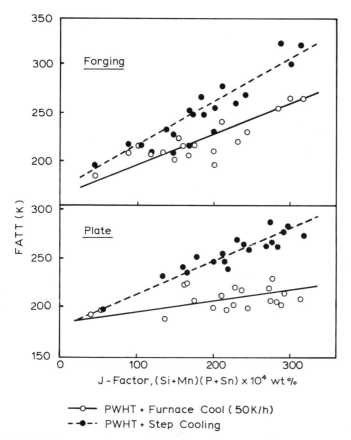

Fig. 7-16. Relation between J factor and fracture appearance transition temperature (FATT) before and after a step cooling of $2\frac{1}{4}$Cr–Mo steels (adapted from Ref. 31).

analyzed to give estimated transition temperature changes, from which run/retire decisions can be made.

7.3.2. Alternative Blade Materials

Steam turbine manufacturers all over the world are considering replacing the conventional AISI Type 403, 12 Cr stainless steel LP turbine blading material by alternative alloys, principally the titanium alloys,[30,36] and also 17-4 PH stainless steel and a modified AISI Type 403 stainless steel.[8]

A survey was conducted by Battelle Columbus[36] to summarize the worldwide status of titanium alloy blading in the LP steam turbine. Two main characteristics of titanium make it attractive as a steam turbine blading alloy: its high strength-to-weight ratio, which is favorable for longer blades in larger turbines, and its corrosion resistance. However, titanium blades will be more costly. It is estimated that a row of titanium blades normally costs about 1.25 to 1.5 times that of a steel row. If the last stage of a large LP turbine were completely bladed with titanium blades, one estimate is that the cost differential of the overall turbine would be only about 0.25 to 0.5% higher than with steel blades; therefore at 0.5%, the overall cost of a 1200 MW turbine would be increased by about $300,000.[30]

Titanium blading appears close to being introduced into production steam turbines, particularly in western Europe, and is in standard production in the Soviet Union.[36] In the U.S., titanium blades are being evaluated for last stage (L) and next-to-last stage (L-1) rows on a limited basis and are routinely used for closing blade application.[3,30,36,37]

Titanium alloys are being proposed for the L and L-1 blading of the LP turbine, because these blades are the most highly loaded, develop the most power, and are the major sources of forced outages in both fossil and nuclear turbines.[37] The potential performance improvements, based on laboratory experimentation, are in the areas of general corrosion resistance, fatigue strength, and erosion; a probable problem area is the low damping capacity exhibited by titanium.

Up to the present, the titanium alloy, Ti–6Al–4V (Ti6-4), has been the primary blading candidate. It is a two-phase alloy, the two phases, α and β, resulting from the vanadium addition. The existence of this two-phase structure, and the addition of aluminum as a strengthener of the α-phase, account for its high strength. The mill-annealed condition has been considered adequate for demonstration purposes.[30,37] Jaffee[30] recently compared the important properties of the Ti–6Al–4V alloy (Ti6-4) to those of the standard 12Cr steel blading material. The following draws liberally from his discussion.

Corrosion. The last stages of the LP turbine will be mostly subject to chlorides, usually on the acidic side, since most of the NaOH will have already condensed in earlier stages of the turbine. Data indicate, however, that titanium is passive over the entire range of pH anticipated in the steam turbine (i.e., TiO_2 film is stable). Also, as shown in Fig. 7-17, the resistance to crevice corrosion is also high, at least to ~350 K; at higher temperatures, Ti would be susceptible to crevice corrosion in acidic solutions. In the blade itself crevices are not present, but the blade-root and blade-shroud attachments have recesses where crevice corrosion could occur as a result of acidic conditions. Also shown in Fig. 7-17 is the remarkable effect that a small amount of palladium has on reducing the susceptibility of titanium to crevice corrosion.

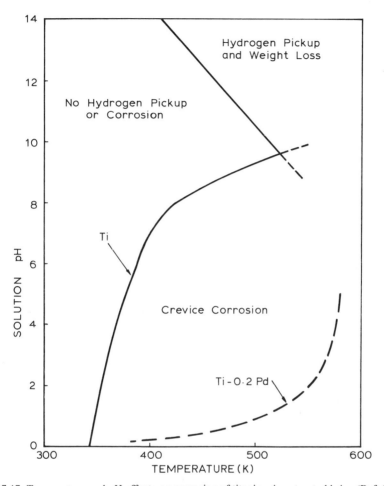

Fig. 7-17. Temperature and pH effects on corrosion of titanium in saturated brine (Ref. 30).

Fatigue Strength. The design criterion for turbine blading is high-cycle fatigue (HCF). The fatigue properties in steam and sodium chloride solution are shown in Fig. 7-18. The fatigue endurance limit of 12Cr is reduced drastically in 1% NaCl solution, whereas Ti6-4 has essentially the same fatigue endurance limit in air, 3.5% Nacl, and steam (low and high oxygen); see Fig. 7-18a. Taking density into account, the endurance limit of Ti–6Al–4V in NaCl is an order of magnitude greater than that of 12Cr steel (Fig. 7-18b).

Once a crack of appreciable length has started, the number of stress cycles on a turbine blade are so high that failure soon results. Thus, an

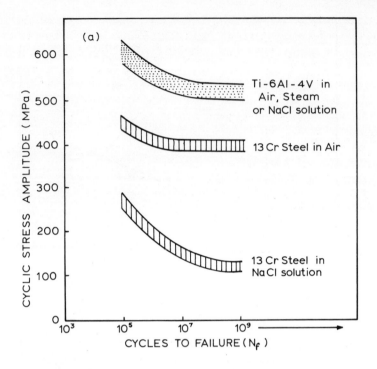

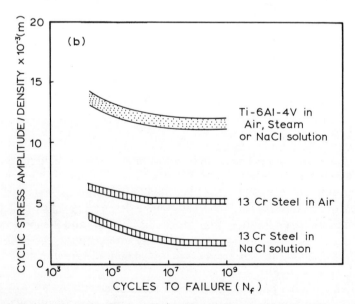

Fig. 7-18. Fatigue properties ($R = -1$) of turbine blading materials in various media: (a) laboratory data; (b) data compensated for density (adapted from Ref. 30).

important design criterion is the initiation of cracks, which is intimately related to propensity to pitting. Also important is the threshold stress intensity required to start fatigue crack growth. The rate of growth at the threshold is sufficiently low that it might be possible to remove cracked blades during inspection at scheduled outages, e.g., a safe-life approach. However, this is unlikely, because the turbine is not opened for very long intervals. The da/dN data for titanium compared with 12Cr steel is shown in Fig. 7-19. The lower titanium bound of $\Delta K_{th} \simeq 4$ MPa$\sqrt{m}$ was obtained with the basal planes parallel to the crack growth plane, whereas the higher titanium $\Delta K_{th} \simeq 8$ MPa$\sqrt{m}$ was obtained on a texture-free material. After rolling high in the α–β field, the normal texture for titanium is to have the basal planes transverse to the rolling plane. Therefore a titanium blade would have its basal planes perpendicular to the edgewise crack growth direction, the favorable direction, as reflected by the higher ΔK_{th} values. The 12Cr steel generally has lower ΔK_{th} values than the titanium alloys when normally processed. Thus, according to Jaffee,[30] in the event that corrosion pits are formed in Ti6-4 (which apparently do not occur under turbine conditions) or there are small scratches or other stress risers on the surface, the titanium alloy would require about twice

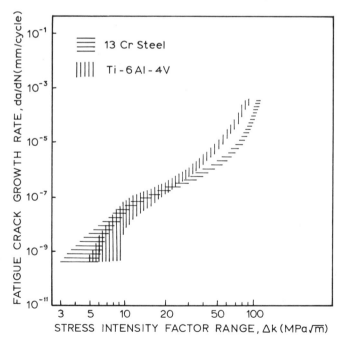

Fig. 7-19. Fatigue crack propagation ($R = 0.1$) in NaCl solutions for turbine blading materials (adapted from Ref. 30).

the stress intensity as 12Cr steel to initiate crack growth. However, once a crack starts growing in Ti6-4 it propagates at the same rate or faster than in 12Cr steel (Fig. 7-19).

Damping Characteristics. Damping is a major area of uncertainty in the application of titanium blades as a substitute for martensitic steel blades like 12Cr. Under conditions in blades, which include steady and alternating shear stresses, the difference in damping between 12Cr steel and Ti–6Al–4V is about 20:1.

Damping is important in blades because the vibration amplitude at resonance increases as the damping capacity decreases, and the shear stress developed in the blade is inversely proportional to the damping capacity. Thus, for a given exciting torque, higher stresses are developed in a material with lower damping. However, Jaffee[30] points out that materials damping is only part of the picture. Total blade damping consists of materials damping, mechanical damping, and aerodynamic damping. Although the relative contributions of these factors will vary from turbine to turbine, the contribution of materials damping to total damping in measurements made on steel blades in an operating turbine showed that the materials damping contributed less than 50% of the system damping. Jaffee,[30] therefore, concludes that "even if the materials damping of titanium blades was discounted completely, the stresses developed would be well below the fatigue strength of Ti6-4 blades. Since the fatigue strength of Ti6-4 is substantially greater than that of 12Cr, this appears to be a relatively small penalty."

Actual experience with operating blades has shown that the low damping capacity of titanium is not a major deterrent to its application as low-pressure steam turbine blading. No blade failures have been reported[36] in 17 cases of experimental titanium blades having been run in operating turbines or in operational tests for periods of up to 14 years. Most of the experimental blades were relatively short, up to about 0.6 m long, however, and so more critical tests are now being conducted. A turbine being built by Brown Boveri will have 1.2 m titanium last-stage blades. Measurements of stresses and stress/strain amplitudes in a retrofitted titanium L-1 row in a 580 MW Westinghouse turbine are planned to be conducted over the next several years.[8,30,37]

Water Droplet Erosion. As noted in Section 7.2.3, long last-stage blades are subject to impingement by water droplets with supersonic velocities. In steel blades erosion by water droplets is resisted by coating with a hard material like Stellite No. 6 (Co–Cr–W–C) or by heat treatment of the leading edge. The latter treatment leads to such high strength levels that these edges are more susceptible to corrosion fatigue. Water impingement tests on 12% Cr steel, Stellite, and a titanium alloy indicate that titanium's resistance to cavitation damage exceeds even that of Stellite.[38] However, experience in operating turbines indicates that this may be somewhat optimistic.

Rust et al.,[37] for example, recently reported on Westinghouse experience with a complete row of Ti alloy steam turbine blades, installed in a fossil plant over six years ago. The blades were given a complete inspection in September 1976 after 44 months of operation, including a fluorescent penetrant inspection for cracking and a visual inspection for erosion damage. No indications of cracking were found. In the erosion check, the Ti6-4 and the 12% Cr stainless steel blades were compared visually. The steel blades had Stellite 6B erosion shields in the leading edge of the airfoil. Ti6-4 blades were found to be only slightly more eroded than those with Stellite material attached to the 12% Cr stainless steel blades.

Manufacture. In general, manufacturing processes to produce titanium alloy components are more complex. Forging of titanium steam turbine blades does not present a problem. Most configurations that can be forged in steel can also be forged in titanium despite the added difficulty of working at lower temperatures and surface contamination.[37] However, the alloy is not as forgeable, nor as machinable as the standard 12% Cr stainless steel blade material. Comparatively, a lower temperature is required for titanium to avoid a metallurgical structure transformation which results in lower ductilities—this, in turn, means more energy is required to forge a given amount of material; and when forgings are large, a larger press is required.

During forging, oxygen from the air diffuses into the material, causing brittleness on the surface. A protective glass coating provides partial protection. The coating also acts as a die lubricant. Nevertheless, even with coatings, there is usually some contamination that must be removed by pickling. Milling speeds must be reduced substantially, and tool geometry must be altered to eliminate galling. Fixturing must be more rigid due to the increased flexibility (lower modulus) of titanium. Grinding, most notably manual grinding, can result in overheating, surface contamination, and structural changes at the surface.

Ti6-4 is welded with difficulty.[37] Two factors are involved: a propensity for weld embrittlement by contamination and a lack of heat penetration. Due to the rapid pickup of oxygen from the air during welding, it is essential that the entire weld area be blanketed with inert gas. This requires gas shield fixturing which adds to welding time and costs. Care must also be taken that all weld preparations, weld wire, and equipment are absolutely free of dirt and other contaminants. The alloy exhibits lower thermal conductivity than the standard blading steel. As a result, heat penetration is limited during welding, and fusion zones are shallow. This can be overcome in most cases by an altered weld-joint geometry avoiding butt joints and other configurations requiring deep fusion zones. The properties of welded Ti6-4 are good despite the difficulties in achieving an uncontaminated weld. Strengths are comparable with those of the base material, though reduced ductilities are the rule.[37]

In summary, based on this good early experience with Ti6-4, it is expected that retrofitting of titanium L and L-1 blades in existing turbines will increase in the future, providing the much needed operational data.[30] In particular, data on the vibrational characteristics of operating titanium blades will establish why titanium blades have performed so well in experimental units despite their low damping capacity. The contributions of longer titanium blades to the power output and thermal efficiency of large turbines also need to be quantitatively demonstrated. Analyses[39] indicate heat-rate improvements of ~2% for nuclear units (1% for fossil) for increased last-row blade lengths of 30 to 40%. For the longest blade length studied in this analysis (+40%) the total steady stress level of the titanium blades is ~90% that of the existing steel blade. Lastly, the indirect benefit of titanium blades to reduce stress-corrosion cracking of discs and blade-root attachments needs to be demonstrated in both fossil and nuclear turbines. The substitution of titanium for steel blades in regions of turbines where the initial condensates are deposited from the steam is obvious. However, the question of how high a temperature titanium blades can be safety operated without hydriding remains unanswered. It is at least 350 K but it may extend to 450 K.

In the material development area, Jaffee[30] sees two opportunities: first, the optimization of Ti6-4 fatigue properties through heat treatment. For instance, it has been shown[40] that the 10^7 cycles fatigue strength in air can be raised as high as 680 MPa by heat treating hot- or warm-worked Ti–6Al–4V such that it recrystallizes into the microduplex equiaxed α–β condition, with grain sizes of 1 to 2 μm. This condition appears to have the highest fatigue strength.[40] For the second, Jaffee points out that Ti6-4 is not necessarily the optimum titanium alloy for steam turbine blading applications. The alloy was selected for evaluation because of the extensive data and experience bases derived from aircraft jet engine compressors. However, he notes that the LP turbine condition differs considerably in design criteria and application from the aircraft compressor application, and therefore special development and optimization of titanium alloys for steam turbine applications are warranted.

References

1. M. Wahlster and R. Schumann, A Contribution to the Electroslag Remelting of Large Forging Ingots, in: *Proceedings of the 4th International Symposium on Electroslag Remelting Processes*, Tokyo, Japan, 1973, Japan Iron and Steel Institute.

2. R. H. Koppe and E. A. Olson, Nuclear and Large Fossil Unit Operating Experience, EPRI NP-1191 (September 1979).
3. C. H. Wells and F. E. Gelhaus, Structural Integrity of Steam Turbine Rotors, in: *Conference on Structural Integrity Technology*, Washington, D.C., May 9–11, 1979, J. P. Gallagher and T. W. Crooker, eds., American Society of Mechanical Engineers.
4. J. M. Hodge and I. L. Mogford, U.K. Experience of Stress-Corrosion Cracking in Steam Turbine Discs, Proc. Inst. Mech. Eng., **193**, 93–109 (1979).
5. S. H. Bush, A Reassessment of Turbine-Generation Failure Probability, *Nucl. Safety*, **19**, 681–698 (1978).
6. L. D. Kramer and D. Randolph, Analysis of TVA Gallatin No. 2 Rotor Burst. Part 1: Metallurgical Considerations, in: *1976 ASME-MPC Symposium on Creep-Fatigue Interaction*, R. M. Curran, ed., Metal Properties Council Publication MPC-3.
7. D. A. Weisz, Analysis of TVA Gallatin No. 2 Rotor Burst. Part II: Mechanical Analysis, in: *1976 ASME-MPC Symposium on Creep-Fatigue Interaction*, R. M. Curran, ed., Metals Properties Council Publication MPC-3.
8. EPRI Steam-Turbine-Related Research Projects, Electric Power Research Institute, EPRI NP-888-SR (August 1978).
9. C. H. Wells, Reliability of Steam Turbine Rotors, EPRI NP-923-SY, Project 502, summary report (October 1978).
10. H. C. Smith and G. S. Hartman, Manufacture of Large Generator Rotor Forgings over 135 Metric Tons, presented at the International Forgemasters Meeting, Terni, Italy, May 5–9, 1970.
11. I. Roman, C. A. Rau, Jr., A. S. Tetelman, and K. Ono, Fracture and Fatigue Properties of 1Cr–Mo–V Bainitic Turbine Rotor Steels, Electric Power Research Institute EPRI NP-1023 Research Project 700-1, Tech. Rep (March 1979).
12. J. R. Weeks, Stress Corrosion Cracking in Turbine Rotors in Nuclear Powered Reactors, Brookhaven National Laboratory, BNL-NUREG 22689-R, informal report (June 1978).
13. F. F. Lyle, Jr., A. J. Basche, H. C. Burghard, Jr., and G. R. Leverant, Stress Corrosion Cracking of Steels in Low-Pressure Turbine Environments, paper presented at Corrosion '80, Chicago, March 1980.
14. R. E. Sperry, S. Toney, and D. J. Shade, Some Adverse Effects of Sress Corrosion in Steam Turbines, *J. Eng. Power*, 255–260 (April 1977).
15. D. Weinstein, BWR Environmental Cracking Margins for Carbon Steel Piping; First Semi-Annual Progress Report, July 1978 to December 1978, General Electric Report NEDC-24625 (January 1979), EPRI Contract No. RP 1248-1.
16. B. L. King and G. Wigmore, Temper Embrittlement in a 3% Cr–Mo Turbine Disc Steel, *Metall. Trans. A*, **7**, 1761–1767 (1976).
17. C. J. McMahon, Jr., Problems of Alloy Design in Pressure Vessel Steels, in: *Fundamental Aspects of Structural Alloy Design*, R. I. Jaffee and B. A. Wilcox, eds., Battelle Institute Materials Science Colloquia 1975, Plenum Press (1976).
18. B. W. Bussert, R. M. Curran, and G. C. Gould, The Effect of Water Chemistry on the Reliability of Modern Large Steam Turbines, *J. Eng. Power*, 1–6 (1978).
19. O. Jonas, Identification and Behavior of Turbine Steam Impurities, Paper 124, Corrosion 77, NACE, Houston, 1977.
20. G. A. Clarke, T. T. Shih, and L. D. Kramer, Final Report Research Project: EPRI RP502, Task IV, Mechanical Properties Testing, Reliability of Steam Turbine Rotors (March 1978).
21. C. E. Jaske and H. Mindlin, Elevated Temperature Low-Cycle Fatigue Behavior of $2\frac{1}{4}$Cr–1Mo–$\frac{1}{4}$V Steels, *Symposium on $2\frac{1}{4}$ Chrome–1 Molybdenum Steel in Pressure Vessels and Piping*, Metal properties Council, Second Annual Pressure Vessels and Piping Conference, Denver, September 1970, ASME Publication G11 (1971), pp. 137–210.
22. R. M. Curran and D. M. Wundt, Continuation of a Study of Low-Cycle Fatigue and

Creep Interaction in Steels at Elevated Temperatures, in: *1976 ASME-MPC Symposium on Creep-Fatigue Interaction*, ASME Publication G113 (1976) pp. 203–282.
23. R. M. Goldhoff and H. J. Beattie, Jr., The Correlation of High-Temperature Properties and Structures in 1Cr–Mo–V Forging Steels, *Trans. Metall. Soc. AIME*, **233**, 1743–1756 (1965).
24. F. E. Gelhaus, private communication, Electric Power Research Institute, Palo Alto, California (September 1979).
25. J. S. Takhar, R. V. Collins, J. E. Shaefer, C. D. Bucska, and J. Saez, Run/Retire Decision on a 26-Year-Old LP Turbine Rotor Based on Boresonic and Material Test Results and Fracture Mechanics Analysis, paper presented to American Power Conference, April 1979.
26. Moisture Erosion: We begin to learn, *Electr. World*, 46–49 (February 15, 1972).
27a. H. R. Tipler, The Role of Trace Elements in Creep Embrittlement and Cavitation of Cr–Mo–V Steels in: *International Conference on Properties of Heat-Resistant Steels*, Vol. 2, Paper 7.4, Dusseldorf (May 1972).
27b. H. R. Tipler, The Influence of Purity on the Strength and Ductility in Creep of Cr–Mo–V Steels of Varied Microstructures, *Philos. Trans. R. Soc. London, Ser. A*, **295**, 213–233 (1980).
28. L. R. Cooper, Advanced Technology for Producing Large Forging Ingots by Central Zone Remelting, presented at the International Forgemasters Meeting, Paris, France, April 20–25, 1975.
29. R. A. Wood, Status of Electroslag Processing for Production of Large Rotor Forgings, EPRI FP-799, TPS 77-721, final report (July 1978).
30. R. I. Jaffee, Metallurgical Problems and Opportunities in Coal-Fired Steam Power Plants, *Metall. Trans. A*, **10**, 139–164 (1979).
31. Y. Murakami, J. Watanabe, and S. Mima, Heavy Section Cr–Mo Steels for Hydrogeneration Services, paper presented at Workshop on High-Temperature Hydrogen Attack of Steels, May 24–25, 1979, Electric Power Research Institute.
32. G. M. Spienk, Reversible Temper Embrittlement of Rotor Steels, *Metall. Trans. A*, **8**, 135–143 (1977).
33. C. J. McMahon, Jr., S. Takayama, T. Ogura, Shin Chen Fu, J. C. Murza, W. R. Graham, A. C. Yen, and R. Didio, The Elimination of Impurity-Induced Embrittlement in Steels. Part 1: Impurity Segregation and Temper Embrittlement, EPRI Report RP559 (July 1980).
34. J. M. Capus, The Mechanism of Temper Brittleness, in: *Temper Embrittlement in Steel*, ASTM STP 407, pp. 3–19, ASTM, Philadelphia (1968).
35. A. Joshi and D. F. Stein, Temper Embrittlement of Low-Alloy Steels, In: *Temper Embrittlement of Alloy Steels*, ASTM STP 499, pp. 59–89, ASTM, Philadelphia (1972).
36. R. A. Wood, Status of Titanium Blading for Low-Pressure Steam Turbines, EPRI AF-445, TPS 76-641, final report (February 1977).
37. T. M. Rust, B. B. Seth, and R. E. Warner, Operating Experience with Titanium, *Met. Prog.*, 62–66 (July 1979).
38. S. M. DeCorso, Erosion Tests of Steam Turbine Blade Materials, *ASTM Proceedings* **64**, pp. 782–796 (1964).
39. W. G. Steltz, Turbine Cycle Performance Improvement through Titanium Blades, EPRI AF-903, Project TPS 77-746, interim report (September 1978).
40. G. Lütgering, Influence of Texture and Microstructure on Properties of Ti–6Al–4V as LP Turbine Blading Material, Electric Power Research Institute, EPRI Project RP 1266-1, Progress Report, May 5, 1979.

8

Future Trends in Nuclear Materials

A nuclear power plant is clearly a demanding proving ground of a material's resilience. Materials possess a memory and damage is cumulative. Abuse early in life is rarely annealed out; even if it is, the accompanying microstructural changes are rarely acceptable for long lifetimes. Thus, for application in a hostile environment, and there is probably none more demanding than that of a nuclear reactor, materials must be fabricated with care, components must be constructed with care, and systems must be operated with care. These requirements place great demands on all concerned, from the forgemaster to the reactor operator. The reliability and safety record of the nuclear industry shows that this challenge is being met, but, with the increasing need to extract more energy from available systems, further improvements are called for.

As was noted in Chapter 1, the incentives are clear cut; the significantly lower fuel costs associated with the operation of a nuclear plant (factor of 3 or more over fossil) mean that these plants will mostly operate in a base-loaded mode (i.e., noncycling), while the expensive fossil fuels are restricted for use during peak demands. Achievement of the corresponding 75 to 80% capacity factor (currently attained by only a few plants) for all nuclear plants will clearly require significant improvements in overall component reliability. For the majority of cases, this mandates an improvement in materials reliability.

Considering the efforts being undertaken to improve overall material reliability and performance, this book could well have been titled, "The Search for More Forgiving Materials." If I had chosen that title, then this final chapter would have to be called "The Search Goes On" because it is evident that, despite the use of a wide range of materials, truly forgiving materials have yet to be demonstrated for many key applications. On the positive side, however, there are encouraging signs that several highly reliable component-specific materials will be qualified in the next five to ten years. In this final chapter, I will present a personal viewpoint of the trends in material development that are evolving from the R&D efforts described earlier, and comment briefly on related technologies that must be improved concommitantly for our goals to be realized.

8.1. Overview of the Materials Problems

Table 8-1 is an attempt to summarize the important aspects of the preceding chapters—components, materials of construction, problems, and causes. It is intended to illustrate two important points: (1) many different classes of materials are already in use in nuclear systems; and (2) the causes of most of the problems are similar in nature in that they involve environmentally assisted subcritical crack growth (i.e., stress-corrosion cracking or corrosion fatigue).

It is generally recognized that materials which contain no impurities or alloy additions are immune to stress corrosion. However, most engineering materials are dependent upon critical alloy additions to develop properties such as low-temperature strength, toughness, creep-rupture strength, and general corrosion resistance; hence, engineering materials are susceptible to stress corrosion.

What examination of Table 8-1 tells me is that, for many components, our priorities might have been misplaced. It appears somewhat irrelevant to have materials that possess design code-approved high strengths, toughnesses, and creep resistance, if they are susceptible to SCC at stresses of the order of one-half to one-third of their yield strength! The experience described in the preceding chapters carries the important message that a compromise is necessary; that is, to maintain adequate strength, toughness, and overall corrosion resistance, while substantially raising the threshold for stress corrosion. This requirement seems to me to demand that the overwhelming trend in materials development should be toward lower-alloy materials and proper control of minor elements.

Perhaps an even more compelling reason for the trend to be away from "higher" alloys is the scarcity or likely scarcity of the alloying elements. Chromium is the major problem now, but manganese, tin, and nickel supplies will be future problems.[1] Currently, the U.S. imports essentially all (91%) of its chromium and about 71% of its nickel. If a large number of nuclear reactors are constructed, the extensive use of austenitic stainless steels would consume a high fraction of these imported element, and, as supplies go down, costs will escalate. Therefore, metallurgists are going to be asked to do the same job with less expenditure and fewer alloy additions.

Thus, the problem boils down to one of defining high-strength, stress-corrosion resistant ("forgiving") materials that contain minimal amounts of chromium, nickel, manganese, etc. The balance of this chapter will explore what progress is being made toward this goal.

8.2. Specific Materials Developments

There is a diversity in the level of "needs" for material design, and the nuclear system spans the entire range. At one extreme are those applications for which the material requirements are readily defined by an existing design or application. This case generally is associated with a static or evolutionary system-design process (e.g., components of the LWR). At the opposite extreme are those applications where either changes in existing design or totally new directions make it difficult to define or anticipate alloy needs. In this case, the system-design process is fluid, in many cases synergistically changing with alloy design (e.g., the first wall of the CTR).

Rarely have brand new materials been "dreamt up" for the nuclear system; to my knowledge only the Zircaloys fit that category. Regardless of end purpose, generally the material (metallic alloy or ceramic) evolves by successive modifications of the composition, thermomechanical treatment, and/or the fabrication process of an *established* material. Either materials having desired properties for a specific application, or one material for a multitude of applications, is produced. Of relevance to this discussion are, first, the selection of "base" materials, and, second, the directions that are being taken to optimize that material. Not too surprisingly, one finds that opinions differ on both initial selection and the directions for improvement. These different philosophies should become evident as I review trends in the following classes of materials: (1) low-alloy (C, Mn, Mo, Cr) steels; (2) "stainless" steels; (3) nickel-base alloys; (4) "new materials" (for nuclear plants)—titanium; and, finally, (5) "barrier" materials.

8.2.1. Clean (Low-Alloy) Steels

High-strength, low-alloy (HSLA) steels are used extensively in the construction of the large nuclear components—LWR reactor vessels and steam generator shells (C, Mn steel), LMFBR steam generator shells ($2\frac{1}{4}$Cr–1Mo), and turbine rotors (Ni–Cr–Mo–V steels). The properties of

Table 8-1. Nuclear Systems: Component

Reactor components	Key materials
1. Core:	
Fuel rod (LWR)	Zircaloy/UO_2
Fuel rod (LMFBR)	316 SS/$(U-Pu)O_x$
Control rod (BWR)	304 SS/B_4C
Fuel assembly { Duct (LMFBR) / Channel (BWR) }	316 SS / Zircaloy
2. Pressure boundary:	
Vessel (LWR)	Low-alloy steel
First wall (CTR)	Candidates: 316 SS, ferritic SS, Ti alloys or V alloys and/or Ceramic (graphite, SiC)
Piping (BWR)	304 SS
Piping (PWR)	C steel
3. Heat exchanger:	
Steam generator (PWR)	C steel/Ni alloy
Steam generator (LMFBR)	$2\frac{1}{4}$Cr–1Mo steel and/or Inconel alloy 800
Condenser (all)	Cu–Ni alloys
4. Turbine:	Ni–Cr–Mo–V bainitic steels, 12Cr SS

Problems and Materials-Related Causes

Major problems	Primary causes
Cladding perforation	Stress-corrosion cracking (SCC)
Wastage	Corrosion, hydriding
Marginal economics: low breeding ratio and doubling time	Fuel properties and clad creep/swelling and corrosion
Cladding perforation/B_4C leachout	Stress-corrosion cracking (SCC)
Bowing and dilation	Creep/swelling Creep/corrosion
Demonstrate 30-year integrity in presence of small cracks	Radiation embrittlement and corrosion fatigue
Requirement for absolute integrity under high-cyclic stress, neutron flux, and temperature conditions	High-cycle fatigue, creep/corrosion fatigue, and radiation damage Thermal shock and radiation damage
Cracking in weld heat-affected zone (HAZ)	Intergranular SCC (IGSCC)
Cracking	Corrosion fatigue
Tube and tube support distortion and cracking	General corrosion and SCC
Possible rupture of tubes	General corrosion
Possible tube cracking	SCC
Tube failures	SCC, erosion, and/or wastage
Rotor bursts, disc cracking, blade cracking	Environmental fatigue, temper embrittlement, IGSCC, corrosion fatigue, or SCC?

these steels are determined by the following factors:

(a) The use of alloying elements such as niobium, vanadium, and titanium, which promote both grain refinement of the austenite and precipitation hardening in the ferrite
(b) The ability to process the austenite phase by hot deformation in a controlled manner to produce the required microstructure (e.g., bainite or martensite)
(c) The trend toward improvement of weldability by reducing the carbon content, in addition to control of the alloy content
(d) The trend toward control of inclusion content using improved casting practices

Experience with LWR pressure vessels (Chapter 4) and steam turbines (Chapter 7) points to the last two areas for future developments. Strength, toughness, and weldability do indeed increase as minor-element levels decrease, but, more importantly, resistance to stress-corrosion cracking and fatigue cracking also increases. Therefore, the ingot preparation techniques that maintain low carbon (<0.22 wt.%) and sulfur (0.01 to 0.015 wt.%) levels and virtually eliminate phosphorus, antimony, silicon, etc., must be made standard practice worldwide for vessel and turbine rotor production.

As was described in Chapter 7, the technology to produce "clean" steel ingots is available, but the steelmaking industry must be convinced that the cost-benefits are sufficiently favorable to convert from the conventional melting practices. Not only must the utilities be prepared to pay the higher price for clean steels, but the current and future demands for these materials must be sufficient to justify the steelmakers' capital expenditure on new equipment.

A good case in point is the slow introduction of the electroslag remelt (ESR) technology. Figure 8-1 shows qualitatively what has been realized for a long time: that, due to a much better ingot-to-forging yield than conventional casting practice, the production of large forgings by ESR will require roughly 33% lower raw ingot weight.[2] ESR has also been shown to reduce overall impurity and defect content. Nevertheless, apart from a pilot plant facility built in West Germany in the late 1960's (capable of producing 170,000-kg ingots) and some small units operational in the U.S. and USSR, there has been no move to introduce this technology for large-scale ingot production.

Cost is of course a major consideration (ESR is 20% more costly than conventional practice),[3] and scale-up to the ingot sizes required for integral rotors of the new generation of 1000 MWe plus nuclear plants is a problem. But combinations of conventional casting practice and ESR

Future Trends in Nuclear Materials

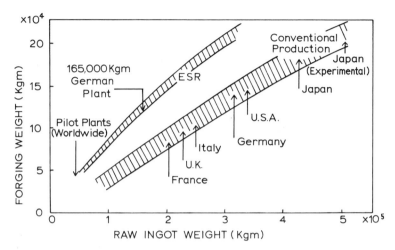

Fig. 8-1. Raw ingot and forging weights for turbine rotors, comparing electroslag remelt to conventional casting practice (adapted from Ref. 2).

(e.g., electroslag hot top or ESHT), which have no apparent scale-up problems and lower costs, are still limited in commercial application, in this instance to two European steelmakers.

What seems to be needed here is an unequivocal demonstration of the benefits in reliability and performance that will accrue from increased steel "purity." Laboratory studies are the primary source of data; clearly, in-service statistics are urgently required. This should be remedied in the near future by the new industry–government cooperative evaluation programs that will place clean steel components out in the field for detailed, long-term testing. The data must, however, be fed back to the steelmakers, along with the message that utilities are willing to pay for increased reliability, if clean steels are to become an accepted standard.

8.2.2. "Stainless" Steels

The austenitic stainless steels, Types 304 and 316 in particular, are in widespread use in nuclear systems (Chapters 3–6). Their selection is predicated on high-temperature strength and creep resistance, and overall corrosion resistance in water and liquid sodium. Their failing is their high susceptibility to stress corrosion cracking, whether sensitized or not, in water containing O^-, Cl^-, OH^- and SO_4^{-2} ions. Consequently, although these alloys will remain primary alloys for sodium environments, there is a growing trend away from this class of steel for applications involving expsoure to steam/water environments [e.g., steam generator materials (Chapter 6)].

The replacement materials that, in my opinion, represent one of the soundest responses to the two problems discussed at the outset of this chapter are the ferritic "stainless" steels. This is a class of alloys containing 9 to 13% Cr, with small additions of various strengthening elements such as Mo (1 to 2%). Generally, they contain only trace amounts of nickel. These steels are conventionally used in the normalized and tempered condition for high-temperature application, and, although this treatment can result in either a martensitic or bainitic structure, they are often referred to as "martensitic" stainless steels.

This class of steel has been employed successfully in such high-temperature applications as steam turbines, jet engines, and gas turbines, and a commerical 12Cr–1Mo–0.3V alloy (Sandvik HT-9) has been used in Europe for steam pipelines and fossil-fired boiler components for over ten years. Strictly speaking, therefore, they are only new in the sense of their potential use for nuclear components.

For nuclear applications, the benefits of the ferritic stainless steels over the lower-alloy ferritic steels (e.g., $2\frac{1}{4}$Cr–1Mo) which should be exploited include: significantly higher strength, increased carbon stability, improved resistance to thermal shock, and enhanced waterside corrosion resistance. The characteristic higher yield strength, lower thermal expansion, higher thermal diffusivity, and improved resistance to caustic and chloride SCC render the ferritic stainless steels also superior to the austenitic stainless steels and nickel-base alloys for certain ex-core applications. In addition, their radiation damage resistance can be utilized in in-core applications.

Thus, this alloy class may well approach being a near-term multipurpose, forgiving material for nuclear plant construction. Consider the applications for which preliminary data appear very promising.

1. LMFBR Core Materials (Chapter 3). Zero- to low-void swelling, high resistance to radiation creep, and apparent absence of helium embrittlement effects, combined with good compatibility with sodium, now make the ferritic stainless steels a leading candidate for duct (and possibly cladding) materials in the next generation of LMFBR cores. The French are already committed to a 11Cr steel for the Super Phenix ducts, while the U.S. are pursuing a developmental 10Cr alloy.

2. CTR First-Wall Materials (Chapter 5). The "radiation properties" exhibited in the LMFBR studies have to make these alloys a leading candidate for the first wall of either tokamak or ICF reactors. They are a late entry into the U.S. CTR alloy development program, and, to my knowledge, no CTR-related data exist. Nevertheless, based on the potential for reduced helium embrittlement and lower thermal stress generation (which are the two key CTR first-wall problems) one should expect significantly longer wall lifetimes at higher wall loadings than can be expected from the austenitic steels.

3. *Steam Generator Components (Chapter 6).* The U.S. has evolved a 9Cr stabilized alloy (0.1C–9Cr–1Mo, V, Nb) that appears to combine the needed high-temperature mechanical properties (fatigue, creep rupture, etc.) with general corrosion resistance, and, importantly, high SCC resistance considered necessary for LMFBR steam generator application. Other important needs such as fabricability and weldability also seem to be in hand. Inasmuch as the steam generator is a pressure-boundary component, Code acceptance is necessary (for the CTR first wall also, but the urgency is not there) and much work remains to obtain all the necessary data. Unfortunately, based on experience with conventional pressure-vessel steels, it could take up to ten years to gain formal Code acceptance.

Interest in the ferritic stainless steels for LMFBR steam generator application is worldwide, with each country pursuing development of its favorite alloy. The French are developing the data and fabrication experience with a 9Cr–2Mo–Nb–V alloy (Vallourec trademark EM12), the British have several alloys in the 9–12Cr range under study, while the Japanese are considering a 9Cr–2Mo steel (HCM9M). The 12Cr–1Mo, W, V, Ni Sandvik HT-9 alloy is also still popular.

Application of ferritic stainless steels in PWR steam generators is evidently still an open question. They have been introduced as support-plate materials in new steam generator designs both in the U.S. and in France, but early laboratory experience is not encouraging. There appears to be no move to substitute these steels for the nickel-base alloy tubing.

In reviewing the current enthusiastic activities on the ferritic stainless steels, one major concern emerges. We are so busy seeking the optimum alloy composition in this class that, rather than converging on a common alloy composition for a specific application, each country (and even groups within a country) is developing its own alloy. Given the limited resources, and the enormous body of data that is required to fully qualify a material, it is clear that, unless some convergence or coordination of efforts is accomplished, it will be a long time before any of these materials see wide-scale acceptance and application. I would therefore advocate a coordinated worldwide effort to first select the optimum ferritic steels for application (say in radiation and in steam/water environments) and then to provide the necessary laboratory and field data to qualify the alloys. Such an effort could cut the normal qualification period in half at a fraction of the normal cost to an individual country.

8.2.3. Nickel-Base Alloys

By nickel-base alloys I refer, of course, primarily to the Inconel series of alloys that are popularly used in regions of high corrosivity—e.g., in the LWR vessel as sleeving (Chapter 4), in the LWR core as control rod cladding (Chapter 2), and in the steam generators as tubing (Chapter 6). It is the latter use that employs the largest quantities of the alloy and so this is where we will explore the trends in nickel-base alloys.

There are at least three, and probably more, schools of thought on the future directions for nickel-base alloys. There are those who would make the Inconel 600 alloy work (for the PWR) by optimizing the microstructure. Then, there is a second school that quite clearly prefers the higher chromium Inconel "Alloy" 800 (for PWR and LMFBR) and are working to optimize its properties also. The third school is looking at alternative higher-chromium alloys (for PWR and maybe LMFBR) such as Inconel 690 and SCR3. It is important to note that these three schools are not restricted to specific countries. While Inconel 600 can be assigned to the U.S. interests, Alloy 800 has supporters in both the U.S. and Europe, and the higher-chromium alloys are being actively considered in the U.S., Japan and France.

From my review of the available literature, I must conclude that all of these alloys possess strength properties and general waterside (and sodium, if necessary) corrosion resistance that are adequate for the intended application, and, therefore, resistance to SCC should be the major factor determining their selection. However, it is difficult to provide a ranking with respect to SCC susceptibility inasmuch as each alloy is under development and the data reviewed in Chapter 6 showed the sensitivity of SCC to microstructure and impurity content. Historical data would exclude Inconel 600 and possibly even Alloy 800 in favor of Inconel 690, but such a decision would neglect the more recent improvements in these alloys derived from new heat treatments. Therefore, I propose to argue a case for selecting the alloy on the basis of the second part to the original problem, namely, the scarcity of alloying elements, and trust the metallurgist to find a processing treatment that will provide the necessary SCC resistance.

Accordingly, my choice for further consideration for PWR steam generator tubing is Inconel 600 (16% Cr) in the improved microstructural condition developed, for instance, by the 24-h, 977 K heat treatment. For LMFBR steam generator tubing, I propose Alloy 800 (21% Cr) due to its better high-temperature strength, but only for the near-term, pending the introduction of Code-qualified ferritic stainless steels. I do not see how we can proliferate the use of the higher-Cr alloys ($\geq 25\%$); rather, such materials should be restricted to special applications for which there are no obvious lower-alloy substitutes.

8.2.4. New Materials—Titanium

There are, no doubt, other materials that could be considered new to the nuclear industry, but, in my opinion, in this category the one with the most promising future is titanium.

Experience with titanium in nuclear components is, admittedly, limited, and only commercially pure titanium and the Ti–6Al–4V alloy, which possesses average properties, have been tested. Nevertheless, the results from condenser tube (Chapter 6) and steam turbine (Chapter 7) performance are encouraging and justify further detailed consideration of titanium.

Titanium poses no resource problems, and an established industry already exists. Consequently titanium alloys are a sound solution to the problem of identifying materials with high SCC resistance but low strategic (i.e., Cr, Ni) alloy content. Further, inasmuch as only the Ti6-4 alloy has been used in nuclear components to this point, there is a great deal of room for improvement of the corrosion and strength properties that are important to nuclear component reliability.

The "window" available to the nuclear metallurgist can best be illustrated by reference to the advances made in titanium alloy development in the aircraft industry. The major effort here has been to advance the temperature capability of titanium alloys through alloying, processing, and heat treatments. Just how far they have moved up from Ti6-4 can be seen in Fig. 8-2.[4] The maximum temperature for acceptable creep strength has been doubled over Ti6-4 by the development of a β-form of Ti–6Al–2Sn–4Zr–2Mo (Ti6242). Even further increases in high-temperature creep strength are projected for the TiAl intermetallic alloys. Oxidation resistance and resistance to fatigue crack propagation have experienced concomitant improvements.

Clearly, nuclear plant environments (e.g., in the turbine, condenser, or steam generator) differ considerably from aircraft compressors and so special development and optimization of titanium alloys are warranted. One major concern in the nuclear system will be the temperature limit imposed on the use of titanium by the problem of hydriding; this could be as low as 550 K, which would eliminate LMFBR steam generator application, for example. Because of this and other requirements the direction(s) may well not be the same as indicated in Fig. 8-2. Nevertheless, the technology developed over the last two decades should facilitate

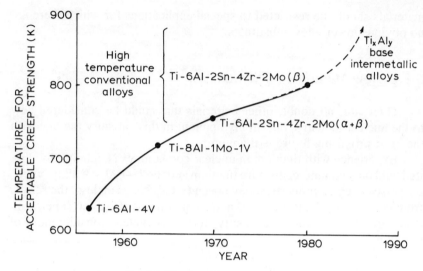

Fig. 8-2. Titanium alloy trends (Ref. 4).

development of titanium alloys for uses in the nuclear system beyond those currently planned.

That titanium alloys may achieve the status of multipurpose nuclear materials in the future is also apparent from developments in the CTR community (Chapter 5). Several design studies and material evaluations have identified titanium alloys (specifically Ti6-4 and Ti6-2-4-2) as candidate first-wall materials. The major properties are, however, generally different from those touted for titanium use in fission-reactor balance-of-plant components, and in this respect, titanium can be compared to the ferritic stainless class of alloys. Titanium alloys exhibit a high resistance to radiation damage (both by atomic displacement and transmutation products), equalling or exceeding that of the ferritic stainless alloys. When coupled with high fatigue resistance and low thermal stress generation, this translates into greatly improved first-wall lifetimes for operating temperatures at or below 770 K. Additionally, titanium is a low-activation material, which property should substantially reduce the long-term (>10 year) radioactivity relative to a Type 316 stainless steel (or Ni-base alloy) wall.

The question of hydrogen effects is always raised in connection with titanium, and, for CTR application, tritium absorption and hydriding will be the major issues. However, this may not be as big a concern as some believe because hydrogen should diffuse rapidly through or out of titanium at 670 to 770 K.[5] Other areas that should receive more attention in the titanium development program are long-term phase stability and segregation under irradiation, which may necessitate some alloy optimization work, and general corrosion and corrosion fatigue resistance in water and liquid metals.

8.2.5. Barriers and Coatings

It is not new to the nuclear system to use a coating or a barrier to improve corrosion and/or erosion resistance, but my general impression is that the use of this approach is on the increase. Clearly, the move to barriers, coatings, or duplex systems to contend with corrosive/erosive environments is a logical one, and, if the integrity of the "layer" can be assured, it is a highly effective means of gaining the improved component reliability without the expense or complexity of fabricating the entire component from the barrier. In many cases this latter approach would be undesirable anyway.

As indications of future trends, new uses of barriers will be discussed in this section, rather than the conventional use of stainless steel weld overlay cladding—now extended from LWR vessels to BWR piping—and Stellite coatings on valves and turbine blades.

8.2.5.1. Fuel Rod Cladding Barriers

There is no near-term replacement for the uranium–plutonium fuel cycle in fission reactors, nor is there the justification to switch from Zircaloy cladding for the LWRs (Chapter 2) or from austenitic stainless steel cladding for the LMFBRs (Chapter 3). Hence, the challenge to the metallurgist is to improve what is already a highly reliable fuel-cladding system. How can one reasonably expect to maintain a fuel rod failure rate at 0.01% or less, while responding to the call for a flexible, high-capacity-factor core? The fission product barrier may well provide the answer to this dilemma, and give us the truly "forgiving" fuel rod design.

Uranium dioxide and Zircaloy are initially compatible, but the result of the fission process rapidly renders the two materials incompatible. The same can be said, to a lesser degree, for the stainless steel–mixed uranium–plutonium oxide system. Thus, the trick is to keep the corrosive fission products from the cladding, and what better way than to coat the inside surface of the cladding with a fission product getter or barrier.

The approach that has the highest likelihood for success for the UO_2–Zircaloy fuel rod is the pure zirconium barrier that is coextruded, $\sim$.08 mm thick, with the Zircaloy tube. My reasons for this choice are that: (1) pure metals are inherently resistant to SCC; (2) a thick barrier has more chance of survival than a thin layer (such as in the 5–10-μm thick copper or siloxane remedy); and (3) a single material fuel cladding component has a better chance of passing the strict safety scrutiny.

For the future generation of fast breeder fuel rods, the requirement for thinner cladding and the likelihood of prolonged fuel pellet–cladding contact due to higher smeared fuel densities and swelling-resistant cladding alloys may necessitate the use of fission product barriers in this system also.

This potential need has been recognized in the LMFBR industry, and small-scale experiments have been implemented to test various fission product getters. Stainless steel appears to have a distinct advantage in this environment in that it is immune to SCC; fission product attack is simply grain-boundary penetration and general wall thinning. Hence, the getter can be a thin coating (of Ti, Nb, V, Cr) on the inside cladding wall or can even be dispersed within the fuel. The only requirement of the getter is to control the oxygen partial pressure to keep the fission products away from the cladding. Whether this more straightforward state of affairs will persist to the longer-burnup, higher-power fuel of the advanced breeder fuel designs remains to be seen.

8.2.5.2. Fusion First-Wall Coatings

Helium blistering is a serious erosion process for all fusion components in contact with the plasma, but for the first wall in particular due to the requirement for absolute integrity (Chapter 5). Thickness losses of about 0.09 mm per year can be estimated based on laboratory studies on Type 304 stainless steel, for example, and these losses are likely to be enhanced in a real system by the vaporization of the blister skin. All metals studied seem to exhibit this phenomenon. Work is in progress to find ways to reduce the helium blistering process, and one of the most promising approaches is the use of ceramic liners or coatings.

The ceramics such as carbon (graphite) and silicon carbide possess other properties that make them attractive for use on the "plasma side" of fusion devices; for example, they are "low-activation" or low Z materials and, hence, will minimize plasma poisoning. In fact there are some who believe that a first wall could be built entirely from ceramics. However, the property that is likely to be exploited first is the helium erosion resistance that is characteristic of ceramics because of their inherent porosity. Results indicate that helium reemission rates are significantly higher in graphite and, to a lesser extent, in high-density silicon carbide. Both materials have well-established industry bases, and coating technology is advanced. Thermal and mechanical properties of these materials are also conducive to long lifetimes at high operating temperatures.

Therefore, a program to develop silicon carbide and carbon for fusion first-wall coatings should be easily justified. Beyond this application, however, I forsee problems in the use of these materials because of their inherent brittleness. Nevertheless, if one includes in the selection criteria the virtually infinite raw material resources, then the larger-scale R&D effort that would be required to extend their use is justified.

8.2.5.3. Tritium Barriers

The high tritium concentrations or partial pressures that provide the high recovery rates desired from the breeding blanket of some fusion reactor designs may also cause high rates of permeation loss to the environment. To keep permeation losses to an acceptably low level in a liquid-lithium breeding blanket, the partial pressure of tritium must not exceed 10^{-10} Torr above the lithium.[6] This is an incredibly low partial pressure to maintain for a product in an industrial situation and, to date, no way of doing this has been found.

An obvious way to reduce some permeation losses is to use hydrogen-impermeable heat exchangers (Chapter 6). This can be done by either dropping the coolant operating temperature (and efficiency) or by adding additional heat exchange loops; but the addition of new materials to the heat exchangers represents a more efficient method. Some of these new materials are barrier metals (copper, tungsten), barrier oxides (oxidation films, BeO), and metal getters (titanium, cerium), and, as shown in Fig. 8-3,[6] all might be placed into laminated (triplex) tubing to reduce or eliminate permeation losses. Of the different candidates oxide barriers are far less permeable than the least permeable of metals (tungsten) and should be developed further. In particular, thin film techniques should be developed to produce pore-free films.

8.3. Related Technologies

The improvement in materials reliability through increases in key properties (SCC resistance, radiation damage resistance, strength, etc.) is just one part of the solution to overall component reliability. Material-joining processes, material and component inspection methods (i.e., nondestructive evaluation or NDE), and material performance analyses must all be improved in parallel.

In my strict adherence to materials technology in the nuclear power industry, I recognize that I have neglected these three related, and equally important, technologies. Clearly, each could justify a separate book in-

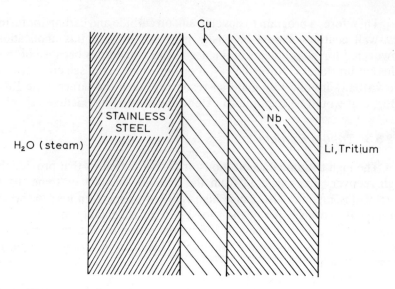

Fig. 8-3. Schematic wall cross section of a possible triple heat-exchanger tube designed to reduce tritium escape via the permeation mechanism (Ref. 6).

asmuch as the disciplines are currently just as dynamic. The developments of particular interest to me, and which offer potentially significant improvements in material reliability, include the following:

(1) Welding techniques that (a) reduce the number of passes (and hence the number of thermal cycles and resulting HAZs), and (b) optimally redistribute the resulting internal stresses. In particular, for the former problem, resistance welding techniques using a homopolar generator offer a cost-saving solution, while for the latter case heat-sink welding (HSW) and induction-heating stress-improvement (IHSI) welding look promising.[8]

(2) Analytical techniques that facilitate the interpretation of NDE signals, enabling the operator to both discern true cracks or defects from the general background noise and to establish their dimensions (e.g., the concept of adaptive learning).[9] In this respect, the materials that I singled out for future development lend themselves more readily to NDE inspection, since it is a known fact that the reduction of slag inclusions, particles, and precipitates of the same or similar size to crack defects substantially increases the ability of conventional eddy current and ultrasonic inspection capabilities to "pick up" the true cracks.

(3) Continuous monitoring of the performance of major components to provide the plant operator with assistance in anticipating (or projecting) problems so that they can be avoided or a least remedied quickly. I am

aware of one such system (dubbed the power shape monitoring system or PSMS)[10] in operation, which utilizes reactor core data to "drive" models for fuel rod behavior to track the mechanical condition (fail/no fail) of the fuel in LWR cores. The total cost of this system's hardware is equivalent to the replacement power costs for one day's outage if oil has to replace nuclear (i.e., $300–500K). On this basis there appears to be ample justification to broaden the scope of this type of system to include other major problem areas. The individual elements are available—materials behavioral models; temperature, stress, and corrosivity sensors; and fast, compact computer systems. What is required is development work to integrate the plant operational data and computer software and hardware. Examples of near-term applications should include the automatic tracking and control of water chemistry in LWRs (and LMFBRs) to avoid unnecessary corrosion of steam generators and major piping, and automatic sensing of turbine performance to minimize catastrophic failures of rotors, discs, and blades.

8.4. Closing Remarks

The materials scientist working in the nuclear industry should take pride in achievements to date and in the promise that the developing materials holds for the near future.

It is not "wishful thinking" to expect that the progress being made on the problem components in the LWR will result in 75–80% capacity factors (already enjoyed by some plants) as an industry-wide average within the next five to ten years. Nor is it out of the question to see similar capacity factors posted for 1000 MWe LMFBRs if they were to be built; indeed, the French might even achieve this goal during this period if the Super Phenix construction adheres to schedule.

Clearly, the biggest challenge to the materials scientist is the fusion reactor. The fundamental question of whether there is a material that can contain the plasma and retain its integrity over the time span necessary to make fusion economically viable still remains. I am optimistic that we can meet this challenge also. The ferritic stainless steels, vanadium alloys, titanium alloys, and ceramics offer the hopes that such a forgiving material will emerge ten years or so from now when countries begin to move seriously toward commercial fusion power.

References

1. J. P. Hirth and A. H. Clauer, Concluding Agenda Discussion—Critical Issues, in: *Fundamental Aspects of Structural Alloy Design*, R. I. Jaffee and B. A. Wilcox, eds.,

Battelle Institute Materials Science Colloquium, Plenum Press, New York (1975), pp. 647–661.
2. M. Wahlster and R. Schumann, A Contribution to the Electroslag Remelting of Large Forging Ingots, in: *Proceedings of the 4th International Symposium on Electroslag Remelting Processes*, Tokyo, Japan, 1973, Japan Iron and Steel Institute.
3. R. I. Jaffee, Metallurgical Problems and Opportunities in Coal-Fired Steam Power Plants, *Metall. Trans. A,* **10,** 139–164 (1979).
4. F. L. VerSnyder and M. Gill, New Directions in Alloy Design for Gas Turbines, in: *Fundamental Aspects of Structural Alloy Design,* R. I. Jaffee and B. A. Wilcox, eds., Plenum Press, New York (1975), pp. 290–227.
5. R. H. Jones, B. R. Leonard, Jr., and A. B. Johnson, Jr., An Assessment of Titanium Alloys for Fusion Reactor First-Wall and Blanket Applications, Electric Power Research Institute, EPRI-RP 1045-3, project final report (1979).
6. R. G. Hickman, Tritium-Related Materials Problems in Fusion Reactors, in: Critical Materials Problems in Energy Production, Charles Stein, ed., Academic Press, New York (1976), pp. 189–220.
7. R. L. Jones, T. U. Marston, S. T. Oldberg, and K. E. Stahlkopf, Pressure Boundary Technology Program: Progress 1974 through 1978, Electric Power Research Institute, NP 1103-SR (March, 1979).
8. M. J. Povich and R. E. Smith, An Overview of Intergranular Stress-Corrosion Cracking in Light Water Reactors, paper presented at American Power Conference, April 1979, Chicago, Illinois.
9. M. F. Whalen and A. N. Mucciardi, Pressure Vessel Nozzle Inspection Using Adaptive Learning Techniques, presented at the First International Seminar on Nondestructive Examination in Relation to Structural Integrity, 5th SMIRT Conference, West Berlin, August 1979.
10. F. E. Gelhaus, D. L. Pomeroy, and M. T. Pitek, Impact of Fuel Damage on Nuclear Power Plant Operations, invited paper, American Nuclear Society Winter Meeting, Washington, D.C., November 12–16, 1978.

Appendixes

Appendix A. Conversion Table from SI Units to Engineering Units.................................... 470
Appendix B. Glossary of Terms and Acronyms 471
Appendix C. Nominal Compositions of Alloys Used (or Candidates for Use) in Nuclear Components 476

Appendix A. Conversion Table from SI Units to Engineering Units

Parameter	To convert from (SI unit)	To (engineering unit)	Use equation or multiply by
Temperature	degree Kelvin (K)	degree Celsius (°C)	°C = K − 273
	degree Kelvin (K)	degree Fahrenheit (°F)	°F = 1.8K − 460
Time	second (s)	hour (h)	2.78×10^{-4}
	second (s)	day	1.15×10^{-5}
Length	meter (m)	inch (in.)	39.37
Area	meter^2 (m^2)	inch^2 (in.^{2})	1550
Volume	meter^3 (m^3)	inch^3 (in.^{3})	61023
Weight	kilogram (kg)	pound-mass (lbm)	2.20
Pressure or stress	Pascal (Pa)	psi (lbf/in^2)	1.45×10^{-4}
	Pascal (Pa)	bar	1.00×10^{-5}
Energy	joule (J)	British thermal unit (Btu)	9.48×10^{-4}
	joule (J)	calorie (cal)	0.239
	joule (J)	watt-second (W-s)	1.00
	joule (J)	foot-lb (ft-lb)	0.735
Linear heat rating	kilowatts/meter (kW/m)	kilowatts/ft (kW/ft)	0.305
Burnup	gigajoule/kg of heavy[a] metal (GJ/kgM)	megawatt-day/metric ton of heavy metal (MWd/MTM)	11.574
Fast neutron damage	displacements per atom[b] (dpa)	neutrons/cm^2 (n/cm^2)	$\sim 2 \times 10^{21}$
	neutrons per m^2 (n/m^2)	neutrons/cm^2 (n/cm^2)	10^{-4}
Stress intensity	megapascal$\sqrt{\text{meter}}$ (MPa$\sqrt{\text{m}}$)	Ksi$\sqrt{\text{in.}}$	0.909

[a] M is normally the SI symbol for mega (10^6); an exception in this book is the additional use of M in burnup units to denote heavy metal.

[b] Fast neutron damage is reported as displacements per atom or neutrons/cm^2 ($E > 0.1$ or 1 MeV) in the literature. The conversion represents an approximation since true conversion must include details of the radiation conditions which are not usually given.

Appendix B. Glossary of Terms and Acronyms

1. Frequently Used terms

Blanket That part of the fission or fusion reactor which contains the "fertile" material. Neutron reactions in the blanket breed the fissionable or fusionable materials.

Breeder A breeder reactor is one which generates more fissionable fuel than it burns up.

Burnable poison A material which absorbs neutrons and, hence, moderates or reduces the heat energy which would otherwise be generated in a reactor core. Burnable poisons are used as control rods which are moved in and out of a reactor for its control and also as fixed rods interspersed with the fuel rods.

Chemical cleaning The use of chemicals to remove corrosion products such as scale and crud that contribute to reactivity buildup in the system.

Cleavage Crack follows certain "preferred" crystal planes.

Crud Generic term for corrosion products in the primary system that circulate with the flow and settle on other components in the primary side, such as the core. These are primarily Fe, Co, Mn, Ni, and Cu-based compounds produced by corrosion of the steel surfaces.

COMETHE A fuel performance code developed by Belgo Nucleaire for prediction of both LWR and LMFBR fuel rod behavior.

Diversion resistant There is currently political and social concern that some of the reactor fuels such as plutonium can be used for the production of bombs. Diversion-resistant procedures will make it more difficult, if not impossible, for small groups of terrorists to divert fissile material from peaceful power production to bombs.

Dopant Addition (metal or ceramic) made to fuel powder prior to pressing and sintering to cause grain growth, enhanced plasticity in-reactor, and/or to trap fission products.

Duty cycle A "catchall term" for the manifestations of the power–time history of the plant on individual components. It could be stress vs. time, temperature vs. time, corrosivity vs. time, or a combination of parameters.

Enrichment When the proportion of fissile material in a fuel is increased, the fuel is said to be enriched.

Equation of state Equation or series of equations that completely describe a phenomenon. Used in the context of E. Hart's model of plastic deformation of metals.

Erosion Material loss due to the removal of surface layers by bombardment with some environment (gaseous, liquid, or radiation).

Fast Neutrons generated by radioactivity or fission have a very wide range of energies, and those with high energy have significantly different properties than those with low energy. It has become standard in the industry to refer to fast neutrons as those with energies above 0.1 million electron volts. The slow ones are called thermal neutrons.

Ferritic "stainless" steels Class of alloys containing 9 to 13% Cr, with small additions of various strengthening elements such as Mo (1 to 2%). Since the general heat treatment yields a martensitic structure, they are often referred to as "martensitic" stainless steels.

Fertile Some isotopes which do not fission or split when bombarded with neutrons will nevertheless capture neutrons and be converted to fissile isotopes.

Fissile Fissile isotopes are those which divide or split when bombarded with neutrons and release large quantities of heat. These are the primary fuels of fission reactors.

Fission The process of splitting or division of nuclei with release of heat energy. Only a few of the heaviest elements are capable of fission.

Fretting The "rubbing" interaction between two adjacent materials, caused by vibration, leads to local deformation and wear of one or both components.

Fuel Performance Code A computerized model of a fuel rod that analytically treats the synergistic interaction of material properties and irradiation effects. Such codes are used extensively in fuel management planning, fuel performance analysis, and licensing.

Fusion The process of the combination of nuclei of the lightest elements with concomitant generation of heat energy.

Hot pressing Compaction into pellet shapes with simultaneous action of pressure and temperature.

Hypothetical accident An extremely low-probability accident that would cause failure of fuel rods in significant numbers if it occurred. The licensee is required to show that his fuel can still retain sufficient integrity so that the core can be cooled and hence the accident terminated. An example of such an accident is the loss of coolant accident (LOCA).

Appendix B 473

Intergranular Phenomenon (e.g., cracking, precipitation, etc.) occurs *at* grain boundaries in material.

Leak-before-break This concept requires that crack growth be stable and that component failure be benign, in the form of a small crack, rather than an unstable large rupture.

Licensing All fuel (new cores or reloads) must receive a license from the particular country's nuclear regulatory body. An important part of the licensing process is the proof (by data and analysis) that the fuel rods will have a very low probability of failure (<1% of the core) during normal operation, and that the core will be able to be cooled even in the event of a low-probability extreme accident.

Load-follow A power plant operated in synchronization with electricity demand; cyclic operation.

MATMOD *Ma*terials *mod*el developed by A. K. Miller to describe the time-dependent inelastic deformation of metals and alloys.

Over-aging Precipitate growth beyond the optimal size and distribution for favorable properties.

Overpower A power increase either planned or inadvertent above the intended normal design power for a fuel rod or a reactor.

Passivation Formation of a tight, adherent, protective (i.e., self-healing) oxide film on the surface of a metal.

Pitting The formation of small (micron size) depressions in the material surface. A form of *local* corrosion.

Power ramp The rate of increase of power generated by a reactor has effects on the fuel which were initially unexpected. Present practice is to control the rate of change (i.e., the power ramp), particularly in the upward direction of the power output with great care.

Reference This applies to an accepted first design of a plant or component.

Sensitization Removal of "passivating" chromium from grain boundaries, leaving them at the mercy of corrosive environments. The material is now sensitized to IGSCC.

Smear density Density of fuel in place within the cladding when both the free volume in the fuel porosity and the volume in the gap between the fuel and cladding are taken into account.

Spherepac A trade name for a fuel rod containing fuel particles derived from the sol–gel process.

"Stabilized" Refers to a microstructure that is resistant to change even at high temperatures (above $0.5\ T_m$) and, therefore, retains its mechanical strength. Generally achieved through formation of stable precipitates at grain boundaries or in the matrix.

Surveillance Intermittent or continuous monitoring of key plant components, e.g., vessel; surveillance capsules contain specimens of vessel steel that are removed periodically to check if any vessel property degradation has occurred as a result of irradiation.

Thermal Neutrons have a wide range of energy. Those having energies below 0.1 million electron volts are called thermal.

(Thermal) aging The time–temperature effects on material microstructure, e.g., precipitation, precipitate growth, and precipitate–dislocation interactions.

Transgranular Term almost exclusively associated with cracking *through* the grains of materials.

Vibrosol A trade name for a fuel rod containing fuel particles derived from the sol–gel process.

VIPAC A trade name for a fuel rod containing particles of fuel, rather than pellets.

Wastage Reduction in metal thickness caused by accelerated oxidation corrosion.

2. Acronyms

AISI	American Iron and Steel Institute
ANL	Argonne National Laboratory
ANS	American Nuclear Standards
ASME	American Society of Mechanical Engineers
B&W	Babcock and Wilcox Company, U.S.
BNFL	British Nuclear Fuels Ltd.
BWR	Boiling water reactor
CE	Combustion Engineering, U.S.
CFR	Code of Federal Regulations
CRBR	Clinch River breeder reactor (U.S.A.)
CTR	Controlled thermonuclear reactor
CVN	Charpy "V" Notch (energy, data, or tests)
DFR	Dounrey fast reactor (U.K.)
DNB	Departure from nucleate boiling
DNBR	Departure from nucleate boiling ratio
DOE	U.S. Department of Energy
DPA	Displacements per atom (radiation damage)
D–T	Deuterium–tritium
EBR-II	Experimental Breeder Reactor II (U.S.A.)
ENC	Exxon Nuclear Corporation, U.S.
EPRI	Electric Power Research Institute
FFTF	Fast Flux Test Facility (U.S.A.)

Appendix B

GE	General Electric
GYFM	General yielding fracture mechanics
HAZ	Heat-affected zone (of weld)
HCF	High-cycle fatigue
HEDL	Hanford Engineering Development Laboratory
HIP	Hot isostatic processing
HP	High pressure (turbine)
IGSCC	Intergranular stress corrosion cracking
IHX	Intermediate heat exchanger
KWU	Kraftwerk Union, AG, Germany
LCF	Low-cycle fatigue
LEFM	Linear elastic fracture mechanics
LHGR	Linear heat generation rate
LMFBR	Liquid metal (cooled) fast breeder reactor
LOCA	Loss of coolant accident
LP	Low pressure (turbine)
LWR	Light water reactor
MIT	Massachusetts Institute of Technology
MPC	Metals Properties Council
NDE	Nondestructive evaluation (sometimes NDT, nondestructive testing)
NDTT	Nil-ductility transition temperature (or ductile–brittle transition temperature)
ΔTT	Shift in NDTT
NRC	U.S. Nuclear Regulatory Commission
NSMH	Nuclear Systems Materials Handbook of USLMFBR Program
NSSS	Nuclear steam supply system
OECD	Organization for Economic Cooperation and Development
OTSG	Once-through steam generator
PCI	Pellet–cladding interaction
PCM	Power–coolant mismatch
PCMI	Pellet–cladding mechanical interaction
PFR	Prototype Fast Reactor (U.K.)
PWHT	Postweld heat treatment
RIA	Reactivity insertion accident
SCC	Stress-corrosion cracking
SI	Structural integrity
TD	Theoretical density—this is the density which would occur if there were no voids and the atoms were packed tightly.
TTT	Time–temperature–transformation (map for Fe-base alloys generally)
USE	Upper shelf energy
W	Westinghouse Electric, U.S.

Appendix C. Nominal Compositions of Alloys Used (or Candidates for Use) in Nuclear Components

1. Steels

AISI type	Fe	Cr	Ni	Mn
Carbon (ferritic) steels				
A501	Balance	4–6	—	1.0
A508/2	Balance	0.35	0.7	0.7
A508/3	Balance	—	0.6	1.3
A508/4	Balance	1.7	3.3	—
A508/5	Balance	1.7	3.3	—
A533	Balance	—	—	1.15–1.50
Low-alloy (bainitic) steels				
1Cr–1Mo–0.25V	96.07	1.0	—	0.85
2¼Cr–1Mo (Grade 22)	95.7	2.42	—	0.49
Ni–Cr–MoV (A469 Class 8)	Balance	1.25–2.00	3.25–4.00	0.60
Ni–Cr–MoV (A470 Class 8)	Balance	0.90–1.50	0.75	1.00
Ni–Cr–MoV (A471 Class 8)	Balance	0.75–2.00	2.00–4.00	0.70
Ferritic (martensitic) stainless steels				
403 (S40300)	Balance	11.5–13.0	—	—
410 (S41000)	Balance	11.5–13.5	—	—
Sandvik, Sweden (HT-9)	Balance	11.5	0.05	1.0
(HT-7)	Balance	9	—	1.0
French (R8)	Balance	9.5	—	2.0
French (EM-12)	Balance	9.0	—	2.0
Combustion engineering developmental alloy (U.S.)	Balance	9	—	1.0
Japanese (HCM9M)	Balance	8/10.0	—	1.8/2.2
Austenitic stainless steels				
304 (S30400)	Balance	18–20	8–10.5	—
304L (S30403)	Balance	18–20	8–12	—
316 (S31600)	Balance	16–18	10–14	2.0–3.0
316L (S31603)	Balance	16–18	10–14	2.0–3.0
321 (S32100)	Balance	17–19	9–12	—
347 (S34700)	Balance	17–19	9–13	—

Appendix C

Composition (wt.%)					
Mo	Si	V	C	S	Other constituents
0.40–0.65	1.0	—	0.10	0.030	0.40 (P)
0.6	—	0.05	0.27	—	
0.52	—	0.05	0.20	—	
0.5	—	0.03	0.23	—	
0.5	—	0.1	0.23	—	
0.45–0.60	0.15–0.30	0.05	0.25	—	
1.25	0.25	0.25	0.33	—	
0.98	0.28	—	0.026	0.009	0.012 (P) and 0.05 (Cu) Nb in stabilized version
0.30–0.60	0.15–0.30	0.05–0.15	0.28	0.018	0.015 (P)
1.00–1.50	0.15–0.35	0.20–0.30	0.25–0.35	0.018	0.015 (P)
0.20–0.70	0.15–0.35	0.05	0.28	0.015	0.015 (P)
1.0	0.5	—	0.15	0.030	0.040 (P)
1.0	1.0	—	0.15	0.030	0.040 (P)
0.55	0.4	0.3	0.20	—	0.5 (W)
0.45	0.65	—	0.1	—	
1.0	0.3	0.35	0.10		0.5 (Nb)
1.0	0.75	0.3	0.15	—	0.5 (Nb)
?	?	1.0?	0.01	—	1.0? (Nb)
0.3/0.7	≤0.5	—	≤0.06	≤0.03	
2.0	1.0	—	0.08	0.030	0.045 (P)
2.0	1.0	—	0.03	0.030	0.045 (P)
2.0	1.0	—	0.080	0.030	0.045 (P)
2.0	1.0	—	0.030	0.030	0.045 (P)
2.0	1.0	—	0.08	0.030	0.045 (P) Ti = 5 × C min.
2.0	1.0	—	0.08	0.030	0.045 (P)

2. Copper-Base Alloys

Type	Cu	Sn	Al	Ni
Admiralty brass (443)	70–73	0.9–1.2	—	—
Admiralty brass (444)	70–73	0.9–1.2	—	—
Admiralty brass (445)	70–73	0.9–1.2	—	—
90–10 copper–nickel (706)	86.5 min.	—	—	9.0–11.0
70–30 copper–nickel (715)	65.0 min.	—	—	29.0–33.0
Aluminum brass (687)	76.0–79.0	—	1.8–2.5	—
Aluminum bronze (608)	93.0 min.	—	5.0–6.5	—
IN-838	82.3	—	—	16.0
IN-848	68.6	—	—	30.0

* R = remainder.

3. Nickel-Base Alloys

Type	Ni	Cr	Fe
Inconel 600	60.5	23.0	14.1
Inconel 690	60.0	30.0	9.5
Inconel (Alloy) 800H	32.5	21.0	46.0
Inconel (Alloy) 800	34.0	21.0	43.0
PE 16	42.3 max.	16.6	34.1

4. Zirconium-Base Alloys

Type	Composition (wt.%)				
	Zr	Sn	Fe	Cr	Ni
Zircaloy 2[a]	Balance	1.5	0.12	0.1	0.05
Zircaloy 4[b]	Balance	1.5	0.2	0.1	—

[a] Zircaloy 2 is also called Zirconium 20.
[b] Zircaloy 4 is also called Zirconium 40.

Composition (wt.%)

Pb	Fe	Zn	Mn	As	Sb	P	Cr
0.07	0.6 max.	R*	—	0.02–0.10	—	—	—
0.07	0.6 max.	R*	—	—	0.02–0.10	—	—
0.07	0.6 max.	R*	—	—	—	0.02–0.10	—
0.05	1.0–1.8	1.0 max.	1.0 max.	—	—	—	—
0.05	0.40–0.70	1.0 max.	1.0 max.	—	—	—	—
0.07	0.06 max.	—	—	0.02–0.10	—	—	—
0.10	0.10 max.	—	—	0.02–0.35	—	—	—
—	0.8	—	0.5	—	—	—	0.4
—	0.3	—	0.7	—	—	—	0.4

Composition (wt.%)

C	Mn	Si	Ti	Mo	S	Others
0.08	0.5	0.2	1.4	—	—	
0.03	—	—	—	—	—	
0.05	0.8	0.5	0.4	—	—	0.4 Cu
0.02	0.64	0.3	—	—	—	—
0.06	0.12	0.24	1.31	3.63	0.004	0.99 Al 0.003 P

5. Titanium-Base Alloys

Type	Composition (wt.%)						
	Ti	Al	Mo	V	C	Sn	Zr
Ti-6242	Balance	6	2	—	<.1	2.0	4.0
Ti-64	Balance	6	—	4	0.08	—	—
Ti-6246	Balance	6	6.0	—	<0.1	2.0	4.0

Index

Accident
  criteria, 95–98, 105–106, 199, 214, 221, 223
  Three Mile Island, 109–112
Anodize, 90
Austenitize, 265, 412
Availability factor, 13–14, 63, 106, 406; *see also* Capacity factor

Barriers, 84–85, 463, 465
Boiling water reactor, 3–6, 16, 54-60, 62–64, 76, 82–85, 86–91, 107, 118, 121, 200, 201–204, 207–209, 216–217, 230, 243–244, 421
Boric acid, 7, 346
Boron carbide, 6, 7, 8, 58, 119–121, 122, 140, 143, 180–184
  mixtures of, 58, 122, 123
  See also Control blade, Control rod
Bow
  of fuel rods, 107–109
  of ducts, 153–154
Breeding
  blanket, 8, 10, 141, 143, 284
  gain, 139
Burnable poisons, 58, 122–124; *see also* Gadalinia

Capacity factor, 13–14, 63, 332, 382, 406, 451; *see also* Availability factor
Carbon, 235, 265, 281, 305–307, 311–313, 314–317, 464
  transport of, 163, 177, 358, 361, 365, 376

Carburization, 163, 177, 179, 250, 259, 261, 361–368, 375–379
Castings, 271–272, 412
Cathode, 337–339
Cathodic protection, 387, 396
Ceramics
  structural, 282, 284, 304–317
  See also Fuel pellet, Plutonium, Uranium dioxide
Channel, 6, 86, 89; *see also* Duct, Zircaloy
Charpy (V-notch energy), 30–32, 218–220, 221, 413–414
Coatings, 281, 304, 316, 464
Condenser, 6, 321, 322, 336, 346, 381–398, 461
Control blade, 6, 58–59
Control rod, 6, 7, 8, 59–60, 119–122, 143, 180–185
Copper
  alloys of, 332, 383–391
  ions of, 339
Core (of reactor), 6, 8, 54–60, 94, 117, 129, 140–143
Corrosion
  in crevices, 46, 243, 334, 335, 371, 419, 422, 442–443
  denting, 334, 335, 337–340, 346, 355
  intergranular, 147–149, 239–242
  nodular, 86–88
  pitting, 46, 368, 385, 387, 392
  uniform, 45–47, 85–86, 89–93, 162–164, 360–361, 369–373

481

Corrosion (*cont.*)
  wastage, 46, 334, 335, 343–344, 374–375, 388, 392
Corrosion fatigue, 36–38, 226–231
Creep, 17, 18, 21–23, 30, 43, 153, 434
  crack growth in, 28
Creep fatigue, 17, 18, 36–38, 245–249, 253, 264, 295–297, 364, 378, 426–428
Creep rupture, 17, 18, 37, 290, 363, 373, 377–378, 413, 424–426, 433–434

Damping, 446
Decarburization, 163, 259, 358, 361–368, 375–379
Design rules, 17–19, 25, 36, 60, 151, 201, 206, 221, 229, 238, 244
Deuterium–tritium (D–T), 9, 10, 12, 288
Dislocations, 23, 40–41, 73, 156, 172
Duct, 139, 143–144, 150, 153–154, 167

Electroslag remelt (ESR), 265, 356, 435–438, 456–457
Erosion,
  by irradiation, 44–45, 279–281, 311–313
  by water/steam, 336, 374, 384–386, 388–391, 391–392, 431–432, 446–447
Europia (Europium dioxide), 184–185

Fatigue
  crack growth, 19, 26, 33, 36, 226–231, 241, 255–263, 298–300, 310, 378–379, 422–431, 445
  environmental, 36–38, 46, 227, 370–371
  high cycle, 33–36, 253–254, 295–296, 344–345
  low cycle, 33–36, 210, 227, 244–245, 264, 295–302, 367, 378, 424–425, 443–445
  See also Corrosion, Creep fatigue, Stress corrosion cracking
Finite element, 207, 213
First wall (of fusion reactor)
  damage, 288–302, 306, 316
  design of, 10, 279–284
  materials, 287–317, 458–459, 462, 464–465

Fission product(s)
  behavior in fuel, 76–77, 147, 178
  release of, 62, 65, 77, 100–105, 119–120, 147–149, 158, 175, 179, 181–182
Flaw(s)
  detection of, 18, 19, 207, 221–222, 243, 430
  evaluation of, 18, 19–21, 207–211, 232, 306–309, 427–430, 466
Fracture mechanics
  general yielding (GYFM), 27–28, 211–214, 255
  linear elastic (LEFM), 19–21, 25–27, 206–211, 219–227, 242, 255, 300, 306–309, 428–429
Fracture surface
  cleavage and fluting, 66–68, 73
  intergranular, 233, 424
  ductile-dimple, 265
Fracture strength, 306–310, 314–316
Fracture toughness
  arrest, 32, 217–220
  ductile, 27, 32, 213–214, 221, 440
  dynamic, 32, 217–220
  static, 25, 30–32, 46, 209–211, 217–220, 222–225, 241, 419, 430
Fretting: *see* Wear
Fuel pellet
  densification of, 98–100, 113, 146
  design changes to, 83–84, 165–167, 178–179, 186–187
  fabrication of, 54–55, 112
  irradiation behavior of, 45, 65, 74–77, 106, 145–147, 158–162, 175–179
Fuel rod
  corrosion of, 85–93, 162–164
  design of, 6, 7, 8, 53–60, 82–85, 107–109, 137–138, 140–141, 144, 152–153, 164, 174–175, 463–464
  failure of, 14, 60–65, 77–82
  performance of, 14, 15, 60–112, 112–119, 127–129, 143–145, 149, 151–152, 175–179, 188–189, 190–191
  safety of, 93–112, 161
Fuel rod codes, 63, 77–82, 83, 104–105, 145, 160–161
Fusion reactor: *see* Magnetic fusion, Inertial confinement fusion, Laser fusion, Tokamak

Index

Gadalinia, 57, 122–124
Grain boundaries
  cracking at, 74, 231–235, 418–419, 424, 434; *see also*: Stress corrosion cracking
  gas bubbles on, 40, 104, 145–156, 288–293, 300; *see also*: Fission products and helium embrittlement
  impurities at, 117, 433–434, 438–441; *see also*: Temper embrittlement
  precipitation at, 250, 341, 350–351, 413
  refining of, 265, 433–434
  segregation to, 231–235, 420, 438–440
  sensitization of, 235–237, 340–341
  sliding of, 30, 251, 253
Graphite: *see* Carbon

Hafnium, 121
Helium, 39, 40, 44, 119, 180–183, 311–313, 330, 464
  bonding with, 175
Helium embrittlement, 39, 40, 150–151, 286, 288–293
Hydriding
  of titanium, 396–397
  of Zircaloy, 61, 87
Hydrogen, 110, 115, 304
Hydrogen embrittlement, 228, 395–396, 422

Inconels (or Inconel alloys), 6, 121, 243, 324, 340–344, 349–354, 356–357, 359–360, 367–368, 372–373, 374, 460
  sensitization of, 340
Inertial confinement fusion, 9, 11–12, 283–284, 295, 331
Intermediate heat exchanger, 2, 8, 10, 321, 328–330, 356–357, 375–379
Irradiation, effects of: *see* Radiation

Laser Fusion, 11, 12, 283–284
Liquid metal fast breeder reactor, 7–9, 13, 15, 29, 39, 41, 45, 137–138, 159, 165, 204–205, 209–211, 244, 256, 263, 287, 293, 324–329, 380, 458–459
Lithium, 10, 12, 283–284, 300, 302, 460, 465
Lithium hydroxide, 91
Lithium oxide, 331

Magnetic fusion, 9–11, 281–283; *see also* Tokamak

Neutron absorber: *see* Control rod, Control blade, Burnable poison, Boron carbide
Neutron flux, 9, 11, 39, 43–44, 70, 91, 155, 222, 279, 287, 311–313; *see also* Radiation damage
Nozzles, 207, 230–231, 243; *see also* Pressure vessel and piping

Operating margin (or Thermal margin), 93–95
Oxide film, 47–48, 69, 85–93, 96–98, 337–338, 346–347, 365, 369–370, 386, 422, 442; *see also* Corrosion

Pellet-cladding interaction (or fuel-cladding interaction), 64–85, 113, 117–119, 158–162, 175–178, 179–180
Piping, 17, 199, 201–205, 209–211, 231–242, 244, 266–271; *see also* Steel, stainless; Steel, pipe
Plasma
  confinement of, 9
  damage from, 45, 279–280, 288
  poisoning of, 45, 280, 304
Plastic flow, 29–30, 73–74; *see also* Strain, Stress
Plutonium
  carbide of, 139, 174–179
  metal alloys, 139, 190–192
  nitride of, 139, 179–180
  oxide of, 112–114, 137, 165–167
  See also Uranium–plutonium (mixed) oxide
Pourbaix diagrams, 346–347
Pressure vessel, 3, 17, 60, 199, 200–205, 219, 222–224, 226, 265–266, 456
Pressurized water reactor, 6–7, 54–60, 62–64, 75, 82–85, 91–93, 94, 106, 118, 123–124, 126–127, 201–204, 216–217, 222, 226, 231, 244, 321–324, 332, 353, 380, 415–416, 421, 460
Pump, 9, 199, 271–272

Radiation creep, 43, 153, 158, 168–172, 297

Radiation damage (embrittlement),
    38–45, 70, 73, 87, 150, 221–226,
    251–255, 261–263, 286, 288–295,
    313–316
Radiation dose, 95
Radiation growth, 41
Replacement power, 63, 129, 333

Silicon carbide, 305–307, 311–313,
    314–317, 464
Silver-indium-cadmium (AgInCd), 60,
    121–122
Sodium, 7–9, 10
    bonding with, 175–176
    effect(s) of, 46, 162–164, 177–178,
        249–251, 258–259, 304, 327—329,
        356, 360–368, 373–375, 375–379
Sputtering: *see* Plasma, Erosion
Steam
    corrosion/erosion in, 96–98, 368–373,
        374–375, 431–432
    purity of, 336, 345–346, 381–382,
        420–421
Steam generator, 2, 6, 8, 17, 231,
    321–328, 330–332, 332–356,
    358–381, 459, 460, 465
Steel
    ferritic, 10, 30, 170–173, 201–202, 222,
        265–266, 268–270, 290–291, 294,
        302–303, 324, 328, 337–338,
        354–355, 356–357, 358–360,
        360–367, 368–372, 374–375,
        458–459
    ingots, 407–409
    low alloys (bainitic), 6, 201, 227, 265-
        266, 412–414, 432–441, 453, 456
    pipe, 6, 199, 201–205, 231–242
    sensitization of, 234–237
    stainless, 6, 8, 10, 24, 39, 40, 55, 60,
        114–119, 121, 137, 143, 150–158,
        162–164, 167, 172–173, 184,
        201–205, 209–211, 231–243,
        244–263, 266–268, 289–290,
        293–294, 302, 328–329, 332, 356,
        367, 374–375, 375–379, 391–393,
        441, 457, 460, 464
Strain
    elastic, 17, 18
    hardening, 24, 213
    inelastic (plastic), 17, 18, 21–24,
        29–30, 38, 74, 151, 177, 180, 212,
        241, 253–254

Strain (*cont.*)
    plane, 27, 32
    radiation induced, 313–314
    See also Plastic flow
Stress, 17, 18, 21–24, 29, 74, 207–210,
    228, 237–239, 296–297, 340,
    358–360, 421, 446
    reduction of, 266–268, 348–349
    rule, 238–239
Stress corrosion cracking, 19, 27, 46, 452
    of copper alloys, 385
    of ferritic steels, 269–270, 371–372,
        416–422
    models of, 71–73, 80–81, 241–242
    of nickel alloys, 334, 340–343,
        349–354, 359; *see also* Inconels
    of stainless steel, 117–119, 121,
        231–243, 267, 341, 392
    of titanium alloys, 396, 442–443
    of Zircaloy, 65, 68–73
Structural integrity analysis, 206, 214,
    296–300, 423, 428–429
Structural integrity (SI) plan, 19–21, 206
Swelling
    of ceramics, 119, 147, 149, 165–166,
        177–178, 180–185, 311–313
    due to fission product, 45; *see also*
        Fission products
    of metals, 41, 150, 155, 157, 167–168,
        172, 293–295

Temper embrittlement, 413, 418–419,
    427, 438–441
Texture, 70, 73
Thermal aging, 250–251, 259–261, 362,
    378
Thermal treatment, 350–351, 409,
    412–413
Thorium, 124, 187–191
Titanium
    alloys of, 291–293, 294–295, 302,
        441–448, 461–463
    metal, 292, 332, 346, 383, 393–397
    modified steels, 167–174, 290, 294, 302
Tokamak, 10, 281–282, 295, 330
Turbine
    design of, 6, 407–412, 461
    failures in, 405–406, 414–432

Upper shelf energy, 30–31, 42, 214–225,
    356, 414; *see also* Charpy energy;
    Fracture toughness

Index

Uranium, conservation of, 124–129
Uranium alloys, 190–192
Uranium dioxide
  behavior of: *see* Fuel pellet, Fuel rod
  pellets of, 6–8, 54–55, 111, 141, 149
Uranium–plutonium (mixed) oxide, 8,
    112–114, 137, 145–150, 165–167

Valve, 199, 271–272
Vanadium, 300–301
  alloys of, 292, 294, 303–304
  carbide, 413

Wear, 344; *see also* Erosion
Weibull modulus, 306–309, 314–316
Welds (weldments and weld repair), 201,

Welds (*cont.*)
    203, 205, 227, 230–231, 231–242,
    243, 263, 265, 266–268, 271, 332,
    391, 395, 447–448, 466

Zircaloy
  cladding, 6, 16, 53–61, 84–85, 453,
    463–464; *see also* Fuel rod
  corrosion of, 46–47, 85–93
  creep of, 89, 109
  embrittlement, 96–98
  growth of, 109
  oxidation of, 85–98, 110–112, 115
  plastic deformation of, 24, 73–74, 116
  stress corrosion cracking of, 64–73; *see also* Stress corrosion cracking